T0121294

Also by James Shreeve

The Sex Token *(fiction)*

Nature: The Other Earthlings

Lucy's Child: The Discovery of a Human Ancestor *(with Donald Johanson)*

The Neandertal Enigma

THE GENOME WAR

THE GENOME WAR

How Craig Venter Tried to Capture the Code of Life and Save the World

JAMES SHREEVE

Ballantine Books New York

2005 Ballantine Books Trade Paperback Edition

Published in the United States by Ballantine Books, an imprint of The Random House Publishing Group, a division of Random House, Inc., New York.

Ballantine and colophon are registered trademarks of Random House, Inc.

Originally published in hardcover in the United States by Alfred A. Knopf, a division of Random House, Inc., in 2004.

Grateful acknowledgment is made to Oxford University Press for permission to reprint an excerpt from *Faust, Part 1* by Goethe, edited by David Luke (OWC, 1987). Reprinted by permission of Oxford University Press.

Library of Congress Cataloging-in-Publication Data
Shreeve, James.
The genome war : how Craig Venter tried to capture the code of life and save the world / James Shreeve.
p. cm.
ISBN 978-0-345-43374-9
1. Human Genome Project. 2. Venter, J. Craig. I. Title.
QH431.S5577 2004
611'.01816—dc21 2003047580

www.ballantinebooks.com

Text design by Iris Weinstein

147028622

To
Walton Shreeve
and
Phyllis Heidenreich Shreeve
(1922–2000)

Dass ich erkenne, was die Welt
Im Innersten zusammenhält

I'll learn what holds the world together
There, at its inmost core.

—GOETHE, *Faust*

CONTENTS

PART THREE

THE GENOME WAR

PROLOGUE

On the morning of June 26, 2000, a short, elderly man walked up the pedestrian lane along the east side of the White House grounds and joined a couple of dozen people already gathered outside the guest entrance, under the shade of the willow oak trees that line the lane. It was the beginning of a sultry Washington day, one of the rare ones in that mild summer, and the man's shirt was already damp from perspiration. "That I should live to see this day is beyond belief," he said, to no one in particular. "I was there in the beginning, you know. I was there before DNA."

Norton Zinder, professor emeritus at Rockefeller University, had been a molecular biologist since before 1953, when James Watson and Francis Crick made the essential discovery that the shape of the deoxyribonucleic acid molecule, or DNA, enabled life to happen. He had made some important contributions back then, too, and much later, in the late 1980s, he had helped Watson organize a government science program called the Human Genome Project. Its goal was to reveal the innermost secret of life: the entire code, spelled out in the language of DNA, for the construction and maintenance of a human being. Like the others waiting in the shade, he had been invited to attend the president's announcement that the human genome had at last been deciphered.

Zinder's gratitude for having lived until this June day was not entirely rhetorical. He had recently suffered a stroke. While he was recovering, he had tried to mediate a globally watched conflict that had very nearly turned the Human Genome Project into the uttermost embarrassment in science and was well beyond his ability to resolve. But the attempt had wrenched sleep from him, and for several weeks he had felt as if he were living on Valium. He was afraid another stroke would kill him before he could see the end of what he had helped begin.

Zinder talked for a while—he was a voluble man, and the stroke had perhaps made him a little more so—about the significance of this day, and recounted his memories of his life in science, which fell out of him in a semi-organized rush, like folders spilling out of a file cabinet. Meanwhile, other guests continued to arrive. Most were scientists. They were all dressed up for the occasion, and from their solemn excitement when they greeted each other, one sensed that they were not accustomed to seeing their colleagues decked out in such finery. At about 9:30, James Watson himself appeared. A line to go through security had formed along the fence, and he was standing in it by himself: a tall, hollow-chested figure in a white suit and a floppy tennis hat. He was staring into the leaves of the willow oaks, his mouth hanging open slightly. Someone asked him how he felt on such a momentous occasion. "It's a happy day," he answered. But he did not look happy. He looked like a man going through the motions of being happy.

In line behind Watson was a much shorter man, dressed in an expensive dark suit and a light blue shirt with a white collar. It was the kind of business-class shirt that a scientist would never wear, even to an event at the White House. His name was Tony White. He was the CEO of PE Corporation, and he had just flown in on his private jet. He did not look any happier than Watson. Other than that, there couldn't be more contrast between the two men. Watson's face was long, bleached, and spotted, and he seemed to be composed mostly of limbs. White was round and compact, like a cannonball. His face was red, his collar was tight, and he was peering around nervously, as if he'd been invited to a party hosted by an enemy. Somebody asked him too how he felt on the occasion. "I don't know how you go from blatantly trashing each other to being all lovey-dovey overnight," he said. "I've had nothing to do with this." He pulled out a cell phone. "Where is everybody?" he grumbled into it. "Maybe there was some other way of getting down here that I wasn't informed about." He snapped the phone shut and shoved it back into his pocket.

Watson and White had arrived at the White House from different worlds, and they remained oblivious to each other. They existed on either side of a wall that has traditionally divided science into two camps: the basic research conducted in university labs and nonprofit institutions like Watson's Cold Spring Harbor Laboratory on Long Island; and the applied science of pharmaceutical companies, biotechs, and other corporations aimed at developing a marketable product, often using the results of academic research as a jumping-off point. There is of course some traffic back and forth, and people on both sides of the wall believe they are serving the public good. But the two camps obey different codes of conduct and reward excellence in different ways. The currency of academic excellence is recognition—publications, honors, and the esteem of colleagues, with the highest accolade being the Nobel Prize. The currency of success in commercial science is currency—lots of it, in some cases. Watson and White epitomized these separate worlds. The only reason the two men were in line together on that June morning was that another scientist, who was at that moment inside with the president, had tried to excel on both sides of the wall simultaneously, which violated everybody's rules. His name was J. Craig Venter. Whether or not he had succeeded was an open question. But he had certainly succeeded in pissing a lot of people off, Watson and White among them.

A science policy administrator, Kathy Hudson, arrived carrying a fresh copy of *Time* magazine. She passed it around and people leafed through it, laughing and admiring the pictures in the lead story. On the cover of the magazine, two scientists stood shoulder to shoulder, one a little behind the other. Both men were dressed in white lab coats. (Journalistic protocol demands that lab coats be worn by scientists during photo shoots; otherwise we might not be able to tell them from other people.) The scientist on the right was Francis Collins, chief of the government's Human Genome Project, who was also inside with the president. The photo showed a man in his late forties with a large square face, thick brown hair, a neat mustache, and a dogged set to his mouth. The scientist on the left was J. Craig Venter. He was bald, with upturned eyebrows and an oval face tapering down to a mouth that flickered up at the corners, as if he were trying to suppress a grin. His face was dramatically bifurcated by the photographer's lighting, the right side aglare and the left in shadow.

Two years earlier, in May 1998, Venter had announced that with backing from PE Corporation (then known as Perkin Elmer), he was going to form a private company to unravel the human genetic code and would complete the project in three years instead of the seven more years estimated to be needed by the publicly financed Human Genome Project. By making the human code available to the world so soon, he hoped to greatly accelerate the pace of biomedical research and thereby save the lives of thousands of people who would otherwise die of cancer and other diseases. He also hoped to become famous, well loved, and very rich. It was a big gamble on all counts. Nothing like the particular scheme he was proposing had been attempted before. If it were broken down into its various technical components, most of them had never been attempted before, either. All of these untested elements would have to work seamlessly together or the whole enterprise would fail. If it did work, it would be a scientific achievement of huge importance. But even then, few people really understood how it would make Venter's proposed company any money. Those who knew Venter were not surprised that he was going to try anyway. "Craig likes to do high dives into empty pools," one colleague said of him. "He tries to time it so the water is there by the time he hits the bottom."

A couple of weeks after Venter's spectacular announcement in 1998, I asked him to let me observe the progress of his undertaking as it unfolded. If the proposed enterprise did manage to succeed in beating the government's Human Genome Project to the finish line, I told him, it might make a compelling book. I didn't mention that there was at least as good a story to be told if it crashed and burned. At the time of his announcement, Venter was president of a nonprofit research group in Rockville, Maryland, called The Institute for Genomic Research (TIGR), which he had founded in 1992 and which was dedicated to pure science. The company he was organizing to carry out the new project would not be. Simply by announcing his intentions, in fact, he had thrown down a gauntlet to perhaps the greatest concerted undertaking in the history of academic science, an endeavor that was universally regarded to be true and good. Some people said he had made a deal with the devil. Others thought of him as the devil himself.

Venter responded to my proposal in a way that I would learn was typical: he invited me to go sailing. He owned a yacht named *Sorcerer,* and asked if I'd like to join him and some friends for a weekend race off the coast of Nantucket. He told me to meet him at the Montgomery County Airpark in Gaithersburg, where a chartered plane would fly us up to the island. I was waiting for him when a beefy blue SUV drove up, bearing the license plate TIGR 2. Venter tumbled out and beamed at me—a brow-lifting, eye-brightening grin that did look just a trifle satanic. There wasn't any malevolence in the expression, but the irrepressible delight in his smile, and the way it sharply lifted his eyebrows and the lines about his eyes, gave his face a netherworldly impishness, as if our meeting there on the tarmac, the plane noisily revving up, were a prelude to some disobedient romp of mischief. He was bald as the devil, too. Bald men often seem acutely embarrassed above the brow. In contrast, Venter's freckled pate seemed a completion rather than an absence, as if his scalp had purposely emptied its follicles to reduce friction with the elements through which he moved, like the shaved head of a swimmer.

"Hey!" he said, with a rising inflection, more the way a surfer might greet a buddy on the beach than the way a distinguished scientist greets a writer. "Hop in! On the flight up, I can tell you how we're going to change the paradigm of medicine."

Venter did not look like a typical scientist, either. He was wearing a light green linen shirt, blue jeans that looked fresh out of the launderette's cellophane, and brilliant white tennis shoes that flashed like beacons with each step. He is five-foot-eleven, but he seemed taller, perhaps because of that uplifted eagerness in his face. Scientists do not all wear tweed jackets, cardigans, or thick glasses, of course, but most of the ones I have met share a downward orientation somewhere in their body language—a stoop, a slouch, or something less tangible, an indefinable holding back. It comes from spending a lot of time crawling around in mental caves. Modern life science—and genomics, Venter's field, was so modern that most people, including other scientists, had never heard of it—is driven forward by reductionism: a strategy of discovery based on the belief that the more you can divide natural phenomena into their constituent parts, and those parts into subparts and so forth, the more you can learn about how nature really works. Looking at a person will not tell you as much about what makes him tick as if you examine his heart and brain and other organs—or, better yet, the specific cell types in those organs, or, farther down in the cave, the interaction of proteins in those cells or the intimate varying structure of the proteins themselves.

A good scientist never forgets that the purpose of studying ever more reduced parts of an organism is to understand the organism as a whole. But to push knowledge forward, you move downward into the biological hierarchy as if deeper and deeper into a system of tunnels, looking for tiny gems, nuggets of new information, buried in the rock. The farther down into a problem the scientist has gone, the harder it is to back out and explain to the people on the surface what he or she has been doing. When scientists are out of the cave and away from work, the urge to get back in again pulls on some vulnerable part of their physical aspect—their posture, for instance, or their gaze or the set of their mouth.

Craig Venter had spent time mining the mental cave, too, chipping away under the beam of a tiny flashlight on a minuscule portion of the vast, pitch-black tunnel system where the genetic secrets of the human brain are hidden. But he was impatient with the pace of life in the cave, and a decade earlier he had found a faster way to find genes. His new method had paid off handsomely, in its contributions both to basic scientific research and to his wallet. It also made him a lot of enemies. His announcement in May 1998 had made new enemies and galvanized the old ones into unified opposition. It was the exceptional hubris of the plan that riled them. Venter wasn't just trying to capture more than his share of gems from the cave. He was going after the cave itself.

"We're going to be on the forefront of *everything,*" he told me on the plane. "We're going to need to build the fastest computer in the world, with data production orders of magnitude bigger than anything else. We're thinking on a different scale. Just doing the human genome and stopping there is way short of what can be accomplished."

He took out his wallet and pulled out a plastic card. The words U.S. DEPARTMENT OF GENETIC IDENTITY were written across the top of the card in an imposing font. Embossed below the photo of a young man was one of those holographic information-bearing chips you sometimes see on credit cards. The chip, Venter explained, contained the man's complete genetic code. Based on that information, he said, the individual would know that if he smoked he had a 37 percent chance of developing lung cancer before he was sixty. So he wouldn't smoke. He would know that he was among the third of the population for whom aspirin helps to prevent heart disease. This would be crucial knowledge, because the chip also told the bearer that he carried a defective form of a gene called APOE, which put him at a much higher risk of heart attack and stroke. He would orchestrate his diet and lifestyle to minimize his risk. Should he wish to inquire, the chip would also tell him his chances of going mad or committing suicide. It would tell him what time of day he was most likely to be productive, what kind of cologne would best accent his natural body scents, and, perhaps, whether his new girlfriend would be a good match for a long-term relationship.

"It's just a mock-up I use in lectures," Venter told me, slipping the card back in his wallet. "But this is where we're headed. As people start to get genotyped, they'll be able to take control of their lives. Our database will allow them to know their future."

On Nantucket, we found *Sorcerer* near the end of the marina's long main pier. Claire Fraser, Venter's wife and his scientific colleague at TIGR, was already aboard, along with some of their other guests. Venter introduced me to Peter Barrett, a businessman in his early forties who was moving from Perkin Elmer to become the chief business officer for Venter's new company. The yacht was a sleek, dark blue eighty-two-foot sloop originally built for the owner of Lands' End, the clothing mail-order house, and had cost Venter more than a million dollars. He had bought it a few years earlier, soon after a previous discovery had made him rich virtually overnight. While in fact he was not much wealthier than many other academic scientists who were surprised to encounter venture capitalists eager to shower them with money in return for the commercial rights to their discoveries, Venter did not behave with his money the way an academic is supposed to behave. He spent it lavishly on things he enjoyed, like fast cars, big parties, and big houses, and most of all on *Sorcerer.* In addition to its original price tag, the yacht cost him another $300,000 a year to maintain, including salaries for a full-time two-person crew. Claire, his wife, liked sailing, too, but she had a little perspective. "We'd be rich," she told me, "if it weren't for that boat."

The next morning was dazzlingly brilliant, as if the yacht owners had pooled their resources and hired the sun to show off their boats in the best possible light. The twenty-five entries paraded out of the harbor, hulls gleaming. Start times for the racing yachts were staggered over an hour, loosely determined by how much emphasis their designers had put on "racing" and how much on "yacht." The summer before, *Sorcerer* had surprised a dozen larger boats to win the transatlantic Atlantic Challenge

Cup, a victory that had not been unnoticed by the Nantucket race committee. We started thirty-five minutes behind the first boat out and immediately raised the spinnaker. It bore the image of a sorcerer wearing a tall pointed hat, a full white beard, and a Cheshire cat smile. The spinnakers of the boats out before us were arrayed like colorful splotches on the horizon. "I wonder how they feel, watching this wizard crawling up their stern," Venter said.

The course was a simple run to a mark seven miles to the northeast, then back again beating upwind. *Sorcerer* was fast, and by the time we rounded the mark we had made up half the distance between us and the leading boat. But *Sorcerer*'s heavy hull and rigging design were a burden when tacking into the wind. Though we passed one boat after another, we couldn't quite catch the leader, who crossed the finish line ten seconds in front of us. As the gun sounded, I looked at Venter. He was as competitive as anyone I'd ever met, and I thought he might be disappointed with our second-place finish. But the thrill of the moment had stretched his face back in a mask of spent, hysterical delight, like a child at the end of a roller-coaster ride.

On that day in August, Venter was already three months into a race with far greater consequences. It would be a far closer finish than he ever imagined. But even if he had better calculated the capabilities of his opponents—and the intensity of their determination to destroy him—Venter would still have gone sailing that day without a second thought. It was just too much fun to miss. That was another thing that bothered his enemies. Even in the worst times in the coming two years, there was a part of him that was having more fun than they were, and they knew it.

PART ONE

CHAPTER 1
MAY 1998: "YOU CAN DO MOUSE"

On May 8 of that year, three months before the Nantucket race, Nicholas Wade, a veteran science writer for the *New York Times,* entered the lobby of the St. Regis hotel on Fifth Avenue. The day before, he had received a call from a public relations representative of the Perkin Elmer Corporation in suburban Connecticut, offering him an exclusive story on an exciting development. Wade was leery. Public relations people often overestimate the media interest in their company's announcements. In the brash, upstart world of biotechnology, moreover, Perkin Elmer was an unglamorous player—a maker of instruments, not news. But the PR rep mentioned that Craig Venter was a player in the new enterprise. Wade knew that Venter would not be involved in anything unglamorous. He agreed to meet with Perkin Elmer's executives over breakfast.

Wade crossed the lobby and squeezed into an elevator just as its doors began to close. A slightly built, mild-mannered Englishman in his fifties, the *Times* reporter attracted little attention from the dark-suited businessmen already in the elevator. In the Perkin Elmer suite on the fourteenth floor, he was introduced to CEO Tony White and two other company executives. One was Peter Barrett. The other was Michael Hunkapiller, head of Perkin Elmer's Applied Biosystems division, near San Francisco. Wade knew him by reputation. Largely unknown outside the biotech world, Hunkapiller was a legend within it. In the late 1980s, he had co-invented a machine that could automatically sequence DNA—that is, read out the order of a short stretch of chemical letters in the genetic code. Since then his technical genius and business acumen had made him the linchpin of an ongoing effort to develop better, faster instruments for sequencing DNA and speeding up other biotechnical processes.

A lavish breakfast buffet had been set up on a sideboard. Craig Venter was not in the room, but his voice greeted Wade from a speakerphone sitting on a coffee table. "Hey, Nick," he said, in a disarmingly mild tone. "Thanks for coming in so early. There's something we wanted you to be the first to know. Are you sitting down?"

Wade took a chair and opened his notebook. First Venter talked, then Hunkapiller, then Venter again at greater length. Tony White offered an occasional comment, in a broad southern drawl. No one touched the breakfast. Wade sat erect, furiously taking notes. He left an hour and a half later, certain that he had the lead story for the coveted front page of the Sunday edition of the *Times.*

"Genome" is not a pretty word. Even when you say it in a normal tone of voice, you sound like you're mumbling. It has so recently come into common usage that until 1997, almost a decade after the Human Genome Project began, Microsoft Word's spell-checker assumed that anyone writing about the enterprise had made a typo and corrected it to "the Human Gnome Project." In 1999, bioethicist Arthur Caplan of the University of Pennsylvania was invited to address a meeting of state legislators who were puzzled over the issue of human cloning. Caplan asked the lawmakers if they knew where their genome was located. Roughly one third answered that it was in the brain, and another third thought it was in the gonads. The others weren't sure.

In fact, two copies of your genome—one contributed by each of your parents—are spooled on the twenty-three chromosomes inside the nucleus of every one of your cells—brain, gonad, bone, skin, guts, muscle, mucus, and every other kind of cell, except for red blood. The spool is fantastically compact. Each copy is a double-stranded molecule of DNA only 79 billionths of an inch wide, but which, stretched out, would run almost six feet in length. Upon this attenuated thread—

imagine a clothesline running the length of the United States, then back again—lie the chemical instructions that have informed the development of your body and brain from the moment you were conceived. The individual units of instruction are called genes. They are composed of strings of the four chemical bases of DNA: adenine, thymine, guanine, and cytosine—abbreviated A, T, G, and C by scientists. The cell's machinery forms the letters into a series of three-letter words that combined give the recipe for the construction of a specific protein: an enzyme that helps you digest a tuna sandwich, an antibody molecule marshaled to fight off an infection, a receptor protein in your brain that helps you read and understand this paragraph. There are a lot of extra letters in the genome, sloppily referred to as "junk DNA," which do not spell out protein recipes but may serve some other purpose, perhaps vital, perhaps not. The whole human genome contains about 3 billion letters, and is often compared to the text of a book. If you decided to read the book aloud and recited one letter every second, it would take you eleven years to get to the end.

Even if someone had eleven years to spare for such a project, a simple recitation of the DNA letters would give no hint whatsoever of the way the whole genome works to create and operate a human life. Instead of thinking of the genome as a book, imagine it as a piano keyboard. Each piano key represents one gene. If you press down on a key, you hear a single note: the protein that the gene expresses. If you press the key again, you will hear the same note again, monotonously, every time the key is played. But with a piano keyboard, you can do much more than play lots of individual notes. You can combine the notes to make music. Just so, our various cell types play upon the long, thin keyboard of the genome: they combine notes, playing some genes together as chords, tripping several together in a phrase, gathering bundles of notes to create the complex and wonderful effects that find expression in our biological being. Just as a pianist doesn't play all the piano keys in every piece, only some of the genes get played in the cell types of each organ. Sonata in the Key of Kidney. The Heart Fantasia. Variations on the Theme of Brain.

Beautiful music, all of it. But think what can happen to a piano sonata if an important key on the piano sticks, or sounds the wrong note when struck. Such a flaw will ruin every passage in which that key is played. In some cases, it will destroy the music entirely. In the United States, one child out of every four thousand is born with cystic fibrosis, which is caused by a defective gene on chromosome 7. Children with this particular stuck key have abnormally thick mucus in their lungs, leaving them vulnerable to repeated infections that erode the lungs' tissues and eventually the ability to breathe. Most will die before their thirtieth birthday. In another gene, nothing more than a substitution of a T for an A causes sickle-cell anemia. Huntington's disease, a slow, inescapable meltdown of the brain, occurs because a gene near the top of chromosome 4 contains a series of repeated stutters on the letters CAG, playing them over and over like a scratched recording. The patient goes mad and inevitably dies.

Single-gene alterations account for some three thousand to four thousand other inherited diseases. Hard as these defects are to track down, they are by far the easiest targets for gene hunters. Most diseases, including such common killers as cancer and heart disease, stem from disruptions in the interaction among several genes and between genes and the environment. To find their causes, you have to first know what all the keys on the piano are. You have to know the whole genome. This knowledge will not lead directly to a cure for cancer and other killers, but by 1998 even the scientists who had originally opposed the Human Genome Project as ill conceived and not cost-effective were utterly convinced that its fruit would be well worth the $3 billion investment of taxpayer money. The major pharmaceutical companies were betting that it would lead to new drugs worth a whole lot more.

After leaving the St. Regis, Nicholas Wade tried to reach Francis Collins, the head of the government's genome project, whose official title was director of the National Human Genome Research Institute (NHGRI), one of the National Institutes of Health. Collins, more or less against his will, was on a plane headed from Newark to Dulles Airport, where he would catch a flight to Los Angeles. He had originally been scheduled to fly straight from Newark to LA, where he was due to give a lecture the next morning. But the evening before, he had gotten a phone call. The mere sound of Venter's voice on the line gave Collins a shiver of apprehension.

"Francis, I think you need to know about something we're about to announce," Venter said. "We have to meet with you right away."

"Who is 'we'?" Collins asked.

"I can't tell you that," Venter replied. "I don't mean to sound coy. I'm just not authorized to talk about it yet."

The two men had known each other for over a decade. When they were first introduced, Craig Venter was an obscure forty-year-old researcher in the National Institute for Neurological Diseases and Stroke, also part of the NIH. He had not attended college until after service in Vietnam, and thus had begun his career in academic science relatively late. Francis Collins was several years younger but well on his way to becoming perhaps the most famous gene hunter in the world. Soon he would claim a share of the credit for the discovery of five disease-related genes, including two of the most important ones yet found, those responsible for cystic fibrosis and Huntington's disease. In 1994, his success took him to one of the most prominent scientific posts in the country. At the age of forty-three, he was enticed by the secretary of health and human services to leave his large, heavily funded laboratory at the University of Michigan and come to Bethesda, Maryland, to assume leadership of the Human Genome Project, then in its fourth year. The job change meant a cut in pay and much less time for research, but Collins would be in charge of a $200 million annual budget earmarked for an historic enterprise. The decision was a no-brainer. "There is only one human genome program," Collins said when he took the post. "It will only happen once, and this is that moment in history. The chance to stand at the helm of that project and put my own personal stamp on it is more than I could imagine."

Now, four years later, Collins still saw himself as the captain of a great ship moving steadily toward its destination. It would be easy to imagine him at the helm, eyes trained on the horizon off the bow. A gangly six-foot-four, he had a long, broad face with features that competed for attention—nose and ears commandingly prominent, an ample but neat mustache, and sharp blue eyes magnified slightly by oversize glasses. He was in the habit of combing his hair forward, which gave him a kind of folksy, unpolished look. While outwardly easygoing—he often rode a motorcycle to work and occasionally played electric guitar with other scientists in a middle-aged rock band—there was a deliberate resoluteness in the line of his mouth and in the way he drove home a point with his chin up high.

The Human Genome Project was a command requiring a great deal of confidence and political will. To keep the money flowing into the program from Congress, Collins needed to constantly reassure lawmakers about the virtue of an enterprise whose costs were huge and whose payoff was distant and abstract—a perilous combination, especially in a Republican Congress. But overseeing how the money was spent required an even firmer grip on the helm. Big Science attracts big egos, and those leading the laboratories funded by NHGRI were some of the biggest around, all competing for the largest possible slice of Collins's considerable pie.

The managerial challenge had been woefully obvious at a contentious meeting of the leading project scientists in Bethesda the preceding December. The pilot sequencing projects were almost over, the full-scale attack on the human code was about to begin, and it was time to take a hard look at what the final phase of this fifteen-year project was going to cost. Since the project's inception, the price of the sequencing of the human genome had been estimated at around $1.5 billion, or about fifty cents for each of the DNA letters in the 3-billion-letter sequence. But would it cost more? Could it be done for less? Collins had called the meeting to decide the question communally. Unfortunately, the various genome centers were all competing with one another for the millions his institute was about to distribute to finance the sequencing—a situation made worse by the presence of some outside scientists who were likely to be sitting on the grant-review panel deciding on who got how much. Before long, an inverted bidding war broke out, as one scientist after another lowballed his cost estimates to show how he could sequence a base pair cheaper than the previous speaker. The numbers were extremely speculative, since DNA sequencing on a large scale had yet to be tried and the technology to do it was still evolving. The tone began to get nasty. Cries of "You're cooking your books!" and "You're lying!" bounded off the walls. The cost of human code hit rock bottom near the lunch break, when one scientist from a small genome center in Texas confidently declared that he could sequence DNA at the rate of eleven cents per base pair. Over the break, however, people began to realize that they were goading one another in a dangerous direction. If the Human Genome Project's work could be done so cheaply, what justification would they have for maintaining, much less increasing, the program's budget from Congress? The debate resumed, but now the price of a base pair began to rise. People voiced their concern about the new technologies: How can we be sure they are going to work? Do we really want to run the risk of cutting back on quality to save money, when the code of human life is at stake? Thirty cents per base pair was more realistic. No, better make that forty. Maynard Olson of the University of Washington, a passionate advocate of high-quality standards, declared that no sequence of human code should be called "finished" unless it contained no more than a single incorrect base pair in a string of half a million. That level of accuracy might take more time, and if amortized properly the cost might be as much as twenty *dollars* a base pair. The price would go down as technology advanced, but in the meantime, Olson declared, any sequence that did not cross that threshold should not be counted as done at all. "People will forgive you for being slow," he warned, "but they won't forgive you for being sloppy." But none of the other genome center leaders, and especially those running large operations that had invested most heavily in automated equipment, could possibly come so close to perfection. At the end of the day, the scientists soberly agreed that the current estimate for the cost of sequencing a base pair should be . . . fifty cents, where the bidding had started. The meeting then adjourned.

Presiding over such a scene as this ("the low point of the Human Genome Project," according to one of the participants), Francis Collins could only struggle to keep order and hide any panic he might be feeling behind his inextinguishable smile. For the most part, the problem lay not with him or his squabbling generals but in the philosophy of the Human Genome Project itself. One of its major arguments was that only the best and brightest in biomedical research should be in charge of something so essential to the understanding of human life and human disease. The way to ensure excellence was through NIH's competitive grant-based funding system, the mechanism driving the most spectacular discovery machine in the history of the world. But sequencing DNA is not discovery-driven science. Developing the tools and insights to capture the human code certainly takes great intelligence, even genius, but it is not experimental research in the traditional sense, where hypotheses compete to explain how nature works, leading ever nearer to the truth. It is more like a massive construction project, closer to building the pyramids than to finding a cure for cancer. But what was under construction was the biological essence of a human being. What single entity could be trusted with such a precious charge? The undertaking *had* to be collective. Given the decentralization built into the design of the project, the progress that had been made by the end of 1997 was respectable. But more than one scientist left that December meeting shaking his head, convinced there was no way the program would deliver a completed genome by the project deadline of 2005, much less anytime sooner.

In contrast to the politicking and infighting was the majesty of the enterprise itself. In Collins's words and under his leadership, the Human Genome Project had taken on an almost messianic quality, a noble, historic undertaking whose importance was "bigger than splitting the atom" and "dwarfed going to the moon." He referred to the genome in his public lectures as "the Book of Life." A part of such rhetoric was expediency, of course: if your program is dependent on taxpayers' money, lots of it, you must define your mission in the loftiest possible terms. But Collins believed devoutly in his own propaganda. He was a medical geneticist by training, and he was convinced that in time knowing the code would relieve untold suffering. He was also a man of faith, a born-again Christian. For Francis Collins, the human genetic code was part of the unimaginably immense knowledge of the Creator. Uncovering its sequence would provide a tiny, magnificent glimpse into the nature of God's mind. There were people within the genome community, Craig Venter among them, who had begun to rail at the slow pace of the Human Genome Project, but in Collins's view such an obligation should not be rushed. When you are uncovering the text of the Book of Life, your efforts must "stand the test of time."

Since the two men met in the late 1980s, Venter's career had taken a very different path. He, too, had made a mark for himself in the laboratory, first at NIH, where he had perfected a process that greatly accelerated the discovery of human genes. But rather than earn the respect of the exalted elite of molecular biology, the commercial success of Venter's method and his brazen public statements had provoked and appalled them, and the more exalted they were, the more appalled they tended to be. Then in 1995, three years after he left government science to found The Institute for Genomic Research, he announced another stunning achievement: the first revelation of the entire genetic code of a living organism, the bacterium *Haemophilus influenzae.* After this important but commercially unrewarding research, many academics forgave Venter his former transgressions, and TIGR was one of six genome centers receiving support for human sequencing from NIH. But Francis Collins did not trust him. After Venter's mysterious phone call, the government genome leader was preparing himself to be greatly and painfully appalled.

Reluctantly, Collins agreed to change his Los Angeles flight to accommodate a meeting in a private room of the Dulles Red Carpet Club, the business-class lounge of United Airlines. He called some of his staff at NHGRI and asked them to attend, and when he arrived they were already there, looking apprehensive. Venter walked in with two other scientists a few minutes later. One of them was no surprise: Mark Adams, a molecular biologist in his early forties, had been a mainstay in Venter's labs for ten years, the two working so closely together that people joked that they were joined at the hip. The other scientist was Michael Hunkapiller—"Craig's *mystery* guest," as Collins later phrased it, with an emphasis on "mystery" that managed to sound jaunty and disapproving at the same time. It was well known that Hunkapiller's Applied Biosystems had a potentially groundbreaking sequencing machine in development. As soon as Collins saw him come into the room, he understood what all this was about. Venter quickly dispelled any doubt.

"Francis, we want to give you a heads-up on an announcement we're about to make, so you don't get blindsided in the newspapers," Venter began. His tone was quiet, almost deferential, but his words were not. "We don't think people want to wait another seven years for you to finish the genome. Perkin Elmer and I have teamed up to form a new company. Our goal is to do it ourselves, using a couple hundred of Mike's new sequencing machines. We are going to make the genome free and available to everyone, same as you. The main difference is, we estimate we'll be done in 2001, four years ahead of your schedule."

Venter then described how this extraordinary feat was to be accomplished. The strategy being used by the government to conquer the immensity of the human genome was to break it down into manageable chunks, then puzzle out the letters in each chunk using DNA-decoding technology already on the market. Venter's enterprise would have two key advantages: the arsenal of Hunkapiller's new, much faster decoding machines, and a blitzkrieg sequencing strategy called "whole-genome shotgun." Essentially, the technique would blast the genome into tens of millions of tiny pieces and, after the DNA letters in each piece were spelled out on the machines, reassemble them all in the correct order using enormously powerful computers and algorithms.

While Venter talked, Collins sat rigidly, his face betraying none of the emotions roiling his thoughts. The initial shock, like being kicked hard from behind, lasted only a moment. Taking its place was a sort of galled incredulity, like that of a pregnant woman informed that someone else plans to carry her baby and deliver it to term in three months instead of nine. The notion was both offensive and preposterous. Even if the new machines could be built in time and worked as well as anticipated—and the history of automated DNA-decoding technology was riddled with revolutionary breakthroughs that proved to be flops—the whole-genome shotgun technique was grossly inadequate to the monumental task at hand. Venter had used the method to decipher tiny bacterial genomes, but the best experts in the field of computational biology had already shown that the human genome was far too large and complicated to be accurately assembled in such a quick-and-dirty fashion. Yet Collins was deeply apprehensive, too. When the congressmen who controlled NIH's budget read about Venter's enterprise in the newspaper, they would not understand that the technique was unworkable.

"Craig, you were at the meeting in Bermuda when Phil Green showed why whole-genome shotgun won't work," Collins said. "Given the size and structure of the human genome, it's logically impossible."

"We don't know if it's going to work until we try it, and neither do you," Venter replied. "Who knows, we could fall on our ass. That's one reason why we strongly support the continuation of the public genome program. We want to coordinate our efforts with yours and collaborate in any way possible. There is plenty of work to go around for everybody. The mouse genome, for instance, is just as important for research and biomedicine as the human, and having them both would be infinitely better than having just one or the other."

Francis Collins had never seen Craig Venter in a modest mood. But what Venter said next left him at a loss for words. "So while we do the human genome," Venter continued, in an offhand, casual sort of way, "you can do mouse."

On the following Sunday morning, Eric Lander picked up the *New York Times* lying on his front porch in Cambridge, Massachusetts. Lander was the director of the Genome Center at MIT's Whitehead Institute and a prominent player in the Human Genome Project. A former Rhodes Scholar and Princeton valedictorian, at forty-two Lander was well into his third career. Previously he had been an economics professor at the

Harvard Business School, and before that a highly respected mathematician. He had learned molecular biology more or less in his spare time, but well enough to win a MacArthur award before he was thirty.

Lander had returned just the night before from a weeklong trip to Israel, where he had taught a seminar on genomics. Groggy from jet lag, he put the newspaper on the kitchen table and made some coffee. Then he saw the headline. "Scientist's Plan: Map All DNA Within 3 Years," it read. The lead noted that if Venter's enterprise was successful, it "would outstrip and to some extent make redundant the Government's $3 billion program to sequence the human genome by 2005."

Lander was surprised by the development but not really shocked. "There you go again, Craig," he said to himself. As he read on, it was the slant on the news, not the news itself, that roused him angrily from his drowsiness. According to Nicholas Wade's article, the Human Genome Project was halfway through its fifteen-year course, but only 3 percent of the genome was sequenced. It sounded as if the project was way behind its own schedule, which in Lander's opinion was grossly unfair, since a huge amount of the project's funds had been used for the preliminary work needed before the actual sequencing could commence. A few paragraphs later, the reporter hinted that if Venter's collaboration with Perkin Elmer worked, Congress might consider closing down the government program altogether. "Right," Lander thought. "And when Craig issues a press release saying he's cured cancer, we should shut down the National Cancer Institute."

Most disturbing of all was the response attributed to Francis Collins. Though the HGP leader was not quoted directly in the article, Wade implied that both Collins and Harold Varmus, the director of NIH, were considering abandoning the human genome to Venter and switching the public program's focus to the mouse genome and those of other mammals. Lander did not believe it; Collins and Varmus would never capitulate so easily. He thought of the 170 scientists and technicians who worked in his lab waking up and reading the same news. With his family still asleep, he got dressed in a hurry and left for the Whitehead Institute, hoping to reassure anyone who wandered in reeling from the shock.

A few hours later, in Berkeley, California, a bearded, cherubic geneticist named Gerry Rubin padded down his drive in the hills overlooking San

Francisco and picked up his own copy of the *New York Times.* Rubin was a half step removed from the human genome community. Although Collins's National Human Genome Research Institute funded his research at Berkeley, he worked not on human genes but on the DNA of the fruit fly *Drosophila melanogaster,* the same little insect that dances around your bananas on a summer afternoon. The fruit fly had been a mainstay in genetics for a hundred years, and over that time drosophilists like Rubin had contributed more to the understanding of the fundamental nature of inheritance than anyone working on human genes. The reason is simple: DNA works basically the same in all organisms, and fruit flies are a lot easier to study than human beings. If you disable the function of a fruit fly gene and insert a functioning human counterpart in its place, the fly will likely run just as well as if you'd taken a tire off a Mercedes and put it on a Ford. But unlike humans, flies mature and breed within a week of their birth, which makes it easy to follow patterns of inheritance through successive generations. They also don't complain when you need large quantities of their DNA and you grind them up in a blender to get it.

The drosophilists formed a relaxed, far less competitive community than that of their colleagues who buzzed around the fatter fruit of the human genome. Like Eric Lander, Rubin noted the slighting of the public human genome program in Wade's article, and he was sure it was going to cause trouble. But he knew Craig Venter, and could not help admiring what he was attempting to do. Rubin had taken time off from his regular research to head the Drosophila Genome Project, an international effort to sequence the fruit fly's genetic code, and he was well aware of how complex a task tackling an entire genome was. His project was about a fifth completed and was scheduled to take another three years. The human code was thirty times larger than that of the fly, and Venter was promising to complete it in the same time frame. If anyone could pull off such a technical triumph, he thought, it would probably be Venter. Imagine what he could do with *Drosophila*!

Back in Cambridge, Robert Millman, a patent attorney, read Wade's story the following afternoon. He was sitting with his long legs up on the desk in his office at Millennium Pharmaceuticals, a biotech firm that Eric Lander had co-founded, located just a few blocks from the Whitehead Institute. Millman's job was to file gene patents and in other ways secure intellectual property rights for the company. Whatever you are imagining a patent attorney to look like sitting at his desk, throw the picture away. Millman was a tall, gaunt man in his mid-thirties with a scraggly red beard and long red hair tied back in a ponytail, his eyes deeply set and very bright. He looked less like a lawyer than a Viking seer, except that Viking seers do not wear loose-fitting black bowling shirts with "Hoss" inscribed over the pocket and "Ace High Plumbing (Home of the Royal Flush)" embroidered across the back. That day, Millman was complementing the bowling shirt with a pair of bright yellow pants, lime-green-and-blue striped socks, and crimson Hush Puppies.

Whatever you think a patent attorney's office looks like, throw that image away, too. Millman had crammed his tiny windowless space full of "cool stuff." Crowding the bookshelves was a tiny butter churn, a miniature lard boiler, a perforated funnel, and a collection of other inventions from the late nineteenth century, when patent applicants were required to submit a scale model to the Patent Office. From Millman's ceiling hung a couple of ghoulish puppets with long pointed ears and pointed teeth. One had a bag over his shoulder and was grinning slyly, like a thief. The other was dangling beneath a hang glider. Millman liked to hang glide, too.

In his extraordinarily long-fingered hands was a collection of news articles relevant to the biotech industry. Millennium Pharmaceuticals' librarian delivered them to his office every day at noon. Wade's article was on the top. Millman read the story through, noting the concerns expressed in it by some scientists and ethicists over the notion that a single private company might gain possession of the entire code of human life. He took some notes for a memo to his boss on how the announcement might affect Millennium's business strategy. Then he flipped to the next article in the stack. Before he started reading, a thought flashed through his head. Wouldn't it be amazing to patent the entire human genome?

Cool.

The *Wall Street Journal* and other major dailies published their own stories that day on Venter's announcement. In Tucson, M'Liz Robinson, a recent MBA from the University of Arizona, circled a copy of the *Jour-*

nal's article with a red magic marker and placed it squarely in the middle of her husband's desk, where he couldn't miss it. Eugene Myers was a mathematician at the university. Working with a little-known medical geneticist named James Weber, Myers two years earlier had proposed that in theory the human genome could be assembled by the whole-genome shotgun technique. When Weber first aired the theory at a meeting of the HGP, it had been very badly received. A few weeks later, he and Myers submitted a paper for publication in *Genome Research.* The journal accepted the paper, but published it with a damning rebuttal written by computational biologist Phil Green of the University of Washington, a highly influential member of the HGP community. Green pointed out a long list of flaws and shaky assumptions in the Weber-Myers approach. In short, their paper was toast as soon as it came into print. This was all old history by now. But Myers had never really gotten over his bitterness about the way the theory had been treated.

The *Wall Street Journal* article made no mention of whole-genome shotgun, but Myers's wife could read between the lines that Craig Venter was going to attempt to put the theory, or something like it, into practice. When her husband came home from work, he went into his office, put on his headphones for some background music, and got directly to work. He noticed the newspaper with the article circled in red but pushed it aside along with everything else on his desk. Two days later his computer chimed with an incoming e-mail. A colleague in Texas was sending congratulations: Craig Venter was going to use the Weber-Myers technique to try to assemble the human code. This time the message got through. Myers scrambled through the papers on his desk, looking for the *Wall Street Journal* article. He, too, could read between the lines. He started to think. His old resentment over the way his proposal had been treated welled up again. But this time there was a mind-arousing edge to his brooding. It got to be late. His wife said good night, but Myers did not answer. He stayed at his desk, thinking, scratching formulas on a pad, tapping keys on the computer until after one. When he woke up the next morning, he called a colleague, Granger Sutton, who worked for Venter at The Institute for Genomic Research. "Hey, Granger," he said. "I hear you guys are going to try to shotgun the human genome. Can I play, too?"

CHAPTER 2
THE SECRET OF LIFE

Ego is a common propellant in science. Its importance is implicit in the tenet of modern scientific method that a field advances when experiments refute accepted theories, not confirm them. The more important a discovery you make, the more ego you will need to combat the battle-tested egos of the people whose ideas were once grand and new but now are old and threatened. The big questions tend to encourage ego growth. Around a very great question—the identity of the first human ancestor, let's say, or what causes AIDS—so much ego gathers and gums up the works that progress might actually be impeded. But when so much is at stake—lives, fortunes, essential truths about the laws of nature—it is probably better to err on the side of arrogance than humility.

A good illustration of where too much humility will get you is the case of Gregor Mendel, the nineteenth-century Austrian monk who is widely considered the father of genetics. Mendel had no awareness of the honor that would be bestowed upon him. He died sixteen years before anyone understood what he had discovered. Born in the tiny village of Heinzendorf in 1822, Mendel was the son of a tenant farmer. He spent much of his adolescence in bed with a mysterious illness that today might be diagnosed as acute anxiety. He was good at mathematics, but after one entire year languishing under the covers watching any hope of attending university drain away, he took the advice of a teacher and signed on to become a friar. It was the only other route available for the son of a destitute peasant in search of an education.

The Augustinian monastery in the Moravian city of Brünn (now Brno, in the Czech Republic) was an ideal haven for a young man with an interest in natural science and little forward momentum in life. But Mendel was expected to perform outside its walls, too, and there he continued to disappoint, twice failing the examination to become a high school teacher. The second time he was so undone by the stress of the exam that he left most of the questions blank. Back to the monastery he went, where the abbot wisely nourished his interests in plant breeding and mathematics, and otherwise expected little in return other than the hope that Brother Gregor's fondness for God would not be completely overshadowed by his apparent fondness for good food. He was thirty-one years old, stout, amiable, and seemingly bound for insignificance.

Mendel developed a deep curiosity about the breeding of plants, especially the common pea species *Pisum sativum.* The abbot supplied him with an ample garden to grow his peas, which the celibate monk took to calling "my children." Mendel observed that in some features, pea plants came in two distinct varieties. Some had tall stems, for instance, and others short. Some had yellow seeds, and others green ones; some had smooth pods, and on others the pods were constricted. One might see the same variation in human beings and other species, too, of course: people can have either blue eyes or brown, some are right-handed and others left-handed, and so forth. Mendel, like a lot of other people, wondered why this was so. Unlike others, he devised an experiment to find out. Eight years and 30,000 pea plants later, he found the answer.

He began his experiment by producing lineages of peas that "bred true" in particular traits: if a pea plant had yellow seeds, for instance, then plants grown from that seed would also have yellow seeds, and so did the plants of the next generation, and so on. Once he was sure that his plants would breed true, he began to crossbreed the types—a tall plant with a short one, for instance, or a plant with yellow seeds with a green-seeded one. The prevailing wisdom at the time was that the characteristics of the parent plants were "blended" in their offspring: a tall and a short pea plant would produce a medium-size one, for instance; a yellow-seeded one and a green-seeded one would produce one with yellowish green seeds. But this is not what Mendel observed. In each trait, the daughter plant resembled one of the parents exactly, while the contribution of the other parent seemed to have vanished. For instance, all of the crossbreeds between yellow-seeded and green-seeded ones had yellow seeds. When Mendel crossed these hybrids with each other, however, he found that the missing trait reappeared in roughly one quarter of the second generation of plants: three yellow seeds to every green one.

The reappearance of the missing trait led Mendel to an insight that should have been galvanic. He reasoned that each parent plant was imparting to its offspring a discrete "factor"—some physical instruction, so to speak, to tell the plant what color seeds to have, or whether to grow tall or short. Each plant carried two of these messages for each character trait, but transmitted only one. When they combined in the new plant, one parent's version was always dominant over the other. If one parent contributed the trait for yellow seeds, for instance, then it didn't matter what the other bestowed on its offspring—the resulting plant would grow up with yellow seeds. Only when both parents happened to bequeath the "recessive" version of the trait would that trait be expressed in the next generation. Three to one.

Mendel had enough ego to know that his experiments proved something very important. Where his confidence failed him was in broadcasting this discovery to the world beyond the monastery's walls. By 1865, he felt he had accumulated enough evidence to present his theory to the Brünn Society for the Study of Natural Science. Forty people attended the first of two lectures, eager to learn about Brother Gregor's hybridization experiments and their implication for agriculture. But the lecture was a disaster—or worse, a non-event. Enraptured by the mathematical logic of his results, Mendel flooded his audience with statistics, numbers and ratios, chances and probabilities. He did not begin by declaring "I have found the secret to heredity." He began instead by observing that "the striking regularity with which the same hybrid forms always reappeared whenever fertilization took place between the same species induced further experiments to be undertaken, the object of which was to follow up the developments of the hybrids in their progeny." His audience was full of dinner and wine, and eyelids began to droop, heads to nod. A couple of hours later, after a tremor of polite applause, the worthies of Brünn gathered their coats and filed out, without a clue that they had just had a glimpse into the logic of creation such as no one had ever been given before.

Mendel was disappointed, but he pushed on nevertheless, publishing his results in the society's journal. He sent copies of his "Experiments in Plant Hybridization" to some of the leading scientists of Europe, including Charles Darwin. He suspected that his theory was of great importance to the understanding of evolution, and he eagerly awaited Darwin's response. It never came. Perhaps if Mendel had put more punch in his title, or talked about the relevance of his experiment in the first few paragraphs instead of how much labor it took to plant all those peas, he would have aroused more curiosity. But the only scientist who responded at all was the botanist Karl-Wilhelm Nägeli in Munich, and he missed the point of what the monk was trying to say. If he'd understood Mendel's argument, he would have realized that some of his own experimental hybridizations confirmed it beautifully. Instead, Nägeli needled Mendel to replicate his work using the hawkweed plant. Unbeknownst to either man, hawkweed happens to follow an aberrant pattern of inheritance. The experiments failed utterly. Mendel gave up, spending his last years as the abbot of the monastery, involved not in science but in an increasingly nasty feud with the authorities over tax policies on monasteries. The copy of his paper sent with such hope to Darwin is still in the Darwin library. In those days, a published manuscript's pages had to be slit apart with a letter opener before they could be read. Darwin never even cut the pages.

They might have been worth the read. Darwin accepted the received wisdom that inheritance was a blending of the contribution from the parents—that's why people look somewhat like their mother, somewhat like their father. But he was uncomfortable with the notion, because it did not sit well with his theory of natural selection, which he had the ego to know was right. The theory of evolution by natural selection argued that those individual organisms innately well-adapted to their environment would be more likely to pass their advantages on to the next generation than those who were less "fit." But if the qualities of each parent were blended together in each successive generation, how could any new advantageous trait take hold in a population through time? The individual with that superior trait would necessarily mate with someone without it, diluting its expression by half in their offspring, then half again in their grandchildren, and so on until it faded away completely.

Darwin was stumped for an answer to this contradiction in his theory. If he had slit open the pages of the treatise sent to him by the humble

Austrian monk, he might have found a resolution to the dilemma. An organism isn't the result of the blending of its parents' essences at all, but the composite result of lots and lots of individual traits inherited from one parent or the other. Mendel called them "factors." We call them genes.

In 1899, no one had heard of Gregor Mendel. A year later, he exploded into a transatlantic phenomenon. In the span of two months in the spring of 1900, three scientists working in three different countries grasped the implications of his paper and cited it in their own. One of their papers caught the attention of English scientist William Bateson, who quickly had Mendel's article translated into English. He ferociously championed the dead monk as the forgotten father of a new science that Bateson called "genetics." He coined the word in 1905, from a Greek word meaning "origin" or "fertile." The word "gene," for the atomlike particle of inheritance implied by Mendel's experiments, came into use a few years later. For the next half century, nobody had any idea what a gene was made of, what it looked like, how it worked, or even if it existed for sure. But given the simplicity of the early geneticists' tools—milk bottles full of fruit flies, low-power microscopes, and some prescient guesswork—what they were able to accomplish was astonishing.

Prescience had a particularly good year in 1902. In *The Lancet,* an English physician, Archibald Garrod, put forth the notion, years ahead of his time, that genes were essentially instructions for making enzymes—the proteins that carry out all of the body's functions. Garrod had been working on a rare, not very serious medical condition called alkaptonuria. One of its symptoms, harmless but horrifying, is that an afflicted person's urine turns black on exposure to air. He knew that the reason for the black urine was an excessive buildup of the chemical homogentisate, and reasoned that some enzyme meant to break down homogentisate wasn't doing its job. Garrod noticed that the majority of his patients with the disorder had parents who were first or second cousins. Garrod knew his Mendel. Could the condition be a recessive trait, like short-stemmed pea plants, inherited only when neither parent contributed the normal, dominant ability to break the enzyme down? If so, then perhaps the function of each of Mendel's "inherited factors" was to make a specific enzyme. Though no one much noticed it at the time, Garrod had founded the field of human genetics and hit upon what decades later would become "the central dogma" of molecular biology:

"one gene = one enzyme (or protein)." On the genomic piano, each key plays a single note.

Just a year after Garrod published his study suggesting what genes do, American graduate student Walter Sutton and German scientist Theodor Boveri independently proposed a theory of where the genes might be found. They noted that chromosomes, the stringy objects inside cell nuclei, usually came in pairs. The exceptions were egg and sperm cells, where the number of chromosomes was reduced to only one member of each pair before fertilization took place. This process of reduction, called meiosis, ensured that each parent contributed only one copy of each chromosome, so the newly formed individual began life with the full complement of two each. Mendelian traits behaved the same way, so it stood to reason that they were associated with the chromosomes. But human beings had only twenty-three pairs of chromosomes. How could all the information needed to make a living being be freighted through to the next generation on so small a number of vehicles? How could all the variation one sees between one individual and another erupt from so simple a scheme?

Thomas Hunt Morgan, a Kentucky-born American aristocrat whose great-grandfather, Francis Scott Key, had composed the national anthem, was one of many scientists who looked on the chromosome theory with a great deal of skepticism. But he eventually proved that it was true, in the process launching a dynasty in genetic research. Morgan was the original drosophilist. He started using fruit flies in his genetics laboratory at Columbia University in 1907 because they took up less space than the pigeons, chickens, starfish, and rats hauled in for his other experiments. When he realized how much easier it was to study evolution and inheritance in flies than in those bulky beasts, half-pint milk bottles full of flies began to take over the lab. It became known famously as "the Fly Room."

In his initial experiments, Morgan's modus operandi was to subject his flies to various indignities—extreme heat or cold, X rays, chemical injections into their genitals, and so on—in hopes of producing mutations whose pattern of inheritance he could then trace through successive generations. In 1910, after two years of inflicting these insults, he and his students were rewarded with a fly with white eyes—an aberration, since the thousands of previous fly eyes he'd stared into were red. They bred the white-eyed fly, which was a male, with over a thousand virgin red-eyed females, and every one of the children had their mother's red eyes. When they bred the progeny with each other, however, they noted that while all the female grandchildren had red eyes, half the males were white-eyed, like the original mutant.

Morgan had also noticed that in male fruit flies one of the four pairs of chromosomes wasn't really a match: one was bigger than the other. We now know these as the X and Y chromosomes. The discrepancy is found only in males, who inherit an X from their mother and a smaller Y from their father; females get an X from each parent. Morgan reasoned that the pattern of inheritance he had observed made sense if a Mendelian gene for eye color resided on the X chromosome contributed by the female parent. All the females of the second generation had red eyes, because at least one of their two Xs was issuing the instruction "color the eyes red," and red was the dominant trait. But for the males, who had only one X chromosome, eye color was a toss of a single coin: red if they had inherited a normal X from their maternal grandmother, or white if their X had been passed down from their white-eyed grandfather. Morgan could point to an X chromosome and say, "There lies the gene that determines the sex of this fly, and there too the one that says its eyes shall be red or white." Genes were real, and the ones for eye color and sex were linked.

The eureka from the Fly Room can still be heard nearly a century later. But it was just the opening chord in a symphony. Morgan and his students began finding other mutants in the milk bottles. One male fly had stubby wings, and that trait too was linked to the X chromosome. It stood to reason, then, that every fly inheriting an aberrant X chromosome from its mother should have white eyes and stubby wings—a double mutant. Weirdly, however, Morgan found that some of the offspring born with an affected X chromosome were only single mutants—white-eyed flies with normal wings, or red-eyed ones with short wings. How could this be? If the X chromosome carried the genes for both traits, how could they split up and go their separate ways in the next generation?

In 1911, Morgan found the solution to the dilemma under the microscope. In most cases, when a cell divides, its chromosome pairs are copied to make four of each chromosome, so that both of the daughter cells can get a full complement of pairs. But remember, the sperm and egg cells get only one set of chromosomes. In the specialized cells that produce the sperm and egg, Morgan noticed a brief sinewy dance taking place between the partners in each pair of chromosomes. The two Xs in a female, for instance, would find each other and intertwine, like two strands of pearls coiling around each other to make a double-stranded necklace. In a flash of insight, Morgan reasoned that this fleeting braid held the answer to his puzzle. Imagine that one strand of pearls was white and the other black. While the two strands were twirled around each other, a stretch of white pearls was swapping places with the black pearls touching it. When the two X chromosomes parted, they were no longer pure white and pure black, but each was a mix of both, and each was the sole X in a new sperm or egg cell. An X chromosome, or any other one, did not stay intact as it passed from generation to generation but was reconstructed each time as a mix of the two Xs the organism had inherited from its own parents. So it was that a fruit fly could inherit a white pearl determining its eye color but a black pearl affecting how long its wings would grow.

The discovery of this phenomenon, called crossing-over, was a big reason that T. H. Morgan won the Nobel Prize in Physiology or Medicine in 1933. Chromosome pairs in all creatures, not just fruit flies, perform this intimate exchange before parting to carry on the lineage. It is the reason why you look somewhat like your brothers and sisters but not exactly like them. Every chromosome in a reproductive cell is a unique mosaic of the traits of the parents, and their parents before them, and so on. The myriad meioses that produced your father's sperm and your mother's egg were each a new act of conjoining between your grandparents, even if they were dead at the time the crossover happened. Just so, if you have children and they, too, have descendants, your own genetic legacy scatters and mixes down in time through a series of tradings between the partnered chromosomes in each new generation.

One afternoon shortly after Morgan had this insight, Alfred Sturtevant, an undergraduate in his lab, took home with him the lab's breeding records. He reasoned that if Morgan was correct about "crossing over," then it might be possible to plot the location of two genes on the X chromosome relative to each other by noting how frequently the two traits were inherited together. If two genes were near each other, then it was unlikely that they would be separated in the reshuffling of meiosis, and they would be inherited in tandem. On the other hand, the farther apart they lay on the X chromosome, the more often they would be separated during crossover, and the more often the offspring born from the mating of a fly showing two traits would have only one trait and not the other.

Sturtevant pored over the inheritance records of the white-eyed trait and four other mutations the lab had found on the X chromosomes. Through the night, he worked out mathematically how often each was inherited with every other one. By dawn the next day, he had produced a linear arrangement of the genes. It was the first genetic map. Morgan later called it "one of the most amazing developments in the history of biology." Sturtevant was nineteen at the time.

Over the next five years, using the sole experimental method of crossing fly with fly and tallying up the traits of the progeny, Morgan and his students laid the groundwork for the subsequent understanding of heredity. Hermann Müller, for instance, began a quest to show that genes were not abstract inferences but solid objects—just like the pearls on the black and white necklace. Later, he contributed the profound discovery that individual genes could be artificially mutated in the laboratory by bombarding fruit flies with ultraviolet light or X rays. This proved that they were not static entities but had some kind of internal structure that could change. Müller believed that his transformation of fruit flies would lead eventually to the controlled transformation of the human species. It hasn't yet, but it did set the stage for the understanding of cancer as a transformation in genes.

Hermann Müller was a brilliant scientist, but he did not have the gene for social grace. Unlike the other famous students of Morgan, who stayed with him for their entire careers, Müller fell out with his mentor and in 1920 moved to the University of Texas. There, suffering from a collapsing marriage and the conviction that others were stealing his research results, he attempted suicide. He left Texas later for Germany, arriving just as the Nazis were coming to power. After watching them smash to pieces the laboratory of his employer because of his tolerance of Jewish scientists, Müller headed east again to the Soviet Union. An ardent socialist, he unfortunately arrived just in time to witness the rise to power of the rabid antigeneticist Trofim Lysenko, who preferred having his opponents tossed in prison or shot to refuting their arguments with experimental results. Müller wandered on, participating in the Spanish Civil War and teaching in Scotland while he waited for some university in his native America to invite him back home. He was finally given a post at Amherst College, and in 1945 moved to Indiana University. The following year, he won the Nobel Prize in Physiology or Medicine for his work on the mutation of genes with X rays.

Not long afterward, Müller's presence there attracted a brash nineteen-year-old college graduate named James Watson. Among many bright young people of his generation, Watson had been inspired by a tiny, powerful book by Erwin Schrödinger entitled simply *What Is Life?* A physicist, Schrödinger approached that hefty question from the bottom up. The chromosomes, he theorized, must contain "some kind of hereditary code-script" that dictated "the entire development and functioning of the individual in the mature state." Given size constraints, the genes making up that code had to be composed of some kind of large molecule. But *what* molecule? What are genes?

Most scientists were convinced that the hereditary molecule was a protein. But the young Watson believed there was a better candidate. Way back in the 1860s, a young Swiss chemist, Friedrich Miescher, examining the nuclei of pus cells he obtained from the discarded bandages of a local surgical clinic, isolated a new kind of biochemical substance that was neither protein, fat, nor carbohydrate. The chemical was acidic, rich in phosphorus, and made up of very large molecules. He called it "nuclein." It was later renamed deoxyribonucleic acid, or DNA for short.

For most of the first half of the century, scientists considered DNA to be a "stupid molecule," like fat and carbohydrate, composed of chains of four chemical subunits repeated over and over again in the same order, like an immensely long package of Life Savers. Proteins were the smart, active molecules. Their subunits, called amino acids, varied in order. It stood to reason, then, that proteins were the only molecules that could carry the information content of a genetic instruction, in the same way the letters in the phrase "Fire! Run for your lives!" carry essential information, while "Blah! Blah blah blah blah!" communicates nothing at all. In 1944, however, Oswald Avery, a scientist at the Rockefeller Institute in New York, showed that a harmless strain of the *Pneumococcus* bacterium could be genetically transformed into the strain that causes pneumonia merely by picking up a stray bit of DNA from the virulent strain and incorporating it into its own chromosome. Avery was reluctant, however, to claim boldly that DNA must therefore be the stuff of genes, and in his papers he buried this hot idea under cold blankets of qualifying caution. Perhaps he, too, lacked sufficient ego.

James Watson, however, did not. In 1951 he arrived at the Cavendish Laboratory at Cambridge University, a bony, brainy young man with popping eyes, a hefty libido, and the mad hair of someone who sleeps under a log. He had a grant to study proteins, and every intention of going after DNA instead. On his first day at the Cavendish, he met Francis Crick, a thirty-five-year-old scientific wastrel with a quick mind but no status or achievement. Crick, too, had read *What Is Life?,* and he, too, suspected that DNA was the molecule carrying Schrödinger's "code-script." He and Watson hit it off immediately.

The result of their meeting was the discovery of the double helical structure of DNA. In the eighteen months it took them to accomplish this feat, Watson and Crick did not conduct a single experiment. They did most of their work using each other as their primary lab instrument: They talked about the problem at their desks, in the tearoom, over beers in local pubs, and walking around Cambridge. They read papers, sketched out ideas, built models, and watched closely what was going on in a laboratory in London headed by a physicist, Maurice Wilkins, who was already working on DNA. They were especially interested in the work of a woman in Wilkins's lab named Rosalind Franklin. She was skilled in the highly difficult technique of X-ray crystallography, which revealed hints of the three-dimensional shape of a molecule by the way a crystal of the substance scattered X rays across a photographic plate.

Franklin did not care much for Watson and Crick—or for Wilkins, for that matter—and they did not care much for her. Science was considered a man's world in postwar England, and a sharp, stubborn-minded woman who did not bring tea to the gents or spread open her lab notes at their request was a bit of a bother. One day Wilkins, without Franklin's permission, showed Watson a particular X-ray photograph she had produced of DNA.

"The instant I saw the picture my mouth fell open and my pulse began to race," Watson later wrote in *The Double Helix,* his brazen account of the quest to find the structure of DNA. He and Crick had suspected that the structure was some kind of helix, and the picture confirmed it, at the same time providing some key dimensions of the molecule. Within weeks they had refined their thinking even more: it was not a single helix, like a metal spring, but two helices spiraling around each other, like mating snakes. Another essential clue came from the work of Erwin Chargaff at Columbia University. It was known that DNA was composed of phosphorus, sugar, and four nucleotide bases called adenine, thymine, guanine, and cytosine (the As, Ts, Gs, and Cs so often referred to in this book). Chargaff had discovered that these four bases were not arranged in a monotonous repeating pattern with an equal number of each one, as had been thought. But in every piece of DNA, the number of As was the same as the number of Ts, and the number of Cs equaled the number of Gs. Watson and Crick reasoned that this must be because the As on one strand of their double helix were glomming on to the Ts on the other strand. The Cs and Gs likewise were complements.

After more talk and a few false starts, Watson and Crick built a model of DNA that looked much like a twisted rope ladder, the rungs between the ropes formed by the joined base pairs, A to T and G to C. They realized that this simple design would make the copying of DNA from one generation of cells to the next effortless and inevitable. During cell division, the two strands pull apart. The bases on these single strands quickly gather up their complements from the supply of free nucleotides floating around in the cell's broth of chemicals: the Ts on the strands grab As, the Cs grab Gs, and so on, until each separated strand has reconstructed an exact replica of its lost partner to form two new, identical double helices. With equal simplicity, the structure also revealed DNA to be the "code-script" of the gene. The four nucleotides function like letters in an alphabet, their sequence arranged into "words" that the cell can understand and translate into all the proteins needed to make an organism grow and live. Far from being a stupid molecule, DNA is the most brilliant chemical on the face of the earth. On the last day of February in 1953, Francis Crick burst through the door of their favorite pub and announced to all that he and Watson had found "the secret of life."

CHAPTER 3
DOWN BUNGTOWN ROAD

To a molecular biologist, Cold Spring Harbor Laboratory, about thirty-five miles from New York City on Long Island's North Shore, is sacred ground. In the 1940s and 1950s it served as a summer gathering spot for the early prophets of the science—reductionists like Watson who believed that the answer to "What is life?" could be found only by venturing down to the bottom of the cave, where life met the physiochemical bedrock of nonlife. Under the inspired guidance of the physicist Max Delbrück, they probed the genetic interactions between bacteria, the lowest forms of life, and their tiny parasitic viruses, called bacteriophages, which might be thought of as the highest form of nonlife. The celebrated phage course taught by Delbrück nurtured the birth and early development of molecular biology and the societal revolution that flowed from it: recombinant DNA, genetic engineering, DNA synthesis and sequencing, gene therapy, DNA forensics, and the molecular approach to understanding the mechanisms of disease.

The lab is no longer just a summer retreat. But as soon as you turn off Route 25A onto Bungtown Road, the lane running through the center of the campus, you feel as though you'd stumbled into a secular monastery, protected from the noise of the world by rustic stone walls and the long shadows of pines. The lane passes by ponds, meadows, and trickling brooks crossed by wooden footbridges. There are some new brick buildings, but the same old clapboard labs and dormitories from that simpler era still cling to the hillside above Long Island Sound. Flagstone terraces, softened by moss and shade, invite high conversations in low voices. The parking lots are grass.

Of course, there are no crosses, church towers, or other trappings of conventional religion. But science has its icons, too. In a courtyard set into the slope of a hill there is a rectangular brick bell tower. High up on the four sides of the belfry are four granite slabs, and carved into the slabs are the four letters A, T, G, and C, the cardinal points of the molecular compass. In the lobby of the main lecture hall, a golden double helix of DNA rises from the floor like an idol, taller than a man. In the hallways and stairwells hang photographs of the original apostles of the new science: Delbrück himself, Salvador Luria, Crick and Watson, Barbara McClintock, Jacques Monod, Alfred Hershey—magical names to a molecular biologist, beneath grainy, dreamy images of postwar scientists at work and play, forever young and cocksure, their eyes bright from the birth of ideas that will take their older, grayer selves to Stockholm.

At a meeting on this hallowed campus in 1986, an emergency session was called to gauge reaction to a proposal by the Department of Energy to uncover the entire sequence of the human genome. That such a feat would someday be attempted had been seen as a distant possibility for a decade, ever since the British scientist Frederick Sanger had discovered a method to read out the order of base pairs in a small sample of DNA, a few hundred letters at most. By 1986, scarcely a hundred human genes had been decoded. In the oft-quoted phrase of Nobel laureate Walter Gilbert of Harvard, the human genome was the Holy Grail of biology. The benefits would be enormous. Having the complete sequence of the genome would make it possible to locate and identify all of the estimated 100,000 genes guiding the construction of a human body and brain. Instead of years of arduous trial-and-error experiments, researchers looking for genes involved in human disease would have 90 percent of their work done for them already, freeing up time that could be better spent looking for ways to fix the errant gene and bring about a cure.

Most of the scientists at that 1986 meeting protested, however, that the time was not right for such a gargantuan undertaking—and certainly not under the auspices of the DOE. The human genome consisted of 3 billion letters, most of which were so-called junk, possibly without any biological purpose at all. Why waste time and money on such a fruitless enterprise, when the technology to do it efficiently didn't even exist yet? Why not focus on finding just the genes themselves first? From the podium, Gilbert pointed out that at the current rate of progress in DNA sequencing, it would take another thousand years to reveal the order of all 3 billion letters. With a dedicated effort involving "thirty thousand person years," however, he estimated that the job could be done at the cost of around $1 per base pair, or $3 billion. Gilbert's numbers triggered an uproar. The audience, composed of university-based molecular biologists and geneticists, were outraged at the prospect of some government-directed "Big Science" project sucking funds from smaller, worthier, and more intellectually stimulating research programs around the country. David Botstein of Stanford University, a burly enfant terrible, stormed the podium, declaring that the DOE's proposal would "indenture all of us, especially the young people, to this enormous thing like the Space Shuttle, instead of what you feel like doing." The audience applauded wildly.

Based on that initial reception, the DOE's Human Genome initiative seemed likely to be stillborn. In the meantime, however, Charles DeLisi, the agency administrator who had conceived the idea in the first place, had gained the support in Congress of Senator Pete Domenici of New Mexico. With the waning of the Cold War, Domenici was keen to develop new, peacetime uses for the DOE-run national laboratories in his state, including Los Alamos National Laboratory, home of the Manhattan Project. But the academic scientists protested that the human genome was a biomedical challenge for the best and brightest minds in molecular biology and genetics—not, in Botstein's words, a "program for unemployed bombmakers." DeLisi eventually agreed that the National Institutes of Health should take a major role in the project. He was able to enlist the support of NIH director James Wyngaarden and other key members of Congress and the scientific community.

It was not until 1988, however, when James Watson committed to the project, that the tide began to turn. Since his great discovery with Crick, Watson's role in science, first at Harvard University and then at Cold Spring Harbor, had become less that of a frontline researcher and more that of a grand impresario—a "mover of people," in his own words, rather than of molecules. He moved people with the written word, too; his textbook *The Molecular Biology of the Gene,* published in 1965, inspired an entire new generation of researchers. Three years later, *The Double Helix* became a best-seller, its literate and audaciously personal account of the race for the helix transforming him from a well-known practitioner of molecular biology into the celebrated embodiment of the science itself. Thirty years later, the Modern Library ranked *The Double Helix* seventh on its list of the hundred best nonfiction books of the twentieth century.

Watson was politically astute, highly principled, and dedicated to inspiring the best science from everyone around him. He could also be scathingly acerbic, offering his opinions on science and his colleagues with a garage-door lift of his forehead that bugged his eyes out in emphasis, this gone-mad look often followed by an abrupt convulsion of wheezy chortling. Watson's epochal achievement as a young man had bestowed upon him a sublime arrogance; he not only saw himself as deserving of a special consideration when he spoke but assumed that everyone else held the same view. At one scientific meeting, Watson was holding forth at length on a subject in which he was not particularly expert. Finally somebody interrupted and said, "Jim, just because you have a Nobel Prize doesn't mean you have the right to tell everyone else what to do." Watson turned and stared at the man. "I do not have *a* Nobel Prize," he said. "I have *the* Nobel Prize."

The imprimatur of the man with *the* Nobel Prize was a major boost for the struggling human genome initiative. Watson threw all his energy into it. Egging him on was the urgent sense of his own corporeal mortality. "To me it is crucial that we get the genome now rather than twenty years from now," he said, "because I might be dead then and I don't want to miss out on learning how life works."

The Human Genome Project officially began on October 1, 1990, with Watson as its director. Congress proffered an increasing flow of money that would eventually bring the program an estimated $200 million annually for the next fifteen years. But the DOE was no longer in charge. Watson, backed by the National Academy of Sciences and the molecular biological community, had made sure that the National Institutes of Health would be chiefly responsible both for funding and for determining the direction of the project. The DOE, whose innovative administrator had come up with the idea in the first place, faded from being the tail wagging the dog to just the tail. The demotion left behind a bitter taste.

The DOE had envisioned a few large production centers bent on sequencing the genome and developing the technology and computer power needed to make sense of all that information. Instead, NIH set up the National Human Genome Research Institute to dispense funds through the tried-and-true, peer-reviewed grant system to a loose community of academic labs around the country. In addition to the ultimate goal of reading out the human genetic code, smaller genomes would be targeted, too, beginning with bacteria, like the laboratory mainstay *E. coli,* and including more complex research animals such as the nematode worm *C. elegans* and the classic genetics standby *Drosophila melanogaster.* Genome fever had taken root in Europe and Japan, too, and Watson made a concerted effort to internationalize the project. While its scientific critics had worried that the project would learn too little about life, many citizens outside science worried that it would learn too much, leading to a nightmarish world in which individuals would be screened, judged, and classed according to their genetic legacies. To address these concerns, Watson set aside 3 percent of the program's funds to explore the ethical, legal, and social issues emerging from the work; the number had come to him at a congressional hearing, off the top of his head.

Strategically, while sequencing the letters in the human genome remained the ultimate goal, the first five years of funding would be devoted to mapping the genome's immense territory. The two concepts, which journalists often mistakenly use interchangeably, are actually very different activities, just as drawing a map of a region is different from pinpointing the location of every tree, rock, and tuft of grass within the region's limits. If you want to understand a landscape with that kind of exactitude, however, a sensible first step would be to identify some fixed landmarks, such as rivers, mountain ranges, and forests, and draw them on a map. In the molecular landscape, these orienting points consisted of the location of known genes relative to each other—simply more detailed versions of the primitive genetic map Alfred Sturtevant had drawn for the fruit fly's X chromosome back in 1913—as well as short, unique stretches of DNA sequence distributed randomly along the chromosomes. Then, only after you have your general bearings, you might send out different teams to begin exploring particular regions in more detail, drawing a boundary around a certain area and carefully noting the location of every object within it. The decision to "map first, sequence later" had the added advantage that the five years spent mapping would allow some time to develop efficient and affordable sequencing technologies that would be ready when the time came to conquer the sequence itself.

Watson imagined sequencing as grunt work that would eventually be carried out in some kind of factory-like setting, leaving the academics free to ponder more interesting problems, like figuring out how those endless concatenations of letters interacted to make a human being. But things would not turn out that way. The competitive structure of federal grant funding conspired with the nature of the genome itself to spawn a kind of molecular land grab. The genome was divided up into twenty-three natural subunits, the chromosomes themselves, and research labs around the country scrambled to secure funds from the NHGRI to map one chromosome or another, or at least a piece of one. This divide-and-conquer approach was, in the words of one of the chief players, "stunningly inefficient." Essentially, the labs were all attacking the same problems separately and running up against the same surprises twenty-three times over. But few dared propose a genomewide approach to be carried out exclusively in their own lab, for a simple reason: the reviewers of any such grant proposal would be the applicant's colleagues in competing labs, who were warily guarding their own claim against just such hegemony.

The mapping phase was completed, sort of, on schedule by 1995. But there were unexpected setbacks. First, many of the maps were proving too crude in scale to be of much use—a river here, a mountain range there, with vast areas still uncharted. Second, the hopes for a cheap, quick method of sequencing DNA had not been realized. Nevertheless, Francis Collins, who by this point had replaced Watson as director of the program at NIH, heeded an urgent call from the community to move on to the sequencing phase. By this time, however, the notion of a few large "factories" taking over the work had been forgotten—the territorial mentality was too strongly ingrained. Instead, the same labs clamored all over again for a piece of the sequencing pie to work on, even a morsel as little as 1 percent of the total genome. "Nobody had killed the beast, but they were all carving up the hindquarters and front section," said the same scientist who had bewailed the inefficiency of the mapping approach.

In 1996, Collins's National Human Genome Research Institute inaugurated a pilot project to get the sequencing under way. Six American genome centers were awarded initial grants to work on various pieces of the genome: the Whitehead Institute, Washington University, Baylor College of Medicine, the University of Washington, Stanford University, and Venter's group at The Institute for Genomic Research (TIGR) in Maryland. In the meantime, the Wellcome Trust in Britain, at the time the world's largest philanthropy, had underwritten the construction of a brand-new genome sequencing facility near Cambridge, England, named after Fred Sanger, who had first invented a method for reading out DNA base pairs. The Sanger Centre promised it would complete fully a sixth of the human genome on its own.

Mitigating the potentially destructive effects of the competition among the Human Genome Project's participants was an overwhelming sense that they were on a collective mission of incalculable importance to the progress of science and humankind. This communalism was defined and inspired by two unshakable commitments. First, the sequence the program produced would be of the highest possible quality, one without gaps in the order of base pairs, and surely without parts of the genome placed on the wrong chromosome. The risks of such crippling mistakes, which could scar the usefulness of the work for decades to come, were far greater than one might think. Like the genomes of most complex organisms, the human genetic script is padded with thousands of near-identical hunks of code. These "repeats" are scattered throughout the chromosomes and were estimated to take up as much as 30 percent of its space. Logically, the only way to cope with the problem and ensure the correct order of letters was to sequence just a little bit of genome at a time. A method had recently been invented to isolate a 150,000-letter fragment of code by cloning it in a genetically engineered *E. coli* cell called a bacterial artificial chromosome, or BAC for short. If you wanted a highly accurate product, sequencing the genome bit by bit, or BAC by BAC, was considered the only way to go, even if it took past 2005 to complete the whole genome.

If "Quality first!" was the scientific rallying cry of the Human Genome Project, its political credo was "Let it be free!" In this regard, the NIH was bucking a trend. However strong the nostalgia for the academic purity of Cold Spring Harbor in its glory days, in the 1990s getting a return on potentially marketable discoveries was not only allowable practice in university departments, it was a requirement. In 1980 Congress had enacted the Bayh-Dole Act, which obliged researchers who were funded with federal tax dollars to secure intellectual property rights on any commercially valuable discovery before putting the information in the public domain. The rationale was to encourage American companies to step in and develop products, protected by patent rights against foreign competition that might otherwise scoop up the inventions for free.

The Human Genome Project was being funded by NIH and the Department of Energy, so it fell under the Bayh-Dole Act. But the leaders of the program in the United States—with vigorous urging from the Wellcome Trust in Britain—believed that the basic human sequence information was a resource too valuable to belong to any one person, university, or country. The matter was formalized at a meeting of the genome program leaders on the island of Bermuda in February 1996. (The Wellcome Trust, which was hosting the meeting, chose the site not for its balmy weather but to underscore the joint leadership of the program by the United States *and* Great Britain.) Fifty-odd scientists arrived at the Princess Hotel in Hamilton with their summer togs and suntan lotion, and then spent the next three days around a huge table in a windowless room, returning as wan as they'd come. With little dissension—and much of what there was coming from Craig Venter—they decided that all human DNA sequenced by the Human Genome Project should be made freely available and placed in the public domain.

To ensure this, the program leaders later determined that all DNA sequences more than 2,000 base pairs long must be made publicly available within twenty-four hours of their discovery on a massive online genetic database called GenBank, administered by the National Center for Biotechnology Information at NIH. This so-called Bermuda Accord made it a physical impossibility for the discoverer to file a patent on raw human DNA. Strictly speaking, the policy directly contradicted the Bayh-Dole Act, but Francis Collins let it be known that human DNA was a special case. Any grant proposal that did not explicitly spell out its allegiance to the Bermuda Accord would be rejected. "The establishment of this principle was one of the defining moments of the Human Genome Project," he later wrote in the journal *Genome Research.*

It was at this meeting in Bermuda that Jim Weber proposed a faster, cheaper method of getting the genome done than the "BAC-by-BAC" approach. Weber, from the Marshfield Medical Research Foundation in Wisconsin, was not a member of the tightly knit Human Genome Project community; rather, he epitomized the intended user of its product—a gene hunter looking for the genetic causes of disease, whose work would be accelerated by orders of magnitude once the genome was finished. Inspired by the success that Craig Venter was having using the whole-genome shotgun method to sequence bacterial genomes, Weber, with the help of computational biologist Gene Myers at the University of Arizona, had come up with a way to adapt the technique to the far more daunting problem of the human code, and he had joined the meeting in Bermuda to present the details of the strategy in person.

He could have saved himself the air fare. After Weber's presentation, Philip Green of the University of Washington, the genome project's guiding light in computational biology, led an assault on the technical flaws and assumptions that Weber and Myers had made in their assessment of the difficulty of the problem. Given the state of sequencing technology at the time, at best the whole-genome shotgun technique would fall far short of producing a sequence that would satisfy Collins's imperative to "stand the test of time." More likely it would be a shoddy approximation, or might even fail to come together in the end at all, wasting money on a project whose utter worthlessness would not be revealed until its final step was complete. It was not the first time someone had proposed a shortcut to the genome. Weber's presentation only hardened his audience's resolve not to heed such temptations but to go instead for the highest possible quality. The "Bermuda standard" of accuracy codified at the meeting called for no more than a single base pair misspelling in every 10,000—or an error rate of .01 percent throughout the code, "junk DNA" and all. Again, there was little or no opposition. Who was going to argue against excellence?

On Wednesday, May 13, 1998, the Human Genome Project's leaders convened for their annual gathering at Cold Spring Harbor, just off a courtyard in a modern conference room with polished blond oak paneling. They were awaiting the arrival of a man whom, in spite of his contributions to the HGP as director of TIGR, most of them considered a heretic. Francis Collins had invited Craig Venter and Michael Hunkapiller to present their plans to a closed session of the project's leaders. At Cold Spring Harbor, dress is always casual. Picture Francis Collins in a plaid flannel shirt and jeans, in his thick glasses looking a little like an intellectual dude rancher. He was talking to the drosophilist Gerry Rubin, who had covered his amiably lumpy frame in whatever old sweater and slacks first presented themselves to him on waking up. Eric Lander of the Whitehead Institute, bull-thick in the torso, red-haired, permanently wired, was moving impatiently about the room, like a prizefighter before a title bout. The directors of the other big genome centers were there, too: Robert Waterston of Washington University in St. Louis, pale and wiry, and Richard Gibbs, an earnest, good-natured Australian in charge of the genome program at the Baylor College of Medicine in Houston. Another key player, John Sulston, who headed the British genome sequencing program at the Sanger Centre, had not yet arrived. He was coping with the Venter crisis in his own way across the Atlantic but would join the gathering later in the week.

Also missing from the group was the original genius of the Human Genome Project, James Watson himself. Seventy years old, Watson had recently retired after almost thirty years as the director of Cold Spring Harbor but still served as its president. Watson had breakfasted with the others in the dining hall earlier that morning, but he refused to attend a meeting convened to listen to Craig Venter. Since the announcement in the *New York Times* on Sunday, the elder scientist had stopped calling Venter by his own name, referring to him in conversations as "Hitler" instead. At breakfast, Watson called out to Collins across the room. His words were still ringing in Collins's ears: "So who are *you* going to be, Francis?" Watson said. "Winston Churchill? Or Neville Chamberlain?"

Typically, Watson had characterized his successor's situation in the most biting way possible. But the Human Genome Project had indeed been invaded, in a preemptive strike more like Pearl Harbor than Hitler's march across Europe. It was just dawning on Collins, and on everyone else, how vulnerable a target his great ship had become. By the spring of 1998, the Human Genome Project's dedication to "quality first" had become an unquestioned creed, but the job was proving harder than expected. Just a few days before the gathering at Cold Spring Harbor, a news article in *Science,* the flagship scientific journal in the United States, had revealed that not one of the genome centers funded by NHGRI had come anywhere near its projected goals. Lander's group at the Whitehead Institute at MIT had aimed to have 23 million base pairs of sequence finished to the Bermuda Standard but had completed only 9 million. At the Baylor College of Medicine, Gibbs's team had hoped for 15 million and produced 8 million. After two years of sequencing, the Stanford University team had not even reached the million-base-pair mark. In all, just 120 million base pairs had been sequenced, or about 4 percent of the total genome. But no matter: the genome would be done, and it would be done right. "It's going to be a long, hard climb," Philip Green told *Science,* adding that he was "still optimistic we're going to get there by the year 2005."

Unfortunately, the tenacious allegiance to quality had blinded the leaders of the program to a fundamental weakness: for most research purposes, such a stringent standard of accuracy was overkill. In a way, the resolute, mustached Francis Collins resembled Alec Guinness's British colonel in *The Bridge on the River Kwai,* who rouses his fellow prisoners of war to build the best possible bridge they can, having left out of his mental calculus that the bridge is for the use of their Japanese captors. Collins's oversight was to forget that most potential users of the genome—academic biologists looking for genes and pharmaceutical companies eager for information that could speed up drug development—did not need a fancy bridge that would "stand the test of time." They just wanted to get across the river. And now suddenly Venter had appeared, carrying a bunch of Perkin Elmer pontoons.

Venter entered the blond-paneled conference room with Michael Hunkapiller and Mark Adams, a bounce in his step and a lighthearted smile on his face, as if there were nothing more than a trifling misunderstanding to untangle. His self-deprecating charm made him harder to despise in person than in the abstract. Most of the forty or so people seated around the U-shaped table were still in a state of shock. Two of them, however, had known secretly for weeks about Venter's plan. Ari Patrinos, a diminutive, refreshingly genteel Greek American, headed the genome program at the Department of Energy, which made him, in theory if not in practice, co-leader of the project with Collins at NIH. He was Collins's friend and neighbor in a Rockville townhouse development. But Patrinos was a friend, confidant, and supporter of Venter, too. He handled the potential conflict with inviolable discretion. The other scientist in the know was David Cox, who directed the genome program at Stanford University. Venter had invited Cox to leave Stanford and join his new company. Cox had decided not to accept but was still pondering an offer to serve on the company's scientific board.

If the Human Genome Project's byword was quality, Venter made it clear right away that his was *speed.* The new company he was forming with Perkin Elmer's $300 million would be dedicated to DNA sequencing at a rate not previously imagined. With unnerving matter-of-factness he described what Hunkapiller's new machines were capable of doing. The numbers were staggering. Each machine could sequence about 1,000 fragments of DNA a day, with each fragment about 500 base pairs in length. With 230 machines running, the number of base pairs churned out daily would thus be 230 x 500 x 1,000, or a total of 115 million base pairs read out every day, roughly the same amount that the Human Genome Project had produced in the last two years. To put together a "consensus" sequence of the genome, the company would be sampling DNA from the blood or semen of five different anonymous individuals, which meant ten different variations on the human genetic code, since each individual would harbor two versions, one inherited from each parent. The planned finish for the company's genome was 2001, four years ahead of the public program—which might not be finished even by then.

The scientists in the audience were extremely skeptical. Applied Biosystems' machines had not even been built yet, so how could anyone know how efficient they would be? More important, everyone in the room knew that simply lining up machines to regurgitate DNA would not create an assembled genome. You had to break the problem down into mapped bits to cope with the problem of repeats. Otherwise the genome you ended up with would be a calamity, a hodgepodge of misplaced elements, like a man with an arm coming out of his head and eyeballs on his knees. *Everybody knows that,* they thought; *even Craig. So what is his real agenda?*

"We are going to assemble the human genome using whole-genome shotgun," Venter said, with casual deference, as if he were offering an hors d'oeuvre. He explained some details of the approach. They had heard it all before; it was Weber-Myers all over again. They were stunned by the confidence Venter seemed to derive from his own naïveté. One of the scientists at the table was slouching down in his seat with his head tilted back, looking like he was about to erupt in anger or a fit of laughter. When Venter cited a mathematical equation, the Lander-Waterman model, to support his method, the sloucher could hold back no longer. "But Craig, you've completely misapplied Lander-Waterman!" he cried. "I should know! I'm Lander!"

What is your policy on data release? somebody else asked. Are you going to file patents on the genes you find? another called out. How do you expect us to believe that you won't, when there isn't any other way your company can recoup its investment? Don't you know that attacking the human code with a whole-genome shotgun strategy is like throwing cans of paint at a wall and hoping they resolve into a Renaissance masterpiece? Don't you know, or care, that the product of your DNA factory will be a ludicrous knockoff? An eyeballs-on-knees kind of genome, misarranged and riddled with gaps? The scientists did not use these words, of course. They spoke in the argot of their trade, and talked of contigs and closure, BACs and YACs, reagents and reactions. The gist, however, was the same: Venter may have had some success decoding bacterial genomes with whole-genome shotgun, but this was the *human* code—a genome thousands of times larger, a million times harder to solve, and infinitely more weighted with implication, and how dare you tell us how to do it. A few—Ari Patrinos and David Cox among them—listened and thought to themselves, *This is just what we need. Something to shake us up.*

Venter answered the questions one by one. There would be widely divergent opinions later on how gracefully he did so. Cox remembered him as "unusually well behaved, going out of his way not to provoke people." Francis Collins recalled him as "his usual supremely confident self, bombastic, dismissive of the efforts of anyone else." While those recollections contradict each other, they may also both be accurate, not just because two people often see the same individual differently, but because some individuals can project two images simultaneously. They are self-contained contradictions, like holograms whose views toggle back and forth depending on the angle of the light striking them.

"If we can get the human genome done," Venter said, as he'd told Collins, "you could concentrate on the mouse genome. Everybody wins."

"When he said that, I almost punched him in his fucking mouth," one scientist at the meeting later remembered. But, strictly speaking, Venter's premise was not unreasonable. If a private company could assume the task of sequencing the human genome, why *not* redirect the taxpayers' money toward other goals equally important to science and medicine? I am doing this for humankind, Venter seemed to say. Turn the hologram, and I am in it for myself. I am Albert Schweitzer. I am Bill Gates. Flip the hologram faster—I am Bill Albert Schweitzer Gates. I am a scientist. I am an entrepreneur. I am a scientist/entrepreneur. I am the slash between the two. How can you not love me? Go ahead, hate me. You think I care?

"We've been thinking, too, about what a big leap it is to go from bacteria to a genome the size of human," Venter was saying. "It would make sense to try it on some model organism in-between in size first." He threw out a couple of possibilities. Then Hunkapiller took over, and while he discussed the mechanics and chemistry of the sequencing machine, Venter quietly moved down behind the seated scientists and tapped Rubin on the shoulder. "Gerry," he whispered. "Have you got five minutes?" The two men slipped out into the corridor. "What I'd really like to do for the pilot project is *Drosophila,*" Venter told him. "But I want you to be happy about it. I want to do it in collaboration with you."

Months later, Venter confided to Rubin that at that moment in the hallway, he was afraid Rubin was going to hit him. But this was just what the drosophilist had been hoping for since he had picked up his copy of the *New York Times* in the driveway the previous Sunday. His group in Berkeley had been working on the fruit fly genome for two years, with an estimated three more to go. Teamed up with Venter, they could do the job in a third the time.

"Great!" Rubin said. "But only on condition the data is released for free. No restrictions."

Venter did not hesitate. "It's a deal," he said. They shook hands and returned to the conference room. There were more questions. "Will you be honoring the Bermuda Accord?" Robert Waterston asked.

"We haven't decided yet on the timing of the release," Venter said, "but it will probably be quarterly."

"Quarterly!" Waterston responded indignantly. "That's a lot different than overnight."

"We're a *company,* Bob," Venter said patiently—or contemptuously, depending on your angle of view. "We don't have to release the data at all. But if you think about it, quarterly is a lot closer to nightly than it is to never."

Through the meeting, Francis Collins had asked only some technical questions. He was gauging the situation carefully and thinking about what Watson had said. Neville Chamberlain or Winston Churchill? As if there were a choice, with all that was on the line. "Well, you've certainly given us a lot to think about," Collins said to Venter in his stalwart, jovial way. Stiff upper lip. Venter, Hunkapiller, and Mark Adams left. Just before the meeting resumed to digest what they'd said, Collins turned to David Cox and asked him if it was true that he was joining Venter's company. The Stanford geneticist admitted that he was considering an offer to serve on its scientific board. In that case, Collins replied, perhaps Cox, too, should leave the room. Cox refused, and Collins did not make a point of it. But later that day he asked Cox to take a walk down Bungtown Road. At Cold Spring Harbor, it is a tradition to "take a walk down Bungtown Road" when you want to talk without being overheard. The road, which is really more a pedestrian path since so few cars ever use it, runs along a wooded slope above the blue harbor from the laboratory entrance to the director's house. It was a perfectly beautiful spring afternoon, with lilac petals and husks of blossoms sprinkled like confetti on the asphalt beneath the two men's feet.

"It will be sad not having you part of the genome project," Collins said. His implication was clear. If Cox accepted Venter's offer, the funding his lab enjoyed from NIH would cease.

"Why does it have to be one or the other?" Cox said. "The worst thing now would be a race between Craig and the public program, trying to see who can get a genome done faster. What we need is integration of this company into the public program's goals. I could help with that."

Collins shook his head and smiled wistfully, as if regret had overcome him for the ideal world that could not be. But there would be no appeasement. "If you join Craig's board," he said, in patiently enunciated syllables, "nobody in the program is going to want to work with you anymore. You have to choose."

Meanwhile, Jim Watson was talking to Gerry Rubin. "So," Watson said, eyebrows launching up toward his scalp. "I understand the fruit fly is going to be Poland."

Venter did not stay to address the full meeting. The rank and file of the genome program trickled in and registered in the lobby in front of the golden helix, checked into their rooms, then wandered down to the little bar in the basement of the dining hall or congregated in small, stunned clutches on the wide flagstone terrace outside the lecture hall. They spoke of the invasion of "Darth Venter." Many were furious—not just at Venter but at their own leaders' "chicken-shit response to Craig." They saw Collins, Lander, Waterston, and the other principal investigators whispering, taking walks down Bungtown Road, going off to restaurants for private dinners. Rumors went around—they're giving in. They're abandoning the mapping approach. Eight years of everyone's hard work wasted.

Two days later, the mood suddenly shifted. John Sulston of the Sanger Centre had arrived, along with his patron, Michael Morgan of the Wellcome Trust. They were like a platoon of British cavalry coming to the rescue. At a special session of the full assembly, Sulston challenged the audience to look on the Venter situation as a new impetus to get the work done in Jim Watson's lifetime. Face flushed, Morgan rose and spoke with the authority of a man whose convictions were matched by his budget. He announced that in light of recent events the Wellcome Trust was doubling its financial support of the Human Genome Project. He received a standing ovation. Next, Collins reassured the assembly that there never was any consideration of abandoning the human genome and switching to mouse, shaking his head at the mere thought. In the audience, laughter. Philip Green explained how a technical flaw in the whole-genome shotgun strategy would make it impossible to figure out where the repeated sections in the human genome really belonged. So Venter's plan was doomed to failure in any case. The scientists left the lecture hall feeling greatly relieved.

CHAPTER 4
GENESIS

Some people will tell you that Craig Venter conceived the idea of privately sequencing the human genome to get rich, or get revenge, or get a Nobel Prize. Others argue that no matter what he hoped to gain for himself, he saw it as the single most important thing that he could contribute to humankind. The fact is, Venter did not conceive of the idea of sequencing the human genome at all. The idea came into being unexpectedly, at a routine meeting of some Perkin Elmer scientists and executives one late November day in 1997, weeks before Venter was hired to implement it. While the inspiration seemed spontaneous at the time, its form had been gradually taking shape for months, in different minds, the way cooks might work on different parts of a recipe but not know what they were making until it all came together in one pot. Even Tony White, who might be said to be the head chef, had no idea where things were leading until the moment they converged.

Tony White is half Cuban by birth, and half North Carolinian. With his southern drawl and pale complexion you might think he'd invented the Cuban half, just to make himself exotic, except that he is not the sort of person to make himself exotic. Compact, blunt, and devoid of affectations, he has small, penetrating eyes that make one uncomfortably aware of one's own pretensions. White has had only two job interviews in his career. The first came right after he graduated from Western Carolina University. A representative from Baxter Laboratories, a Fortune 500 health care supplier, told him they were looking for someone to run the personnel department in their South Carolina factory. "I'd rather die than be the personnel manager of your South Carolina plant," he responded, which so impressed the Baxter representative that he had White flown up to meet with the company chairman in Chicago. The chairman offered him a job in sales. After a stint in the army, White began at Baxter peddling surgical gloves and intravenous equipment to hospitals. He spent the next twenty-six years rising through the ranks until he became the number two man in the company. And that, he figured, was about as far as a short, blunt, Cuban American with a hick country drawl and no MBA would be allowed to go in that company. It was 1995. He figured it was time for another job interview.

A corporate headhunter sent White to Perkin Elmer Corporation in Connecticut, a teetering wreck of a conglomerate in search of a new CEO. Founded in the 1930s, Perkin Elmer had made bombsights during World War II and a hodgepodge of other products since, including engine coatings, silicon-chip-making machines, and various optical and analytical instruments, including the infamous faulty mirror on the Hubble Space Telescope. The company had also recently acquired the rights to make the bench-top lab devices that performed the polymerase chain reaction, or PCR for short—a simple, virtually foolproof method of copying a small sample of DNA a billionfold, amplifying its signal enough for its sequence to be read and its meaning revealed. Since its invention in 1983, PCR had become a ubiquitous tool in any endeavor involving the analysis of DNA, from curing cancer to identifying human remains and resolving paternity suits. It was PCR that identified the DNA on O.J. Simpson's bloody glove and Monica Lewinsky's stained blue dress.

Even with this lucrative acquisition, however, high costs in Perkin Elmer's core analytical-instrument business had kept the company's growth rate stagnant for years. The company's infrastructure was an overgrown mess, with bickering fiefdoms scattered around the globe. At the time White went for his interview, major investors such as George Soros were threatening to bail out. "Why in the world would you *want* this job?" one member of the interviewing board asked the job candidate. In fact, White had initially accepted the appointment just to brush up his interviewing skills. But on closer examination, he'd concluded that Perkin Elmer had a golden nugget hidden in the dross. Two years earlier, the company had acquired Applied Biosystems, Inc., in Foster City, California.

Mike Hunkapiller is a big man, six-foot-two, with an implacable, granitelike face and muscular arms that he keeps habitually folded across his chest, a posture that wards off casual inquiry. It is hard to imagine him as a shy 135-pound twenty-six-year-old from Oklahoma arriving in 1976 in the laboratory of molecular biologist Leroy Hood at Caltech. His plan was to get his Ph.D. in physical chemistry and return to Oklahoma as soon as possible. Things did not work out that way. Hood put great stock in developing tools to do molecular biology better, faster, and cheaper, and Hunkapiller possessed the right combination of talents in chemistry and engineering to make that happen. After he got his degree, his mentor convinced him to stay on awhile as a postdoc. While still a graduate student he designed a machine that could read out the sequence of amino acids in a protein better than anything else around at the time. Hood got some venture capitalists interested in commercializing the protein sequencer, and in 1981 they founded Applied Biosystems. Within two years, the company also had licenses on technology developed at Caltech that could synthesize short stretches of DNA and amino acid sequences. All that was needed to complete a quartet of instruments for the new biology was a machine that could automatically sequence DNA.

Fred Sanger of Cambridge University had already invented his technique for reading letters of DNA a few years earlier, but it was a manual technique. The difficulty of the problem was much greater than the metaphor of "reading letters" suggests. Unlike letters, the four nucleotides of DNA cannot be distinguished by their shape. Sanger's method required a great deal of ingenuity. He began by preparing a solution containing millions of copies of a small DNA fragment, and divided the solution up into four equal parts. Then he heated the four test tubes, which separated the double-stranded DNA into single strands. To each tube, he added DNA polymerase, an enzyme that uses a single strand of DNA as a template to re-create its missing partner, a complementary string of base pairs. He also introduced a primer—a small synthesized fragment of DNA of a given sequence of nucleotides. The primer fragments would glom on to their complements on the template, which would tell the enzyme where to start the copying process. He also supplied each tube with plenty of free-floating nucleotides: the As, Ts, Gs, and Cs that the enzyme would need as raw material to construct the complementary strings.

At this point in the experiment, the solution in each of the test tubes was exactly the same. Next came the ingenious part. Sanger spiked each tube with a smaller amount of a doctored form of one of the four nucleotides, which had been tinkered with to stop the copying process in its tracks. In the first tube, for instance, there might be a generous supply of ordinary Ts, As, Gs, and Cs, but a few of the doctored dead-end Ts as well. Whenever the enzyme happened to grab one of these killjoys and attach it to the growing strand, the reaction on that particular strand would cease. Thus, after reheating the solution to separate the double-stranded DNA again, Sanger ended up with a collection of single strands of differing sizes in that particular tube, each one beginning at the start of the sequence, and each ending when a killer letter "T" was attached. The same process was going on simultaneously in the other three tubes with the three other DNA letters.

To read the sequence of the entire original DNA fragment, Sanger thus had only to sort the fragments by their size, and read the last letter of each one. This was accomplished with the use of a device called an electrophoretic gel. The instrument is as common in a molecular biology lab as a hammer on a construction site. The researcher pours some heated agarose, a sugary substance distilled from seaweed, between two glass plates separated by a narrow gap. When it cools, it hardens into a clear gelatinous plate. If you looked through it with a microscope, you would see that it is shot through with tiny channels and rivulets, like bone tissue. The scientist squirts his sample into one end of the gel. At the other end is a strong positive electric current. Because DNA molecules are negatively charged, the current draws them inexorably toward the positive end. They have to work their way through the latticework of microscopic channels, and naturally the smaller ones slip through faster than the larger ones. After letting a gel run for a while, the researcher turns off the current and the molecules stop where they are. (One two three, *red light*!)

Using a pipette, Sanger squirted an amount of each of his four solutions at the top of a gel, turned on the electric current at the bottom, and waited for the individual molecules in each of the four columns to migrate down, sorting themselves by size along the way. The fragment at the very bottom of the gel would be the smallest one, consisting of a single nucleotide representing the first letter of the DNA fragment. Next would come the second smallest molecule, consisting of that same first letter, plus one "killjoy" A, T, G, or C that had stopped the copying process in its tracks. The next largest fragment would consist of three nucleotides, then four, and so on. Reading from bottom to top, the order of the fragments on the gel would thus represent the sequence of nucleotides in the whole chain. In the example below, for instance, the first letter in the sequence is an A, followed by a C, a T, and so on, until you reach the slowest, biggest fragment shown, which is a T. The complete sequence is ACTCACGGT:

A	T	G	C
	—		
		—	
		—	
			—
—			
			—
	—		
			—
—			

Using this process, Sanger was able to read out the letters of a stretch of DNA as many as 500 base pairs long. By combining fragments, he had by 1977, after thirteen years of refining the technique, sequenced the complete 5,386-letter genetic code of a virus called phiX174. For his effort, he won the Nobel Prize in Chemistry in 1980 (his second), sharing it with Harvard's Walter Gilbert, who had come up with a similar technique independently, and Stanford biochemist Paul Berg, for his pioneering experiments in recombinant DNA.

While sequencing the complete code of a virus was a marvel at the time, using Sanger's method on the human genome would take 100,000 years, which alone might explain why a lot of very smart people initially thought the Human Genome Project was a very stupid idea. Obviously, a faster, more automated technique was urgently needed. In the early 1980s, a Japanese scientist named Akiyoshi Wada had rounded up Hitachi, Fuji Photo, and several other corporations to sponsor the development of an automated sequencing machine, but it never saw the light of day. By the time Hunkapiller and others in Leroy Hood's lab at Caltech started work on the problem, several American companies, including Du Pont, were mounting their own efforts to design an automated DNA sequencing machine, and so was the DOE. But these contraptions did not work for large-scale sequencing projects, if they worked at all.

Though Hunkapiller had not returned to Oklahoma as planned, a piece of Oklahoma came to him in the form of his younger brother Tim, who soon joined Hood's lab as a graduate student. It was Tim Hunkapiller who had the breakthrough idea. In Sanger's method, the DNA letters were read by the researcher's eyeball, scanning both across and up the gel at the same time, a process that was both error prone and squintingly slow. Tim's idea was to color-code the last letters in each fragment by chemically attaching a different-colored dye to each of the four doctored, killjoy nucleotides: blue for C, yellow for G, red for T, and green for A. As each fragment in turn arrived at the bottom of the gel, the color of its last letter could be read by a detector, which would feed this information to a computer. If the process worked, several samples could be run at once.

With Hood serving as a sort of executive producer, the Hunkapiller brothers worked out the details of the technique. Lloyd Smith, a physical chemist in the lab, managed the development of the project at Caltech, while an ABI scientist, Kip Connell, worked on the optics needed to detect the fluorescently labeled samples. In 1983, Michael Hunkapiller left Caltech to join ABI, and a couple of years later the company had a working prototype of an automated machine. It took twelve hours to run, producing sixteen sequenced "reads" of DNA averaging around 300 nucleotides each. Theoretically, it could sequence more DNA in a day than a single researcher could do in a year.

By the time Tony White interviewed for the CEO job at Perkin Elmer in 1995, ABI was a division of the company, which was supplying 90 percent of the world's automated DNA sequencing machines. Perkin Elmer also owned a hefty chunk of the market in other high-ticket technology for basic molecular research. While the market was specialized, Hunkapiller's enterprise was exploiting its dominant position so freely that its customers grumbled that the abbreviation stood for "Arrogance Beyond Imagination." At Perkin Elmer headquarters in Connecticut, the corporate brass viewed it as merely another conquered property that brought an additional product line to the mix. White did not think they appreciated what they had. When the board member asked him "Why do you want this job?" he did not answer, "So I can tear your company apart and rebuild it into something you won't even recognize as your company anymore." But it was that tantalizing prospect that led him to accept their offer.

He did not waste any time. White immediately cleaved Perkin Elmer into two separate businesses: analytical instruments and life sciences, the latter including ABI and the PCR product line and the former everything else. He then began starving the fat off one to feed the growing muscle of the other. The analytical business had been promised $30 million in new research money; White snatched it back and instead ordered $30 million in cost cuts, through retirements, layoffs, and asset sales. He plowed the cash back into ABI to seed creative development of "skunk work" projects that Hunkapiller's scientists and engineers could sink their hands into without corporate interference. To encourage more forward thinking, senior executives found their cash bonuses replaced by stock options. Meanwhile White went on a controlled buying spree to boost Perkin Elmer's presence in the emerging biotechnological markets. Every decision was designed to make the company lean and hungry again. "Our heart and soul," White told an auditorium full of shareholders a year after he took over, "the fiber of everything we do, is oriented toward growth." George Soros and the other high rollers decided to hold on to their stock and see what White could make of his new company.

In spite of all the cost cutting, White had already concluded that the old analytical instrument business was a dead end. ABI was at the heart of his new vision for the company, but it had an inherent weakness. Hunkapiller's machines and the expensive reagents needed to run them were positioned to provide the raw technological power driving the fledgling "pharmacogenomics" industry—finding new drugs based on genes and tailored to an individual's specific genetic makeup. But it irked White that more nimble start-ups were primed to dig wealth out of the genome, while Perkin Elmer, through ABI, was more or less relegated to supplying their picks and shovels.

The most successful of these new genomic information enterprises was a Bay Area company called Incyte Pharmaceuticals. Incyte had been founded in 1991 to feed the growing appetite of the pharmaceutical industry for information about genes. A successful drug costs about half a billion dollars to bring to market. Most of this investment is not spent on laboratory costs, researcher salaries, or even in the expensive clinical trials it takes to develop the drug. It is consumed instead in chasing down thousands of other potential drugs that turn out to be dead ends. This gigantic waste of time and money is implicit in the traditional, trial-and-error way in which "Big Pharma" looks for drugs, which is akin to trying fistfuls of keys in a lock until one of them happens to open the door. The companies test thousands of compounds randomly to see if they will have a desired biochemical effect. The promising ones then have their molecular structures tweaked by medical chemists to enhance their power and effectiveness. Those that still hold promise go into animal trials to gauge their potential side effects. For every 10,000 molecules originally tested, only about 250 make it this far. Perhaps 5 of those go on to the last hugely expensive stage: human clinical trials. With luck, one of them will make it to the pharmacy shelves.

Not surprisingly, anything that can reduce the number of casualties in a company's drug pipeline is a potential gold mine. The biotechnology industry arose in the 1980s as a way to discover drugs not by the old empirical top-down approach but from the bottom up. Rather than throw keys at a locked door, look closely at the lock's mechanism and see if you can design a key that fits. Most often, the "locks" were cell-surface proteins known to be part of some critical chemical chain reaction that was malfunctioning. If you could identify the protein and figure out its shape and chemical properties, perhaps you could design a "magic bullet" molecule specifically to unlock the door, without disrupting some unrelated chemical pathway that would cause an unwanted side effect. This was exactly how Amgen, the very model of a successful biotech, discovered the drug erythropoietin to treat anemia, and rode that one drug to billions in profit.

By the early 1990s, both biotechs like Amgen and traditional pharmaceutical companies realized that a shortcut to finding target proteins for drugs was to look for telltale clues to a protein's function in the human genetic code itself. Incyte was one of the first to capitalize on this trend. It had been a loyal ABI customer since 1992, and Hunkapiller had a good relationship with Roy Whitfield, its CEO and president, and Randy Scott, its chief scientific officer. But like other clients, they complained about ABI's prices. Then one day Tony White heard that somebody at Incyte was referring to ABI as "just a commodity" that could be replaced if Incyte could develop its own instrumentation in-house or find another supplier. The comment made him angry. If his customers were threatening to move into his space, White figured, he had better move into theirs first.

"I don't remember who made that remark about us being just a commodity," he said later. "I think it was Randy. No, it was Roy. Or maybe it was just one of my nightmares. But if I dreamed it, too bad for them."

Meanwhile, Michael Hunkapiller was having his own little nightmare. ABI's automated DNA sequencers had enjoyed a decade of virtual monopoly of the market. But now a Silicon Valley start-up, Molecular Dynamics, appeared poised to bring out a machine using a process that in accuracy and pure speed could bury ABI's best sequencer. The machine was called the MegaBACE. Hunkapiller told people he wasn't concerned. He understood DNA sequencers better than anyone in the world, and in his view there were inherent design flaws in the MegaBACE that limited its threat. But for the last few months, engineers from Molecular Dynamics, by then owned by Amersham Pharmacia, had been suddenly spending an awful lot of time over at Incyte, which was ABI's largest commercial customer. A disgruntled scientist from ABI had ended up at Incyte, too, bringing with him a brainful of ABI's proprietary secrets. Hunkapiller said he wasn't concerned about that, either. But he rarely let on that things concerned him.

For all its sophistication, the current version of ABI's machine, called the Prism 377, still depended on two notoriously unsophisticated components: slab gels and human beings. Gels have to be handmade by pouring a rapidly hardening acrylic liquid between two plates of glass. It is easy to make a pretty good gel; a novice can do it. But it takes skill to make a perfectly uniform gel every time, without a single, near-invisible bubble anywhere in its viscous surface that could throw the slow descending trickles of DNA off course like a pebble diverting a rivulet of water. Once a gel is dry, another technician mounts it on the machine and squirts the samples of DNA into a row of tiny indentations at its top. This technician, too, is highly skilled, but maybe she is standing on tiptoe to get a good angle with her pipette, and maybe on one day out of twenty her hand is trembling a bit more than usual, or maybe it's a guy who had one more beer than he should have had the night before and forgets, one time out of a hundred, to change the plastic tip on the end of his pipette before dipping it in another sample, corrupting it. These things happen. It's human nature.

Even when all goes right, the DNA samples migrating down through the gel can still wander a bit off course, like track runners edging into their opponents' lanes. It's the nature of gels. The camera shuttling back and forth at the finish line is trying to identify the DNA letter coming down at the end of each lane, but when the samples wobble out of their lanes it doesn't get a good strong color signal, so it doesn't know what to think. Instead of transferring a confident "T!" to the computer's software, it sends the message "Maybe it's a T" or even "Huh?" When it comes time to match up a sequence with other ones to see where they overlap, these uncertain base pairs will make things more difficult, or even impossible. Time and money wasted.

An alternative to slab gels had been in the air for years. Instead of running dozens of samples down a common gel, each sample might be enclosed in a thin, gel-filled capillary tube, where it couldn't possibly wander into its neighbor's lane. The capillary technology depended on the manipulation of incredibly tiny samples of DNA, however, and whenever things get very tiny, the margin of error shrinks in proportion. ABI had already brought to market a machine that employed a single capillary tube, which was useful for small research labs and other facilities where increasing the sheer quantity of sequencing per day wasn't an issue. Hunkapiller also had a team, sworn to secrecy, working on a multi-capillary machine, code-named the Manhattan Project. The production schedule called for it to be ready for customers in two years. By then, the MegaBACE capillary machine might have already stolen a huge piece of ABI's market. Hunkapiller insisted he wasn't concerned about the MegaBACE. But he still paid a visit to his production team. "We need the capillary machine a year earlier," he told them. There was gasping, a throwing up of hands. Some engineers quit on the spot.

Next, Hunkapiller went over to Incyte and let them know that ABI was designing a capillary machine, too. "We'd be willing to work with you on this as a preferred partner," he told Roy Whitfield and Randy Scott, "but not if you're talking to someone else at the same time." Whitfield and Scott saw no reason to limit their choices. Hunkapiller conferred with the new boss of his parent company. He and White decided to make Incyte another offer: How about a merger? Again, Whitfield and Scott declined. Their company was a fast-moving start-up that had just begun to turn a profit, and they weren't interested in becoming just another division of a notorious dinosaur like Perkin

Elmer. If Tony White was going to move into Incyte's space, he would have to find some other way than buying it.

Thus it was that when the Perkin Elmer executives met on an afternoon in November 1997 in ABI's headquarters at the end of a long sandy road on the edge of San Francisco Bay, it was not to discuss how they could make history, or beat the public genome program to the ultimate prize, or mount an enterprise that would "dwarf going to the moon." They were doing business. Tony White had in the back of his mind the hope of expanding Perkin Elmer from a maker of instruments to a formidable user of those same instruments. But the competition in that market was well ahead. Michael Hunkapiller's group had some new technology to present to the board, including the multicapillary sequencing machine under development. There was a competitor driving that project, too.

Present at that meeting were some other key players. One was an Australian mergers-and-acquisitions specialist from Morgan Stanley named Alex Lipe, who with Perkin Elmer's Peter Barrett, who was also at the meeting, had guided White in his flurry of purchases of small biotech firms. The buying spree had bolstered Perkin Elmer's newly aggressive presence in the biotech sector, but Lipe had recently advised White that it was time to stop acquiring and start building from within; coming from an acquisitions manager, the advice sounded trustworthy. Perkin Elmer's last purchase had been a Cambridge, Massachusetts, company called PerSeptive Biosystems, a protein-analysis enterprise started by Lebanese-born wunderkind Noubar Afeyan seven years earlier, when the ink was still wet on his Ph.D. from MIT. Afeyan had sold his company to Perkin Elmer for almost $400 million. The deal had yet to be finalized, and formally speaking Afeyan wasn't yet a Perkin Elmer employee. But he was at the meeting in Foster City, too. Like Lipe and Barrett, he shared White's vision of "moving up the food chain" and getting into the genetic information business. But like them, he didn't really have a clear idea how that was to be done.

It was Afeyan, the newcomer, who first said the words. The head of the multicapillary-machine production team was winding up a presentation on the instrument's design and capacity. There were twenty-odd people around the table, discussing such matters as pricing, costs, and marketing strategy. This was the first time Afeyan had heard that the project even existed. He was taking some notes and idly doing some calculations. "You know, with enough of these machines, we could sequence the whole human genome," he remarked. A few people chuckled at the notion, and the discussion returned to serious topics. But now Hunkapiller was hunched over his yellow pad, scribbling. After a minute he looked up. "He's right," he said.

"Who's right?" asked White.

"Noubar. With two hundred machines, we could sequence the human genome in three years."

Most people in the room hardly knew Noubar Afeyan, but they knew Michael Hunkapiller. He would not interrupt a serious discussion except for something even more serious. Having affirmed the possibility, however, Hunkapiller drew back and became cautious: whether or not the idea was technically feasible, it had to make sense financially. But the discussion was gathering momentum. Perkin Elmer wanted to get into the genetic information business, right? But Incyte and one or two others were already there, claiming patent rights to thousands of genes. But they didn't have the *genome.* Nobody else was even thinking of going after the whole human script, except for the Human Genome Project, of course, which had no bottom line to consider. If you thought of the human genome as an endpoint, in fact it was hard to see any commercial value in it. But what if you saw it as an entrance instead? What if you offered its wild vastness as an open invitation for limitless exploration by anyone who hoped to find a gene, a drug, a cure, a new truth about who we are? These explorers—drug companies and academics alike—would arrive in droves, like miners to the banks of a river rumored to hold gold. Anyone who wanted to could squat down by the river, dip in his hand, and see if he could pull out a nugget. It was a risk, of course. But there was ample precedent for a company creating enormous wealth by first giving away something for free. "Think about it," said Afeyan. "How much would Microsoft have paid to have developed the web themselves and made it available to everybody?"

The whole discussion lasted only ten minutes. It was remarkable not just for what was said but for what wasn't. Nothing remotely like a real business model was generated. There was no talk about how the company would finance the enterprise. There wasn't any mention of getting in the face of the public Human Genome Project; far from being a competitor, the government was ABI's biggest customer. There was no talk about making history or about the future of humankind. Curiously, no one gave a moment's thought to the most intimidating obstacle to the idea, scientifically: how they would actually assemble a coherent genome from the millions of tiny fragments those two hundred machines would belch out. Tony White, who had set the process in motion to begin with, didn't say much at all. But he was hanging on every word. He sensed that he was hearing how he could move up the food chain, capture customers, beat Incyte, and put the name of Perkin Elmer in the mouths of everyone on Wall Street. Just a commodity, indeed! He even had an idea where he'd get the money he needed. "Let's get it over with," he said, when there was a pause in the conversation. "Let's just do it."

And that is where the idea for decoding the human genome in a private venture came from—not from the mind of Craig Venter, like Athena fully formed from the head of Zeus, but from businessmen sitting around a table thinking up strategies to increase their shareholder value. Of course, there were details to be hashed out, a business plan to be developed, financing to be structured, and the hiring of someone who could make it all happen. White and Hunkapiller needed to implant the seed of their idea in a nourishing environment that could carry it to term, birth it, and mother its precocious development. So in that sense, at least, it would be fair to say that the venture sprang from the head of Craig Venter.

"A genius," said Hamilton Smith, one of Venter's closest colleagues, "is someone who recognizes a good idea when he hears one."

CHAPTER 5
THE CODE BREAKER

In 1962, the year James Watson received his Nobel Prize, Craig Venter was sixteen years old. A black-and-white photograph of him that year shows a lean, good-looking kid sitting on a couch, wearing a white T-shirt and a studied smirk. His hair is silky and sun-blonded, and an insolent shock of it hangs over his forehead, beach-boy style. By this age, a lot of future leaders in science have already made their presence known by scoring in the stratosphere on standardized tests, winning national scholarships, or excelling in college while still awaiting their full complement of body hair. The boy in this photograph doesn't give a shit about going to college. He is a proactively bad student. Like Mendel, he flunks tests because he leaves questions blank whose answers he surely must know but for some reason cannot or will not articulate. Sometimes he refuses to take the tests at all. Unlike Mendel, he doesn't retreat to his bed afterward and pull the blankets up over his pain. Judging from this photograph, he masks it with contempt instead.

Venter grew up in Millbrae, California, a middle-class community a mile or so west of San Francisco Airport. When he was in grade school, there was no fence around the airport, so he would ride his bike over and race airplanes. When a plane started to take off, he'd dash out of the brush and pedal as fast as he could alongside, with the pilot shaking his fist and a passenger's appalled face framed in each passing window. When the plane pulled ahead and rose into the sky, he'd hear the siren of the airport police car and take off himself, into the reeds and thickets where they couldn't find him.

Then as now, railroad tracks run through Millbrae, dividing the nicer neighborhoods from the poorer ones. Until Venter was ten, his family lived in a tiny bungalow on the wrong side, underneath the roar of the planes, which were so low you could see the rivets in the wings. His birth, fourteen months after his brother Gary's, had been unplanned, and there wasn't enough to go around for a second child. When things got better, his father, John, an accountant, moved his family to a modest split-level house on the better side of the tracks. By then there were two more children. Gary was the pride of his parents—studious and brilliant, a straight-A student with unlimited prospects. Susie was the only girl, and little Keith, troubled by a hearing problem, the loved and protected baby of the family. Craig was given, or took on, the role of family bad egg—flunking in school, getting in trouble, and making life difficult. There is a videotape condensed from the Venters' home movies taken when the kids were growing up. Maybe Craig wasn't around much when the camera was rolling, or it just wasn't aimed very often in his direction, or maybe something was lost in the editing onto videotape. But he hardly makes an appearance.

"My mother would ask him, 'Why are you such a failure? Why can't you be like Gary?'" Keith Venter remembers. "It still pains me to think about it."

The pop-psych answer to the mother's question might be that Gary was already doing a much better job of being Gary than Craig could ever hope to, so Craig chose to get his parents' attention by failing spectacularly at being Gary instead. He spent half his time trying to stay out of trouble and the other half trying to get into as much of it as he could find. His poor performance in school wasn't due to a lack of ability. When he was required to take an IQ test in the military a couple of years after the surly photo was shot, his score was 142. It wasn't due to a lack of ego, either; even back when he was bringing home Ds in high school, Venter felt he was "destined to do something great." He had other priorities. He was a champion swimmer and a hit with the girls. "When I got to high school," Keith says, "two different teachers asked me, 'Are you going to be a genius like your brother Gary? Or a playboy like your brother Craig?'"

Keith Venter is now an architect at the NASA Ames Research Center just a few miles down Route 101 from Millbrae. The scientists at NASA Ames design rockets and spacecraft; Keith designs the buildings they work in. Though he is slimmer, darker, and more bashful than his brother, with a full head of hair, the similarities are striking. Both brothers move with an easy, placid grace, as if their limbs were lighter than most people's, and they share that expectant eagerness in the muscles around the mouth and eyes, as if they are perpetually ready to delight in what is about to happen—or, in Craig's case, what he can cause to happen. Growing up, Keith was mesmerized by his bad big brother's daring and followed him everywhere. They built forts out of scrounged lumber ("That was my architectural training," says Keith). One time they erected a tollgate across the street in front of their house and charged neighbors a nickel to drive through, until somebody without a sense of humor called the cops. Craig saw the police car coming up the street and slipped back into the house. He reemerged just as the officer was confronting the six-year-old tollbooth agent. "Keith, what are you up to *now?*" he exclaimed. "I'm really sorry about this, officer. I'll make sure this stuff is taken out of the street right away."

"I thought that was *way* cool, the way he did that," Keith says now, without a trace of resentment. "He wowed me. He was always wowing me. He still does."

Craig Venter ached to go fast—in the pool, where he had set regional swimming records; on his surfboard; and on the road after his family had enough money to buy a car. He still drives his Porsche in a controlled delirium, as if each surge of velocity is a dose of methamphetamine. When he was fourteen, he found some plans in *Popular Science* on how to build a simple hydroplane. He got a paper route and mowed lawns to buy plywood and screws. One evening, John Venter drove home from work to find his middle son had turned the garage into a construction site. "Get this crap out of here," he ordered.

"Where else am I supposed to build it?" Craig yelled back.

"You can build it anywhere you damn please, but I need this garage to park the car in when I come home. Is that understood?"

Venter thought for a while and then rigged up a pulley system to hoist the hull of the boat up to the ceiling at night so his father could slide the family car in underneath. His father was appeased, his brother Keith wowed, and the neighborhood girls enjoyed coming in the afternoon to watch Craig, stripped to the waist, hammering and bending sheets of plywood. Somebody gave him an ancient outboard motor. He got some books on engines out of the library and rebuilt the motor. When it was finished, his father helped him load the hydroplane on top of the car and they took it down to Coyote Point south of the airport. They launched the boat with the planes roaring overhead. His family watched from the shore, while Craig zipped around and around with his face drawn up in adolescent ecstasy, half blind from the spray off the bow.

But there were limits even to Keith's uncritical awe. In the summer before his senior year in high school, Craig would climb down a rope ladder at night from the bedroom he shared with his younger brother. He rolled the car down the street to get it started out of earshot, drove over to his girlfriend's house, and climbed through her bedroom window. John Venter found out what was going on and told his middle son that if he did it again, he'd tell the girl's father, who was some kind of gangster. After that, Craig made sure to wad a couple of pillows under his blankets before he swung out the window. It fooled his father for a while, but one night on his routine check he got suspicious and checked under the blankets. Then he got in the bed and dozed off, waiting for Craig to come back. Keith, pretending to be asleep, watched from under his own pillow, horrified. He woke up a couple of hours later to hear the two of them shouting at each other. As promised, John informed the girl's father, who drove over and put a gun in Craig's face. "It was scary," Craig remembers. "But what was worse was having your own father turn you in."

"I know what he means, but Craig deserved it, too," says Keith. "Think about it. She was sixteen years old. What if it was *your* daughter?"

When Gary Venter graduated from high school, he went off to study math and physics at Berkeley. When Craig graduated the next year, barely, he moved into his grandmother's garage in Newport Beach, where the surfing was better. He got a job putting price tags on goods at Sears Roebuck. He soon realized that he'd either have to get some skills or end up being a stock clerk all his life, so he enrolled in junior college. But it was too late: this was 1964, and a semester of junior college wasn't enough to keep him from getting drafted into the army. His father, an ex-marine, convinced him to talk to a navy recruiter first. The recruiter noticed his high school swimming record and promised him a three-year enlistment instead of four and a spot on the navy swim team, which sounded like a great way to serve one's country. But just before he reported to boot camp in San Diego, President Johnson announced that he was escalating the war in Vietnam, and all military sports teams were canceled. Venter now found himself in a crowd of 35,000 other young men behind barbed wire in a place dedicated to breaking the will, watching the planes take off from San Diego Airport for far-off places. After a week, he hatched a plan with another miserable recruit to go AWOL by swimming down a creek that ran through the base and eventually out to the ocean. The night before the planned escape, the company commander announced that he'd heard there were two assholes who were thinking they could swim out, and he'd like to remind them that desertion during time of war was a capital offense. They changed their minds.

Soon afterward, Venter took the IQ test along with the other recruits. His surprisingly high score gave him a choice of service options. He chose to train as a hospital corpsman, because it was the only one of the interesting options that didn't require extra years in the military. Posted to the Balboa Navy Hospital nearby for his medical training, Venter quickly learned a lesson in military survival tactics: make yourself useful to those who can protect you. Pretty soon he was performing spinal taps and liver biopsies and teaching others how to do them. The physicians rewarded his work by hiding him from daily inspections, so he rarely had to put on a uniform and could let his hair grow long again like his surfing buddies. By three in the afternoon his shift was over and he could hit the beach.

Venter received orders to transfer to a naval clinic in nearby Long Beach to manage the emergency room. Suddenly things weren't looking so bad; he'd wanted a skill and the navy was giving him one. The day before he was due to leave, however, an officer on the nursing staff told him to get a haircut, and he responded by telling her to fuck off. Nobody could protect him from the consequences of an outburst that blatantly stupid, and when she filed charges against him for disobeying a direct order he was court-martialed and ordered to report to the brig in Long Beach instead of the emergency room, for six weeks of hard labor followed by a certain posting to Vietnam. By this time, Venter had learned that most hospital corpsmen in Vietnam served as medics in combat, and the Vietcong paid a bounty to any soldier who could kill a corpsman and bring back his insignia.

The night before he was due to report, he looked at where his life was headed and made a calculated decision. His orders to the brig were penciled onto the original copy of his orders, which was taped on the outside of the envelope containing his official papers. He steamed open the envelope and discovered that the other copies of his orders inside still had him reporting to the emergency room. He managed to remove the amended original orders on the outside of the envelope; then he got on his motorcycle and drove to Long Beach.

"I'm sorry sir, but my orders were blown off my motorcycle and the copy on the outside was lost," he told the reporting officer. The guy chewed him out, confined him to barracks for a week, and told him to report to the emergency room as instructed.

Venter served in the emergency room for six months. He received a commendation for his work, and when his posting to Vietnam could no longer be delayed, one of the doctors helped him write a letter volunteering for duty at the field hospital in Da Nang. He spent the first six months in the emergency room there, sorting the salvageable from the dying and trying to stitch torn bodies back together. Double amputations on soldiers with both legs mangled by a mine became a kind of personal specialty. For the second six months, he worked in an infectious disease clinic. "Nobody said, 'Hey kid, where's your M.D.?'" he remembers. "If you could do it, you did it."

Through those first six bloody months in Da Nang, Venter, twenty years old, became dumbstruck by the phenomenon of life from watching boys like himself struggle and die. One day a soldier arrived with a spatter of his intestines on the stretcher around him. Most of the rest had been blown away by a mortar. Yet the guy was conscious—indeed, talking animatedly about how he couldn't wait to get back to Brooklyn to play basketball with his buddies again. He never got there, of course. But he survived for two weeks without any guts, regaling his mostly comatose fellow patients with stories of life on the street. Another guy came in DOA with a small, nearly bloodless wound in his head. Venter took out the dead soldier's brain and found that the bullet had left just a tiny track, with hardly any obvious damage. How could life continue in one and be so easily snuffed out in the other? Why was life so powerful? Why was it so fragile?

His own hold on life was beginning to seem a lot more tenuous, especially during the Tet Offensive, when the hospital was subjected to near-constant shelling. One night when he was working a late shift, shrapnel from a rocket tore up the mattress on the bed where he would have been sleeping. The operating room was better protected than the barracks, and for his last three months in Da Nang, Venter would wash the blood off the operating table at the end of his shift, crawl up on it, and go to sleep, figuring there was a better chance of making it to the morning there. When his tour of duty was over, a life spent surfing didn't seem like an option anymore. Death is too powerful a motivator. He enrolled at San Mateo Community College, afraid that with his past performance in school, he wouldn't be able to cut it. But he wanted to be a doctor.

Besides, he was married. He had met his wife while on shore leave in Australia. Lying on the beach, he noticed a woman out past the breakers, drowning. Or maybe she wasn't drowning, but in any case he swam out and rescued her, and brought her back home with him. The marriage did not start on solid ground. "I wanted sex," he says. "She wanted her green card."

After a year and a half at San Mateo, Venter transferred to the University of California at San Diego. Under the eye of the distinguished biochemist Nathan Kaplan, the ex-navy medic who'd poked in wonder at the tiny killing track of a small-caliber bullet in a soldier's brain began looking at brains at a finer resolution. There was a controversy going on then between some British scientists, who thought the receptor for the neurotransmitter adrenaline worked inside the membrane of a brain cell, and their American counterparts, such as Kaplan, who argued that the receptor was exposed on the outside of the membrane. While still an undergraduate, Venter devised an experiment to help settle the dispute. To the adrenaline molecule itself he chemically welded a microscopic glass bead, much too large to pass through the membrane of a brain cell. The club-footed molecule stuck to the membrane of the brain cell and did its work: ergo, the receptor for adrenaline must be on the outside of the membrane. The work was published in *Proceedings of the National Academy of Sciences,* an unusual distinction for an undergraduate. He decided to go into research. "A doctor can save maybe a few hundred lives in a lifetime," Venter told his newly wowed little brother Keith. "A researcher can save the whole world."

Since Vietnam, he had developed both ambition and a clutching fear that he would die before he could accomplish whatever next step he had set for himself. He raced to get his bachelor's degree and his Ph.D. in five years, then skipped entirely the next stage in a conventional academic career—serving as a postdoc—and took the offer of a junior faculty position at the State University of New York at Buffalo. The morning he arrived, the favorite student of a high-ranking professor was scheduled to defend her thesis. The professor wanted to show her off to the new hire from UC San Diego, so he invited him to sit in. The student made her presentation, and after she left, the professor turned to Venter. "So what did you think?" he asked.

"That was the most mediocre load of shit I've ever heard," Venter responded. People stopped asking for his opinion, but that didn't keep him from giving it.

Venter did not act professorial. He had energy, ideas, and an in-your-face brazenness way beyond his untenured place. He also had slightly crazy blue eyes, a scraggly beard and an even scragglier ponytail, a two-year-old son, and a marriage on the rocks. His wife got a job in Texas and left behind their son, whom he often brought with him to work, changing diapers during class. Claire Fraser, a proper New England girl, the daughter of a high school principal in a Boston suburb, was one of his students, and soon their relationship was an open scandal. His $21,000 salary was pretty good for a young researcher in the mid-seventies, but the baby-blue Mercedes he bought with it didn't blend in well in the faculty parking lot. His wardrobe was monstrous. Academics might prefer tweed jackets and cardigans, but in the world that Venter was more familiar with, polyester was in vogue, and he was a one-man walking fire hazard. To one faculty party he wore bell bottoms made of a fabric that Fraser says "would now be used for carpet pads, with a crease you couldn't erase if you ran over it with an eighteen-wheeler." The pants were white with red roses embossed on them. Over a green shirt spattered with little Disney characters he wore a yellow vest with ragged fringe and huge lapels.

In spite of the pants, once Venter's divorce was final Fraser married him. There was a zone around this odd dervish that few of his colleagues wanted to penetrate, but that made him even more exciting to her. At faculty meetings he was full of bravado and blind confidence, but painfully clueless as to why he wasn't being accepted by the others. "He would go to meetings with these distinguished professor types, and pretty much tell them they were stupid," she remembers. "That's not the way to win friends."

In his research, Venter was continuing the work he had started in San Diego, trying to understand the nature of proteins on the surface of brain cells that pick up chemical signals called neurotransmitters and trigger the cell to fire off an impulse in response. If he could characterize the receptor for the adrenaline neurotransmitter and find its gene, it might provide some clues to how messages in the brain are sent and received, which in turn might shed light on why we think and behave the way we do. In the mid-1970s, a molecule called cyclic AMP was being heralded as a long-sought "secondary messenger" controlling a great number of cellular functions, including adrenaline reception. In Buffalo the universality of cAMP was accepted as dogma, but Venter's work on the adrenaline receptor was explicitly contradicting the idea, and he let everybody know it. Perhaps if he hadn't been in such a rush to make up for lost time, he would have done so with more tact. But probably not.

Though he was attracting plenty of students and grant money, Venter was denied tenure in the pharmacology department. Fortunately the biochemistry department picked him up. He and Fraser were later offered appointments at the National Institute of Neurological Disorders and Stroke, part of the National Institutes of Health, and they moved to Maryland. NIH "intramural" positions were considered plums, because research funding was automatic, without the need to apply for grants. But ten years had gone by, and Venter was still chasing after the identity of the adrenaline receptor gene. He eventually lost that race to another lab. He had done nothing to change the world, as he had meant to do when he came home from Vietnam twenty years earlier. Then he heard about a new enterprise under discussion at NIH. It would be organized on a scale more to his liking, and led by the legendary James Watson.

CHAPTER 6
THIS GUY CAN GET SEQUENCERS TO WORK

Before Vietnam, Venter had lived pretty much in the present; Da Nang did something to him, and from then on he seemed to be living in the future instead. His attitude and his lifestyle were scaled to match what he wanted to become, not who he was. Such a person is bound to have collisions with people in authority, because they will insist on relating to him in his present state, while he has already vacated that self-conception and moved on to larger quarters. It can be hard on a spouse, too. Venter and Claire Fraser adored each other, and they worked seamlessly together in the lab. But Claire had to learn to relate to someone who on the inside was eight steps ahead of the person standing in front of her. He used her as a sounding board for his big, future-sized ideas. A lot of them were less than brilliant, and some were downright stupid. Early in the marriage, she would just tell him so. He would get angry, they would fight, and she'd end up in tears. She soon decided that when Craig asked for her opinion on some brainstorm, she would tell him it was terrific. If it wasn't, he'd figure that out later for himself.

On a practical level, the discrepancy between where Venter was and where he wanted to be made a mess of their finances. He had learned to sail in Vietnam, and in Buffalo he had insisted on buying a speedy little catamaran that they couldn't afford. They could have afforded it after they moved to NIH, but at that point Venter bought a larger boat, which put them back into debt. Then one day in 1989, while on an airplane flight back from a meeting in Japan, he got another big, future-sized idea. He had no notion how much boat it would buy.

At that time, the Human Genome Project was just getting off the ground under James Watson's leadership. The National Institute of Neurological Disorders and Stroke, Venter's institute at NIH, had little to do with the new enterprise. Two years before, however, Venter had read an article in *Genomics* by Leroy Hood, Michael Hunkapiller, and Lloyd Smith, describing their first automated DNA-sequencing instrument, soon to be developed by Applied Biosystems. Several other automated sequencers had already been brought to market but were proving so finicky and difficult to operate that their purchasers had taken to calling them "$100,000 paperweights." Watson himself was disenchanted with the technology. In 1987, Craig Venter did not know anything about sequencing DNA—he was supposed to be analyzing brain proteins—and because he was ignorant, there were no constraints on his perceptions. He could see right away that the future did not belong to people squinting at fuzzy columns on a gel and counting out the 3 billion letters of the human genome with their eyeballs. He asked his lab chief at NINDS for the funds to buy not one but two $100,000 paperweights so he could sequence the code of the adrenaline receptor gene. His boss refused and told him to get back to work on proteins. But Venter had a grant of his own of $250,000 from the Department of Defense, meant to be spent on research germane to combating biological warfare. He contacted Hunkapiller and lobbied to have his lab chosen as a test site for the new machine.

It took some time—the technology was complex, not something you could plug into the wall and expect to work—but with the help of Hunkapiller's engineers and a smart young grad student named Jeannine Gocayne, Venter got the machines running. Soon they were spitting out rainbow columns of DNA sequence. By the beginning of 1989, Venter was barely on James Watson's radar screen. But with the speed possible using the machines, Venter figured that his lab could sequence the human X chromosome, or at least a large section of it that was littered with genes implicated in diseases. He met with Watson in his office at the newly created human genome research center at NIH, and told him what he wanted to do.

"Jim could see we were way ahead of everybody else," Richard McCombie, Venter's lab manager at the time, said later. "We asked for five million dollars. He turned to his people and said, 'This is great, give them the money.' But they told him that it didn't work that way. And Jim says to them, 'What do you mean it doesn't work that way? I run this place. What's the point of running the place if I can't make the decisions? Give them the money for two or three years, and if they do good work give them more. If they don't, we fire them.'"

"The Human Genome Project is going to succeed," Watson told Gerry Rubin not long afterward at a lawn party at Cold Spring Harbor, "because I've got this guy who can get automated sequencers to work."

Icon that he was, however, Watson was working now under the constraints that hobble all federal bureaucrats. The entire NIH budget for the infant genome program in 1989 was $28 million. The university-based scientists who had developed the plan to "map first, sequence later"—and were depending on grants from NIH to start mapping—were outraged that such a hefty chunk of the budget might be diverted for sequencing, especially to an NIH investigator they had never heard of. Watson told Venter that in order to get that kind of money, he would have to submit a proposal, in essence following the procedures designed for scientists outside NIH. The review committee bounced the proposal back, asking for more information. Venter wrote a longer proposal, and it too was rejected, but with an assurance that it might be more favorably reviewed if Venter scaled it to the committee's recommendations. He refused. Perhaps if he had been more politic about it, things would have gone differently.

"There were two reasons why those grants were rejected," said McCombie, who parted with Venter in 1992 and now works at Cold Spring Harbor. "First, we were way ahead of everybody else, and nobody realized it. And second, Craig was an asshole, and everybody realized it."

Venter did not much care what most people thought of him. But what Watson thought of him was another matter. He was in awe of the famous scientist, and thrilled that he seemed to enjoy Watson's respect. Venter wanted to change the world with his research; here was someone who had done just that, and now he was championing Venter's own ambitions. When those ambitions grew suddenly larger on that flight back from Japan, Venter assumed that Watson would support him again. But he had painfully overestimated the great man's esteem.

. . .

Venter's idea was a shortcut way to find genes. The human genome is a very bulky contraption. It contains all the instructions for manufacturing and maintaining a human organism, but it also includes vast quantities of junk DNA, both between and *within* the genes themselves, that does not code for proteins. "Junk" is a misnomer: Although protein-coding genes account for less than 3 percent of the DNA in the human genome, inferring that the rest is worthless is like saying there is no value in the deserts of the Middle East because they are composed mostly of sand and only a little bit of oil. The fact is, we don't know what purposes lie hidden in that alleged junk. We do know, however, that some of it performs the vital function of regulating when a gene is turned on or off. Without those switches, there would be no difference between a liver cell, a brain cell, or a cell in your big toe, and we would all be a dysfunctional chaos of overexpressed protein. The importance of the regulatory regions was one reason that Watson and the other molecular biologists in the Human Genome Project had determined to go after the whole sequence and ignore the pleas of the medical geneticists to go for the genes alone. Venter believed that Watson's strategy was right, in the long run. But in the meantime, why not go for the codes of the genes themselves?

DNA is a lazy molecule. It just sits on the chromosomes, like a general who calls the shots in battle without getting up from behind his desk. When a gene is about to be expressed—translated into a protein—the two strands of the genome's double helix split open where the gene resides, like a zipper opening partway up its length. DNA nucleotides floating around in the cell nucleus glom on to their complements on the exposed strands to form single-stranded molecules called messenger RNA. In the process, the noncoding interruptions of "junk" within a gene's sequence get edited out. Thus the messenger RNA, or mRNA for short, represents a tightly abridged version of a gene, containing only those passages that the cell needs to make a protein. The newly created mRNA molecule, carrying its instructions from the general, scurries out of the nucleus and delivers the message to ribosomes, tiny hamburger-shaped structures in the cell's cytoplasm. It is here that the active business of making a protein gets done. Meanwhile, the two strands of the DNA zip back together, rendering the gene inert until its code is needed again.

The role played by RNA in this process immediately suggests a way to isolate and capture the DNA code of the genes themselves: ambush the messenger and steal its orders from the general. Messenger RNA itself is too fragile to be isolated and manipulated in the lab. But with the help of certain enzymes, biologists can transform the fleeting, single-stranded RNA molecule into a durable double-stranded synthetic one called complementary DNA, or cDNA for short. Theoretically, at least, if you could read out the sequence of all the cDNAs expressed in human cells, you would have deciphered the code of all the human genes. How many genes there are in total was a mystery, but in 1989 estimates ranged from around 50,000 to as many as 200,000.

Craig Venter was not the first to have the idea of clubbing the messenger and stealing its code. The notion had been kicking around since the Human Genome Project had first been discussed in 1986. But getting the sequence of entire cDNAs would be a laborious process, even if one were able to capture enough mRNAs to make the effort worthwhile, which most people doubted. Venter's airborne idea was to pluck out a short segment from a cDNA—just enough to identify the gene's presence and define it as unique from the other genes expressed in a cell at the same time—and spell out the fragment's sequence with one swipe on his automated machines. These little molecular dog tags, three hundred to five hundred letters each, would be enough in themselves to verify the presence of a gene and could later be used to pin down the gene's location on the chromosome, leading medical gene hunters to their quarry.

When Venter returned from Japan he called his lab people together and outlined what he had in mind. "So who wants to work on this?" he asked. Nobody answered. They were all busy on their own projects, and they'd heard a lot of big Venter ideas come and go. But a few days later, a new postdoc showed up in the lab from the University of Michigan. His name was Mark Adams. He looked like a poster-boy nerd: a pencil-thin figure with big glasses, Bill Gatesian bangs, and a birdlike economy to his movements. He'd been hired to work on sequencing the human X chromosome, but the NIH had still not funded that project, so Venter asked him if he'd like to try something a little different. Adams snapped it up. He wasn't a big fan of the "map first, sequence later" credo of the Human Genome Project, which seemed to him like "a slow boat to China," so he was more than happy to try out an approach that could potentially yield a lot of exciting information very quickly. He immersed himself in the chemistry needed to run the sequencers, the molecular biology used to prepare the samples of cDNA to go into the machines, and the computer logic needed to make sense of what came out the other end. He worked hard all day, absorbed and apart from the others, and at precisely 5:27 every day, he packed up his briefcase and left to meet his wife at the NIH Metro stop.

As raw material for the project, he and Venter chose a commercially available collection of unsequenced cDNAs extracted from brain tissue. It was a wise decision. Some Canadian scientists had tried a similar approach on muscle tissue as early as 1983 but had not found any new genes. Human brains, however, are run by a lot of different proteins, which means a lot of different genes are expressed in brain cells. Adams captured the sequence of his first gene fragment in a few days, a dozen more in a week. Venter dubbed the fragments "expressed sequence tags," or ESTs for short. Soon ESTs were erupting like popcorn. By the end of the summer they had collected a hundred. By the following summer that number was approaching 2,000, essentially doubling the number of known human genes. Granted, these short stretches of sequence only identified the *presence* of a gene and said little about what function the gene was performing in the cell. But with a little extra work, their locations on the chromosomes could be pinned down and some hints of their function could be gleaned by comparing their sequences to known genes in other organisms. One of the first genes Adams turned up, for example, was the human version of a gene known to be vitally important in guiding embryonic development in the fruit fly. Even knowing that humans possessed such a gene was valuable information and could lead to a better understanding of its function later on.

Venter was ecstatic. He had veered wildly off course from his approved plan of research, but the risk had paid off. While the Human Genome Project grant committee was still dragging its feet over his X-chromosome proposal, he had already leapfrogged ahead of that idea and found a way to go forward even faster, using his ESTs. Venter wrote Watson to let him know what he was up to, hoping to win his approval and some funding to continue the EST project. This time Watson was unimpressed. He was not against going after the protein-coding regions of the genome, but he didn't want to do it in this quick-and-dirty fashion, or before more progress was made on mapping. He also had much less respect for Venter as a scientist than his eager admirer realized. Watson believed that there were people with great minds like his own, and then there was everybody else. He had already decided in which category Craig Venter belonged.

"When Craig first started talking about ESTs, the quality wasn't that good," Richard Roberts, a colleague of Watson's and a future Nobelist himself, later remembered. "He didn't understand the science that well at that time. As soon as you began to ask him questions, it was clear he didn't understand everything he was talking about. Jim took an instant dislike to him because of that."

Watson did not formally respond to Venter's letter, and Venter was growing increasingly frustrated and letting everybody know about it. His EST method, he told *Science,* was "a bargain in comparison to the genome project." For a few million dollars a year, he said, his lab could find 80 to 90 percent of the human genes, whereas sequencing the whole genome would cost hundreds of millions. It was not the sort of public statement that would endear the parvenu scientist to the genome community. The big worry over getting the genes first had always been the fear that Congress would perceive the project's job as done and abandon the effort to get the whole code. Venter agreed that it was important to sequence the whole genome. But he refused to throttle back his own research to match the slower pace of the community project. He seemed astonished that the community did not see it his way.

The dispute was beginning to draw attention. One day in the spring of 1991, Venter was over in the main administration building at NIH when he stopped and asked someone directions to the men's room.

"You're Craig Venter, right?" the man said. "I've been meaning to get in touch with you."

The man was Reid Adler, a lawyer and the head of NIH's Technology Transfer Office, which meant he was responsible for administering the agency's patent policies. He had read about Venter's gene-finding method in a news article in *Science.* So had other people. Adler had already been contacted by an attorney from the biotech firm Genentech, who had expressed concern over the imminent publication of the DNA sequences Venter was finding. According to the attorney, if Venter's short, identifying fragments of genes were released freely into the public domain, it could make it difficult or even impossible for biotechs like Genentech to gain patent protection on their own gene discoveries. The U.S. Patent and Trademark Office had already ruled that an isolated human gene was valid intellectual property, if it fulfilled the requirements that any other invention had to meet in order to get a patent. One of those requirements was that the invention be "novel": if somebody else had already made public the same discovery, then the subsequent patent claim had to be rejected. Conceivably then, publication of Venter's ESTs—mere fragments of genes—could be construed by the Patent Office as preempting the claim of a company to patent the complete gene, even if the company had invested in the hard work of reading out the gene's whole sequence, identifying its protein, and figuring out what role it played in the body. It would be better for business—and American business in particular—if NIH patented the ESTs first, allowing academic researchers to use them for free and commercial ventures to license them at a fair cost. The logic might seem counterintuitive, but it was logical nonetheless. "You should come talk to me before you publish any results," Adler told Venter. "There's a strong argument for patenting all those ESTs."

Venter was initially opposed to the idea. Putting any kind of restrictions at all on use of the ESTs by scientists would seem to limit their usefulness. But Adler's arguments impressed him. Besides, Adler made it clear that whether Venter went along with the notion or not, NIH had a legal obligation under the Bayh-Dole Act to at least try to patent the ESTs. A lot of people might argue that a mere wisp of a gene sequence from brain tissue, with almost nothing known about its function, was not a patentable product. But that question, Adler argued, was up to the experts in the Patent Office to decide, not NIH. Bernadine Healy, the director of NIH, strongly supported Adler's patent applications. Reluctantly at first, but with growing enthusiasm, Venter agreed to go along with the plan. He let the Human Genome Project office at NIH know that he was proceeding with the patents, and when he didn't hear anything back from James Watson, he forged ahead. The patent application was officially filed in June 1991, just before the research was published in *Science.*

The first time Venter learned how Watson felt about the scheme was the last time he had any delusions that he would enjoy the great man's blessing. The occasion was a Senate meeting called by Pete Domenici the following month. Domenici was the leading proponent of the Human Genome Project in Congress. He was also a Republican, with a deep concern for protecting American business interests. At the time, the Japanese were trouncing the American computer chip industry, and Domenici was worried that the biotech industry might be next. He had invited a number of genome scientists to give their opinions on the status of the program, including Venter and Watson. Some press people had come for the meeting, too. Venter testified that he had filed patents on a thousand new human genes, based on the ESTs he used to identify them. It was the first time most of the people in the room had ever heard of the plan to patent the short sequences. One participant later remarked that when he heard it, he "almost fell out of his chair."

Watson, however, was not surprised. He had come prepared, and with his usual imperious aplomb dismissed the idea of patenting gene fragments as "sheer lunacy." Venter's automated sequencing machines, he said, "could be run by monkeys." Venter, sitting next to him, turned pale.

"You could see the dagger go in," a witness later recalled. "It killed him."

On the elevator afterward, Watson mentioned to some people that he had been "too hard on Craig." But the damage was done. In Venter's lab the next day, it was as if someone had died. To cheer people up, one of the team members appeared wearing a gorilla suit. But it was hard to laugh off the sting. Perhaps Venter should have seen what was coming. Watson had never hidden his opposition to the commercialization of science. In his view, molecular biology was a domain in which great minds made great discoveries. He was not opposed to colleagues having ties to industry—owning company shares, consulting, or sitting on boards, all of which were activities he himself engaged in. But the sense of the primary research being overrun by the profit motive was deeply repugnant to him. Any monkey could make money. Venter had fancied himself deserving of Watson's approval, and the elder man had let him know where he thought he belonged: down with the other lower primates.

In April 1992, Watson, embittered by his disputes with Bernadine Healy over the patent issue, resigned from the Human Genome Project. By then, Venter had left NIH, too. If the government genome effort wasn't interested in his quick way to scoop up human genes, it seemed there were plenty of business people who were. His phone began to ring soon after the publication of his EST paper in *Science* in June 1991. He had only $2,000 in savings, and the offers were tempting. But he had not gone into research to get rich—not that he was opposed to that development. He wanted to do recognizably great things. One savvy venture capitalist understood this. Wallace Steinberg was head of a New

Jersey company called HealthCare Investment Corporation. After some preliminary discussions on the phone, Steinberg came down to Bethesda and made the NIH scientist an offer on the spot: HealthCare would channel $70 million over seven years—an astounding amount for basic research—to a nonprofit institute run by Venter, where he could carry on his science however he pleased. In return, the new institute would give proprietary commercial rights for any marketable discoveries it made to a profit-making venture set up by Steinberg. Venter would own 10 percent of the commercial enterprise's stock. It seemed like a great deal. The meeting took all of fifteen minutes, and the deal was closed with a handshake. Venter thought that he had entered the best of all possible worlds. "It's really remarkable. Not a single document was exchanged," he told a reporter for the *Washington Post.* "It's every scientist's dream to have a benefactor invest in their ideas, dreams, and capabilities."

Venter named his new enterprise The Institute for Genomic Research and, bringing Mark Adams and many of his other NIH lab staff along, set up a facility a few miles north of Bethesda. He had tens of millions of dollars with which to move forward at his own speed, impeded neither by academic politics nor by a commercial bottom line. Steinberg brought in William Haseltine, a prominent AIDS researcher at Harvard, to head the for-profit company, which was to be called Human Genome Sciences. The terms of the relationship between TIGR and HGS were simple enough on the surface. TIGR would search through a variety of human tissues looking for expressed genes, sequence five-hundred-nucleotide fragments on a host of new ABI machines, and compile them into a massive database. Haseltine's HGS would then have six months between the discovery of the gene fragments and their publication to look them over and analyze their possible functions. After that, Venter could publish the information, and academics and other nonprofits would be free to use the data however they wished, on condition that they give HGS first option on commercial rights to anything discovered using the data. The six-month delay wasn't arbitrary or unusually restrictive; in fact it matched NIH's own policy, which at the time, before the 1996 Bermuda Accord, required scientists to release their data within six months of generation. If HGS found a gene to be potentially valuable to medicine, it could invoke a special clause granting it up to another year to explore the gene's commercial value and patent it before Venter could make the data public. University researchers could use the data after its release, as long as they signed a form promising not to exploit it for commercial purposes.

Somehow Venter was naïve enough to think this arrangement would come without a cost. William Haseltine was already notorious for his aggressiveness in science and business. His suits were dark and expensive, and he wore his thinning black hair slicked back in a style more often seen on Wall Street than on a campus quad. His voice had a gentle, seductive lilt to it, and his red lips were fixed in a cold, distancing smile. Haseltine had gotten into the race to understand the AIDS virus early, quickly starting a biotech company to characterize retroviruses. He had subsequently founded a series of other biotechs with Steinberg, at the same time keeping close ties to Harvard and the Dana-Farber Cancer Institute. Along the way he had married Gale Hayman, the jet-setting creator of Giorgio perfume. He had tycoon-sized objectives, and by some reports would use extraordinary tactics to achieve them. According to one investigator, during the AIDS race Haseltine would call up his competitors at three o'clock in the morning, just to keep them off balance. After Haseltine became CEO of Human Genome Sciences, a scientist at Incyte Pharmaceuticals in California—HGS's chief competitor—looked out his window one day to find him standing on the lawn, peering back in at him.

Haseltine and Venter first met for lunch at a fast-food restaurant in Bethesda. In Venter's view, Haseltine's role in the new enterprise was merely to mind the purse strings while he merrily went about identifying and publishing new human genes. But Haseltine had a different plan in mind. "I was running the company de facto from day one," he recalls. "After that first lunch with Craig, on the way back on the plane, I formed an idea. The primary goal was to build a new global pharmaceutical company that discovers, manufactures, and sells its own pharmaceutical products, with a market cap of three billion or greater. That was immediately my goal. I don't care what Craig's goal was. He was just a booster rocket. I discussed that explicitly with Wally Steinberg. He said to me, 'Bill, you aren't going to keep him *around,* are you?' "

Not surprisingly, the two men soon found themselves clashing over their competing agendas. From the discussions with Steinberg, Venter had gotten the impression that HGS would be holding only a few dozen genes out of the public domain for more than six months. Instead Haseltine was invoking the extension clause on virtually any sequence that had a whiff of medical importance, and very little information was coming out of TIGR into the public domain. To the academic scientists, Venter's surprise at this turn of events seemed disingenuous, to say the least. "When he went off with Bill Haseltine, Craig was seen as evil," says Norton Zinder of Rockefeller University, a key figure in the establishment of the Human Genome Project. "The world looked on TIGR as an absolute den of corruption." Nor did it help matters that Venter was so visibly enjoying his sudden wealth, including the purchase of a $1-million-plus house in Potomac, with an indoor swimming pool in a room that alone was larger than his entire former house, and his $1-million-plus dream boat, *Sorcerer.* Haseltine was richer. But no one begrudged him his wealth as they did Venter his.

At TIGR, meanwhile, the EST method was beginning to bear real fruit. In 1994, Bert Vogelstein, a highly respected cancer researcher at Johns Hopkins University, tapped into the TIGR database and quickly found a long-sought gene for a particular type of colon cancer. Other researchers were having success using ESTs to search for genes involved in Alzheimer's disease. TIGR was producing thousands of new human-genome-fragment sequences every month, magnifying the value of its database. Of course, HGS retained commercial rights to anything discovered using the information, including Vogelstein's find. "An academic can use the data for anything he wants," Haseltine said. "The only thing they can complain about is that they can't derive income from it. Well, that's life."

Neither the academics nor other private enterprises, however, were willing to accept life on Haseltine's terms. Soon after Vogelstein's discovery, Alan Williamson, a vice president at Merck, offered to fund a separate, aggressive EST discovery project at Washington University in St. Louis, one of the leading academic centers for DNA sequencing in the public genome program. Unlike TIGR's database, the Merck-funded gene information was made available immediately for use by anyone, without restrictions. Meanwhile, Randy Scott and Roy Whitfield had founded Incyte to exploit Venter's EST method to find even more genes. They envisioned Incyte primarily as a supplier of genomic information: compile a huge and easily searchable database of gene sequences and sell access to it to the pharmaceutical companies hungry for a way to find drug-related protein targets. To hedge its bets, Incyte was applying for patent protection on as many ESTs as it could find. If a drug company or biotech was intrigued enough by something they found in the database to order a physical copy of the gene, great—Incyte would issue them a license giving it a share of royalties on any profits derived down the road on a drug developed from that gene. But it would take many years to see any income from those royalties, if they ever appeared at all. In the meantime, the company hoped to turn a profit on subscriptions to the database alone.

William Haseltine was monitoring Incyte's progress very closely. But his own business ambitions were grander. He didn't want to simply *feed* Big Pharma, he wanted to *be* Big Pharma. Compiling a database of ESTs was just the first step in building his global drug company from the gene up. Once Venter's "booster rocket" had gotten Haseltine off the ground, his company would mine the information for itself, developing, testing, manufacturing, and selling gene-based drugs. This was where the really huge profits lay. But Haseltine estimated that it would take twelve years to bring a gene-based drug to market. In the meantime he, too, needed more cash to keep the operation going. In 1993 HGS signed an exclusive deal with SmithKline Beecham, one of the world's top ten drug companies. In return for allowing SmithKline Beecham access to TIGR's swelling lode of gene data, HGS would get $85 million up front, plus the promise of $40 million more for more genes down the road. Ironically, HGS's deal with SmithKline opened up a market for Incyte, too. As soon as the other big pharmaceutical companies saw that one of their own had secured a private gene mine, they flooded to the only other major source of genomic information. In 1994 Incyte signed Pfizer to a nonexclusive subscription to its database and within a couple of years had made nonexclusive deals with most of the other big drug companies. By the time Perkin Elmer tried to acquire Incyte three years later, Scott's company was actually turning a profit.

Back in Rockville, Venter was watching his best of all possible worlds be torn apart. On one side were the academics vilifying him for withholding data, and on the other Haseltine, who was demanding that he withhold even more. Any semblance of trust between the two men had disappeared. Venter felt that he had been double-crossed into serving HGS's commercial interests. Haseltine in turn suspected that Venter was keeping data secret from HGS. Late in 1993 Haseltine installed his own sequencing operation at HGS to pump up production. Venter accused him of going into competition with his own partner.

"There was no reason for me to compete with him," Haseltine said later, with his cool beveled smile. "I already owned everything he did."

The situation was beginning to push Venter over the edge. One day a business associate of Haseltine's, who had purchased a sizable chunk of HGS stock, invited Venter and his wife to lunch in New York and sent his helicopter down to Rockville to pick them up. Suddenly Venter remembered a clause in his contract with Haseltine that required HGS to continue annual payments on the $70 million pledged to TIGR only if Venter were alive. When the chopper arrived, he was afraid to get in. "Do you know how easy it would be for somebody to push us out?" he told Claire.

Not long afterward, at a meeting in France, Venter began feeling terrible sharp pains in his gut. Rushed to the emergency ward, he was diagnosed with severe diverticulitis, an infection of the intestines often associated with high stress. He was flown back to Washington, where a foot of his colon was removed to save his life. It was a bad time. Instead of healing the world, he was making himself sick and his enemy a wealthy man.

It was around then that he met Hamilton Smith.

CHAPTER 7
THE QUIETER WORLD

On the morning of July 7, 1998, Hamilton Smith drove down from his farm in Howard County and pulled into the TIGR parking lot. His 1987 Mercury Grand Marquis rumbled along the rows of cherry Corollas and silver Civics like an old tug trying to dock in a marina. The car had a long piece of trim missing on the driver's side, exposing a parallel row of rusted holes, as if the car had been strafed long ago. The odometer read 244,000 miles. The radio was playing—the knob had stuck in the "on" position a couple of months before—and a sucking sound was emanating from somewhere deep in the steering column. Smith didn't mind, because he had his hearing aid turned down low. The Mercury was among his most beloved possessions. He was more ambivalent about his Nobel Prize.

Smith maneuvered the car into a spot, gathered up his briefcase, and quietly made his way through TIGR's elegant lobby. He was on his way to pop in on Craig when the receptionist called out to him. "Something came for you FedEx, Dr. Smith," she said.

"I'm sorry," he said. "I missed that."

"A FedEx," she said, louder. "Over in shipping."

Smith collected the package and took it directly to his lab. It was from Gerry Rubin's *Drosophila* group at Berkeley. Packed in dry ice was a small plastic vial. Smith held it up to the light. Suspended in alcohol was what looked like the tiniest shred of a cloud: 500 micrograms of purified fruit fly DNA, about the weight of a couple of grains of sugar. He set the vial in a rack and put on his lab coat. It seemed to transform him. He was a very tall, large-boned man with pleasant features and a deliberating hesitancy in his manner, as if he were considering where to put his body next or what next to consider at all. His thick white hair fell down low and oafish on his forehead. But in the white coat, he suddenly looked like the Hollywood version of a famous scientist. The bumpkin comb job became "a shock of white hair"; the lost look, lost in thought. Missing, however, was any trace in his eyes of an assumption of self-worth. He slouched badly, as if he were trying to avoid the higher stratum of air that he had been called to share with the likes of Watson, Crick, Frederick Sanger, and the others who had achieved the supreme award. "There are Nobels, and then there are *Nobels,*" Watson once reportedly said. Ham Smith knew which category Watson would assign Smith's prize to. What's more, he agreed with him.

He flicked the outside of the vial with his finger and the DNA inside did a little dance. Rubin's people were very good at purifying DNA, and Smith had been assured that this sample, prepared from fruit fly embryos, would be very, very pure. It had better be. Smith enjoyed big, lumpy ironies—which was one reason he liked Venter, who excelled at creating ironic situations. He smiled to think that the success or failure of Craig's whole $300 million enterprise rested on this tiny flocculent wisp of DNA from baby fruit flies, like an elephant balancing on an eyelash. He flicked the tube again. Then he stooped down to steady it against the lab bench and pried open its hinged top. Using a sterile pipette tip, he carefully fished the DNA out of its alcohol bath and got to work.

Venter's new company still didn't have a name, much less a physical plant, but it did have a handful of prospective employees from TIGR. Mark Adams was going to oversee the daily operation of sequencing millions of fragments of DNA and the factory-level biochemistry needed to prepare them for the machines. Once the fragments had the order of their base pairs read by the machines, Granger Sutton, TIGR's computational biologist, would be charged with writing the computer algorithms that would reassemble them into a completed sequence of the genome. Others, under Adams's direction, would then sift through the deciphered code, analyzing it for clues to the location of genes, especially ones with biomedical value. Once things got going, Venter imagined as many as four hundred or five hundred employees engaged in these various stages in the company's pipeline.

But the first step belonged to Ham Smith alone. His job was to transform gooey clumps of DNA—first from Rubin's fruit fly embryos, and later, from human blood and sperm—into a precisely ordered form that could be read by the machines and turned into digital data. To do this, he would break up the raw DNA strands into millions of little fragments and insert each of those little pieces into its own virus molecule. The virus molecules, called vectors, are not infectious pathogens but tame laboratory tools—tiny bits of DNA that have been cut open at a particular point so that a foreign bit of DNA can be inserted, like coupling a new boxcar into the middle of a freight train. A collection of such doctored molecules manufactured for a given purpose is called a DNA library. The tiny pair of chemical "scissors" used to cut the virus at a precise point is called a restriction enzyme. To a molecular biologist, using restriction enzymes is as routine as turning a screw. No one thinks anymore about who discovered them, any more than a carpenter would pause to wonder who invented screwdrivers.

Smith was going to begin by making a typical library of a couple of million DNA fragments—just a dry run, to give Adams and the others downstream some raw material to play around with while Smith figured out how to construct a bigger, cleaner, more perfect library than anybody had ever attempted before. Near perfection was demanded by the sheer scale of the project: the larger the genome being shotgunned, the greater the chance for making mistakes in sequencing and assembling it, and thus the greater the need to minimize errors in the raw material—sequences composed of two isolated fragments accidentally stuck together, for instance. The fruit fly genome was estimated to be thirty times the size of any genome that had been assembled with the whole-genome shotgun technique, and the human genome was some thirty times bigger than that of the fly. At a congressional hearing the previous month, convened to review the Human Genome Project's future in light of Venter's private initiative, Maynard Olson of the University of Washington, an adamant advocate of quality above all else in the public genome program, told the legislators that an attempt to attack the human genome with the whole-genome shotgun technique would meet with "catastrophic problems."

"Based on extensive experience with the assembly of composite human DNA sequences in our genome center and other laboratories, I predict that there will be over 100,000 serious gaps in the assembled sequence," Olson warned. He defined a "serious gap" as one that rendered uncertain the order of the completed sequences on either side of it, calling the validity of the whole assembly into question.

Ham Smith had laughed when he heard what Olson said—not because he dismissed the skepticism as unfounded but because he shared it. *Craig is doing another high dive into an empty pool,* he thought, *and I'm jumping with him.* It seemed enormously funny.

If the project failed, Smith was going to make sure it wasn't on account of the quality of his DNA libraries. After checking the purity of Rubin's sample, he got right to work on the first step, the one that gives "shotgun sequencing" its violent name. Smith dissolved some of the sample in a larger amount of buffer and placed it in a nebulizer—a glorified atomizer, like the ones asthmatics use to spray medicine into their lungs, only enclosed in a chamber. With a hose he attached the device to a tank of nitrogen and turned the valve on the top of the tank. The nebulizer gave a hiss and the DNA solution sprayed out into the chamber through a tiny nozzle. The shearing force generated through the nozzle beat the living daylights out of the DNA; Smith had set the pressure on the nitrogen at fifteen pounds per square inch, which he knew was just enough to shatter the DNA into "shotgunned" fragments of around two thousand base pairs each, which would collect in the bottom of the chamber. Next Smith prepared a solution containing the fragments and ran it through an electrophoretic gel to isolate the bits that were just about two thousand base pairs long. They clustered together in a bright band in the middle of the gel. Dimmer bands below and above it contained bits of DNA that the nebulizer had broken up too coarsely, or too finely. He pressed the razor into the gel, cut out a strip from the middle of the bright band, and tilted it into a test tube.

With all the automated sequencers, the supercomputers, and the humming infrastructure of the billion-dollar business that Venter's enterprise would involve, it is pleasing to think of modest, unassuming Ham Smith here at the beginning, bending in solitude over a lab bench with a single-edged razor blade in his liver-spotted hand, cutting into a strip of jelly. After dissolving the fragments once again, he had a solution full of millions of bits of fruit fly DNA, each approximately the same length. He would have to run that solution out again on another gel, and then another if need be, cutting out the center of the bright band to be sure each fragment was no more than fifty base pairs longer or shorter than two thousand. This was routine "kitchen science," all of it—not the sort of thing your average Nobel laureate wastes time on. But Smith just happened to be better at it than anyone else in the world. And he was having fun. "I love bench work," he once remarked. "I guess it's in my blood. I never get tired of extracting DNA."

Like Venter, Ham Smith had grown up with a precociously brilliant brother a year older than he. His name was Norman. Before Ham was five, his father gave the boys a Gilbert chemistry set. They took it down to the basement and never really came out again. The family moved frequently from one rented house to another, and in each new basement or attic the boys set up their lab and their shop, running experiments, building radios, telescopes, blowtorches, rockets. They rigged up a jet engine from an old carburetor and brewed homemade insecticides. They played music together, too, Norm on violin, Ham on piano. They grew around each other like intertwining trunks of a tree. Norm called his younger brother "Butch." Ham called his older brother "Butch."

"We did a lot of incredibly dangerous things," Ham Smith remembered. "My mother allowed us a lot of leeway. We played with lethal high voltage, phosphorus sticks, sulfuric acids, all the concentrated acids. When I was eight or nine, we got a pound jar of sodium cyanide, to kill butterflies. Enough to coat the tip of your finger would kill you. If you breathed it in, you'd die."

His mother was reclusive, and so were the boys. Norm was extremely shy around other kids. Ham was a little less so, and would sometimes have a friend over to shoot hoops and swap comic books. "After a while my mother would say, 'It's time for Larry to go home now,'" he said. "She didn't want bad influences."

In school Norm skipped two grades. Ham skipped a grade, too, but he never thought he had a tenth of his brother's brains. He was just good at tinkering, while Norm was destined for greatness. In 1947, Norm was a finalist in the prestigious Westinghouse Science Talent Search and went off to study physics at Berkeley. Ham stayed home and attended the University of Illinois for a year, but he missed his brother and transferred to

Berkeley. They shared a room, pushing their desks together to make one larger desk. They didn't go to dances or parties, and women were as remote and unapproachable as aliens. Norm graduated with a physics degree. He could have gone anywhere to graduate school but elected to stay at Berkeley, to be near his brother.

It was shortly after entering graduate school that Norm began to act strangely. Later, when everything was understood, Ham would realize that his brother had been behaving strangely for years, but the behavior was embedded in a life so odd and reclusive already that he hadn't really paid any attention to it. "In boyhood, I thought he was perfectly normal," he remembers. "But then, I wasn't normal myself."

Now it was a different story. Norm complained about hearing a continuous noise, like clattering dishes. His eyes stung. He was so gripped with anxiety that he couldn't teach his undergraduate classes. "Somebody is trying to poison my cafeteria food," he told Ham one day. "We have to find out who."

Ham convinced him to see a doctor, who sent him to a psychiatrist, who diagnosed schizophrenia. There was nothing to be done, and it was only going to get worse. Soon Norm was unreachable. Suddenly Ham was without half of his own being, like a man waking up with only a left side. He had always resented his mother's reclusiveness; now he screamed at her for allowing his brother and him to grow up like badgers in a hole, nurturing their own isolation from the world. At the time, there was a gathering awareness that schizophrenia was a genetic disease. Ham became even more withdrawn, monitoring his own thoughts and behavior constantly for signs that he, too, was losing his mind, all the time overcome with guilt because he was still sane and Norm, the destined one, had been taken into that dark world instead of him. Ham went on to medical school at Johns Hopkins but refused to take the required psychiatry courses. He could not bear the idea of learning anything more about what might happen to him.

What he did want to learn about was molecular biology. This was the late 1950s, and the spark set off by Watson and Crick's discovery of the structure of DNA had exploded into an entirely different way of looking at the nature of life. The double-helical structure of DNA virtually shouted out its twin functions. In cell division, when the two strands split apart like a zipper unzipping, each strand would reinvent the other by the natural propensity of each A, T, C, and G to attract its natural partner. Each daughter cell thus got its own double helix, identical to those in all the other cells in the body. If, on the other hand, the helix separated at just one point along its length, like a cookbook laid open to a particular page, the gene exposed on that open page could be read by the cell and put to use. Easy as cake. But *how* did the recipe become a cake? How did the genetic "words" in a gene get translated into the flour, egg, salt, and chocolate of the protein cake itself? All the cells in an organism harbor within them all the organism's genes. How did a liver cell know to open the cookbook only at the pages needed to make the proteins that do the things that livers do? How did they know not to make brain proteins instead? What would cause some cells to go wild and bake the same recipe over and over again, in the abomination of cancer?

Ham Smith, training to be a doctor, stole time down in the library to read the papers coming out of Cold Spring Harbor—by Watson, Crick, Joshua Lederberg, Alfred Hershey, and the others. When François Jacob and Jacques Monod showed how genes are turned off and on in 1960, Smith carried the paper around in his pocket, reading and rereading it like a poem whose every word spoke to him with utmost clarity. Here were people lighting up the bottom of the biological cave, where nobody had gone before, and Smith thought, *Hey, I* get *this! I see what they did. Come to think of it, I could have done it better.*

In 1967, Smith took a junior faculty position at Johns Hopkins. He had dropped medicine and decided to go into research. Junior faculty members are expected to teach, of course, and serve on committees and perform other administrative duties. Smith did all of this badly. Intimidated by a roomful of students, he would turn and lecture to the blackboard. He was better off left in his lab, tinkering. He soon began work with a bacterium called *Haemophilus influenzae,* nicknamed "H flu." The organism has no relation to the flu virus, but a strain of it can infect children with ear infections, or, more seriously, with meningitis and pneumonia. Like other infectious bacteria, H flu stays virulent by picking up stray scraps of DNA floating around from defunct *Haemophilus* cells, recombining their mutations with those in its own living DNA. In effect, it reshuffles the cards in its genetic deck, making it harder for the host's immune system to guess what hand it is playing. It was this property of genetic recombination that interested Smith. He introduced his first graduate student, Kent Wilcox, to his experiments on the microbe. One day in the spring of 1968 Smith suggested that Wilcox put some salmonella virus into a solution of *H. influenzae* cells.

He has forgotten why he did that; he certainly didn't have a clue about what would happen. But as the next few months would bear out, science can be almost magically serendipitous. Smith had just given a seminar to his department colleagues about research going on at Harvard showing that an enzyme in the bacterium *E. coli* seemed to be protecting it from viral invasion by hacking the invader's DNA up into powerless pieces. The biochemist Matthew Meselson called it a restriction enzyme, because it restricted the virus's attempt to turn that bacterium's genome into a Xerox machine for itself.

"What happened to the salmonella virus in the *H. influenzae* cells?" Smith asked his graduate student, the morning after he'd suggested the experiment.

"It's gone," Wilcox said.

"You mean you can't detect it?"

"I mean it's *gone,*" said Wilcox. "Disappeared. I was thinking, do you think H flu has got a restriction enzyme working, like the one you were telling us about in the seminar?"

Smith didn't think so, but he just happened to have a way to find out. In an unrelated experiment, he was using a viscometer, a contraption of glass bulbs and tubes that measures the density of a solution by timing how fast it passes through a thin capillary tube. It occurred to Smith that it was the perfect way to test Wilcox's hypothesis. They prepared two solutions, one containing an extract of H flu cells and the other H flu cells plus salmonella virus. They ran the solutions through the viscometer. In the first tube, the density of the solution stayed the same throughout the experiment: hardly surprising, but this was the control. In the second, the solution decreased steadily in density as time went on—indicating that the salmonella virus was being chewed up into smaller pieces. Smith made an entry in his lab notebook, dated May 25, 1968. They had a restriction enzyme: a protein in a cell that recognizes and cuts foreign DNA but leaves its own life code alone.

Soon there was an even more exciting observation. Unlike the enzyme in *E. coli,* this one in H flu wasn't just hacking randomly at the salmonella virus: it was cutting it into forty discrete pieces, each one about a thousand base pairs long. Smith did a little math. In a code consisting of four letters—A, T, C, and G—a random string of any six letters in a row would repeat itself approximately every thousand letters. Ergo, the enzyme must be homing in like a guided missile on a particular string of six letters in the salmonella virus's DNA. Smith and Wilcox sent off a paper to the *Journal of Molecular Biology.*

Smith wasn't trying to revolutionize biology over the next year; he was just trying to figure out the identity of those six letters. The experimental designs were ingenious, and whenever genius hit a roadblock, serendipity stepped in. Using an extract of snake venom, he determined that the first letter could be either a G or an A. By the end of 1969 and before the original paper had made it into print, Smith and another graduate student, Tom Kelly, had identified the five remaining letters. The restriction site was a six-letter palindrome: C, A, followed by either a G or an A, then the same three letters in reverse. Smith was able to withdraw the original paper and submit a much more impressive one. He was proud of "the nice little bit of biochemistry" he had done, but did not really appreciate its import. But he had a friend and colleague at Johns Hopkins, Dan Nathans, with whom he often shared ideas and experiments. When Smith finished the work, he wrote about it to Nathans, who was on sabbatical in Israel. After a time, Nathans wrote back—a letter full of science news and gossip. Near the end, he joked that he had "designs on your enzyme," with its ability to split open an invading virus's circular genome. "If it were to work and open up the circle at a specific point," he mused, "this could be useful for many things."

"Useful" is an understatement. When a restriction enzyme cuts the virus, it creates the opportunity to splice in a scrap of some other organism's DNA between the two broken ends, like riveting a few extra links into a bicycle chain. The virus is unaffected. It goes on doing what its gazillion ancestors have been doing for eons: infecting a bacterium and hijacking the bacterium's DNA replication machinery to make copies of itself—clones. Only now, the bacterium is also copying that freeloading scrap of foreign DNA, which is carried along like a strip of toilet paper stuck to the bottom of a shoe. In 1972, Paul Berg of Stanford University used restriction enzymes to create the first "recombinant" molecule. He grafted a foreign gene into a bacterium, obliging the bacterium to produce the protein made by that cloned gene. Soon after, Herbert Boyer and Stanley Cohen perfected this technique and used it to manufacture human insulin by splicing the insulin gene into a bacterium. Thus began the biotechnology industry. Berg got the Nobel Prize for his work. Boyer and Cohen, who had advanced the field by an equal quantum leap, did not. They were commercial scientists.

It wasn't hard to see that once you can slice and dice DNA and clone quantities of just the bits you want, the possibilities are endless: you can manufacture proteins in bulk, engineer crops with built-in insecticide genes, introduce healthy genes into patients who lack one needed to survive—in essence, you can redesign life. But Paul Berg, among others, became deeply concerned about the ethical implications of this awesome new power and blew the whistle on himself. Once scientists started patching bits of life together from different species, where would it end? What would happen if one of these recombinant molecules was unleashed for some nefarious purpose? It was even more likely to happen by accident. People began to imagine anthrax viruses engineered into *E. coli* bacteria and dumped in the water supply, or initially innocuous genetic concoctions mutating on their own and spreading like killer bees.

At Berg's urging, a two-year moratorium on recombinant DNA was imposed. Even when the ban on research was lifted, for several more years anyone wishing to work on recombinant molecules had to navigate through a ream of new regulations. Back at Johns Hopkins, Hamilton Smith's lab was now classified "P3" by the government—a veritable hot zone contained under negative pressure and entered only after donning sterile clothing. He was no longer allowed to use *H. influenzae* bacteria, because they were infectious, so he switched to another strain of *Haemophilus.* Otherwise, his life went on essentially as it had before.

Then one morning in October 1978, Smith's parents were driving down a road in Florida when they heard the newscaster on the radio say that Hamilton Smith of Johns Hopkins University had won the Nobel Prize in Physiology or Medicine. His mother turned to her husband and said, "Gee. I didn't know there were two Hamilton Smiths at Johns Hopkins."

You can think of Smith in his lab at TIGR, twenty years later, as a craftsman alone in his workshop. He had a set of tools, a set of skills, and a task at hand, like a cobbler bent over a shoe. He was cobbling molecules. Shotgunning the fruit fly DNA into two-thousand-letter bits, with the wait time for reactions and some refinements in procedure, had taken about a day. To complete this initial *Drosophila* library, he would later have to fit those "inserts" into a virus, the vector, and then introduce the doctored virus to its bacterial host to be cloned into millions of copies. The vector he was using, called pUC18, comes precut with a restriction enzyme from the biological supply company. Should he have needed one, however, he could have ordered it from among the hundreds offered in the company's dog-eared catalogue lying on the lab bench nearby.

Today Smith's first task was to "polish" the ends of the inserts. Their ends were likely to be frayed—one DNA strand in the double helix a few base pairs longer than the other—so he added a chemical to the solution that chews back the longer strand to match its partner. Next he attached an "adapter" molecule onto the ends to link with the vector, rather like fitting an electric plug on the end of a lamp cord, only this procedure is a great deal more delicate, since the cord and the adapter are only a few dozen atoms wide and there are millions of both swimming around in solution. It was a matter of experimenting to see what works, when to add what chemical and in what concentration, how long to incubate, and so forth. On the opposite end of the adapter, Smith grafted a series of four DNA letters reaching out like a tiny Velcro hook: CACA. Then he turned his attention to the vector. It needed cobbling, too. Most important, its ends had to be tailored to present the corresponding Velcro loop: GTGT. Now the inserts of fruit fly DNA would lock into place, completing the virus's broken circle. He worked carefully, pausing to consider each decision, each squirt of a pipette into a test tube, even though he had done it all thousands of times before. He was aiming at perfection, in order to get close to it. The worst mistake would be a library that contained a significant percentage of so-called chimeras: two fruit fly fragments stuck together by accident and inserted into the same vector. The sequencing machines would interpret them as one continuous sequence in the genome, when in fact they could be miles apart, even on different chromosomes. Chimeras happen. But when it came to constructing his DNA libraries, Smith was determined to make sure they happened hardly at all.

Toward afternoon, when the shadows were lengthening across the lab bench, Venter stopped in and asked him a question.

"What was that?" Smith said.

"I said, how's Gerry's DNA looking?" Venter repeated.

"It's beautiful stuff," said Smith. "We should have a library of a couple million clones good enough for Mark to play around with by the end of the week"—though he knew it would probably be ready sooner than that. If he said so, Venter would prod him to get it done even faster.

"Can you have a library ready for full-scale sequencing in two months?" Venter asked.

"No, but I can have it in three," said Smith, thinking it would take two.

"We need it fast, but we need it to be good. I don't want to put on any pressure, but this whole thing depends on the quality of your libraries. So, we need it good more than we need it fast. But we need it fast."

"By the way, Ham," Venter added, as he turned to leave. "Can I ask why you've got a five-thousand-dollar hearing aid and you keep the volume down so low you can't hear anything?"

Smith pondered this. He'd never really thought of it. "I guess I prefer the quieter world," he said.

When Smith learned that he had won the 1978 Nobel Prize in Physiology or Medicine, he felt like someone had punched him in the stomach. The blood rushed from his brain. He hurried up to his office, locked the door, and put his head between his knees. He would be sharing the prize for the discovery of restriction enzymes with the Swiss scientist Werner Arber, who had set the intellectual framework for the discovery by theorizing that such enzymes must exist, and with Dan Nathans, who had done the preliminary work demonstrating their real value. Each Nobel award is given to at most three people each year. Smith was mortified. Why hadn't Matt Meselson at Harvard been the third? Meselson was the one who had discovered the first Type I restriction enzyme. It was all some horrifying mistake. Then the implications for his life began to set in. There would be interviews. Parties to attend, lectures to give. People would gather around him, eager to hear what he had to say, and nothing would occur to him. He would be expected to lecture to huge audiences, to banter, to be charming. The phone was ringing. Smith reached up and took it off the hook.

The reality proved worse. He was sucked into a vortex of invitations. Companies wanted him on their board. Students flocked in, pleading with him to take them into his lab. The money from the prize and its fallout led to more distractions. He bought some stocks and became obsessed with their rise and ebb. He poked at his science, unable to engage that fertile place in his mind that had always been there before. His grant applications got sloppy but were funded anyway; nobody was going to turn down a fresh Nobelist. But after a while the graduate students began leaving, fed up with chasing hypotheses that led nowhere. His classroom teaching went from mediocre to abysmal; the student evaluations in his undergraduate classes got so bad he simply stopped reading them. Most of the stocks lost money. Then one day in 1989 he suffered a terrible humiliation for a Nobelist: his annual grant proposal to the National Cancer Institute came back rejected.

Smith couldn't blame the reviewers. Everything he had done for years was ordinary. He decided to just ride things out until he could retire. Though he was still admired for his skill as a bench scientist, by the early 1990s his larger role in the scientific community had been reduced to honorary functions, such as chairing meeting sessions where younger, more productive scientists would discuss their new research. It was at such a session that Smith met Craig Venter. At the time, Venter was lost, too.

CHAPTER 8
H FLU

In 1998, when Venter announced he was going to go after the human genome, he and Hamilton Smith had been friends for five years. They did not have much in common, except that each was missing qualities that were overabundant in the other. Venter lit up any room that he entered, as if it were a party waiting for him to arrive before things could get started. Smith wilted in social situations, never sure what to say or where to put himself. But he was a joy to come to; for all his awkwardness, there was something warm and settled underneath, a pull-up-a-chair welcoming in his demeanor all the more wonderful because it was so rare in a scientist of his stature. Venter, for all his confidence, seemed to be in a perpetual state of self-invention, trying on this mask or that one, delighting in double entendres and verbal ambiguities, so that one was never sure whether he was being serious or droll, or purposefully leaving the matter unsettled. He said outrageous things—*Watson got his Nobel by peeking into Rosie Franklin's drawers*—and if the reaction wasn't what he'd hoped, he'd say something even more outrageous: *When it comes to standing on the shoulders of giants, the only way I'd stand on Jim Watson's shoulders is if he were lying face down in the mud.* Smith delighted in listening to his friend go on, and as people laughed and the conversation continued, you could sometimes see him struggling to add some wry retort, but the moment would pass before he could put the words together.

Smith also had a Nobel Prize, one that he thought he didn't deserve. Venter had no Nobel but was doing everything possible to show he deserved one. And both men would have said they'd gladly trade psyches for a while, to get some relief from being inside their own heads.

They had met at a scientific conference in Bilbao in 1993. Venter was at a low point, mired in his battle with William Haseltine over the release of the ESTs he was finding at TIGR. Smith, biding time before he could retire, had been asked to chair a session at the meeting, and Venter was one of the speakers. His talk was about the EST work. Smith knew Venter by reputation: Venter was the guy who was trying to patent human genes and screw everybody else. "I shared the opinion most academics had about Craig," he remembers. "He was the Antichrist. But when he presented his work, I could see that this guy was discovering genes at an incredible rate, leaving everybody else behind."

That evening, Smith wandered into the hotel bar, looking for somebody with a familiar face so he could avoid the embarrassment of eating alone. Venter was waiting for some friends who hadn't shown up. He saw Smith by himself and offered to buy him a drink. They hit it off immediately. The friends arrived, and they all went off to dinner. The party clicked, and things got louder and drunker. Venter was clowning around, making caustic jokes about Haseltine and Watson and flirting outrageously with the prettier members of the party. Smith didn't say much, but he was enjoying himself more than he had in years.

"You know, we'd really like to have someone like you on the TIGR board," Venter told him as they rode the elevator up to their rooms. Smith was too drunk to give the offer much thought, but a week after he returned to Baltimore he received a formal invitation from Venter to join the board, with a respectable remuneration. He drove down to TIGR for a look around and was impressed with the power the institute had assembled to sequence DNA. Venter was equally taken with Smith's quick insights, and a Nobelist, even a forgotten one, would obviously add prestige to the TIGR board. It pleased him, too, that Smith was willing to look beyond his notoriety and appreciate his talents. "You know, we're looking for something new to do at TIGR and I've planned a retreat to brainstorm next week," he told Smith. "Why don't you come along?"

The next day Smith was reading over some notes on an experiment in progress using his favorite organism, *Haemophilus influenzae,* when an idea came into his head. He thought it was the best he'd had in years. He drove down to the retreat and slouched in the back listening while the young TIGR scientists tossed ideas around. When there was a break he raised his hand. "This place is called The Institute for Genomic Research, right?" he said. "Well, how would you like to sequence an entire genome? How would you like to sequence *Haemophilus*?"

It wasn't as if no one had thought of reading out the code of an entire organism before. Fred Blattner, a molecular biologist at the University of Wisconsin, had been generously funded by NIH way back in 1986 to sequence the genome of *E. coli,* a mainstay lab organism. But Blattner was using the manual sequencing technique and wasn't expected to finish for several more years. It took Venter only a moment to see what a delicious opportunity Smith had just proposed. Nothing in his contract with Haseltine and Human Genome Sciences specified that he *had* to sequence human DNA. Hamilton Smith knew more about the genetics of *H. influenzae* than anyone on Earth. With Smith's know-how and TIGR's automated techniques, Venter thought he could beat Blattner to the prize—the sequencing of a complete genome—with time to spare.

TIGR had another weapon in its arsenal, too. Granger Sutton, a young computer scientist from the University of Maryland, had written a program called the TIGR Assembler that quickly compared the sequences of thousands of ESTs each against every other. If two sequences contained the same fifty letters in a row, it was nearly certain that they overlapped in a gene. So the program would stitch them together into a longer fragment. By the same process, these longer fragments could be joined with others into still larger ones, and so on until all the DNA in the gene that codes for a protein was accounted for.

While Smith prepared a shotgun library of H flu DNA, Sutton tried to figure out how to revise the TIGR Assembler so it could handle the problem of putting the organism's entire 2-million-letter code back together again. They applied for a grant from NIH to underwrite the project and in the meantime got started by dipping into the TIGR endowment. Several months later they heard back from NIH: the grant was rejected, on the grounds that the proposed method probably wouldn't work. By that time, they were nearly finished sequencing H flu's genome. Venter defiantly tacked the rejection letter up on his office door.

Not surprisingly, William Haseltine was less than thrilled to find that his booster rocket had taken off in such an unexpected direction. It could not have happened at a worse time. Incyte was gobbling up as many human ESTs as it could find; Merck had just launched its own program to sequence human ESTs and put them in the public domain. Every human gene fragment discovered by those two enterprises was one less for Haseltine's coffers. "Look Craig, we are in a race!" he said. "We'll pay you *extra* for human sequencing."

Venter refused. He was through supplying HGS with human genes to patent. Haseltine was furious. He owned the rights to whatever TIGR produced, but he couldn't dictate what that product would be. In the meantime, he was obliged to continue to hand over millions to Venter's operation to "sequence worms, bugs, and so forth," as he later put it. Venter had to be stopped. Haseltine made no attempt to hide his intentions. At a social gathering of the TIGR board, while waiters passed through the well-heeled crowd with silver trays of finger food, Haseltine took his partner aside. "I'm going to get you," he said.

"He's trying to destroy me," Venter told his wife when they got home from the party. "He's trying to destroy TIGR."

"What did you expect?" she said. "If he can't control you, then he's going to try to get rid of you."

Though he couldn't stop Venter from switching his focus to the H flu genome, Haseltine still believed he had a contractual right to whatever TIGR produced, including an organism's whole genome. *H. influenzae* was a pathogen, after all, so knowing its genetic code could very well prove useful in combating it. Equally important, the microbe shared many genes with human beings, and information about those genes could be applied directly to figuring out their roles in the human organism. As the project neared completion, HGS's lawyers made it clear that TIGR was prohibited from publishing the complete genome of *Haemophilus influenzae* before disclosing it to HGS and giving the sister company time to file a patent. But Venter was determined to make the discovery public as soon as possible. His lawyers argued that the six-month review period had begun for each individual H flu sequence independently on the day it was submitted to HGS. By this reasoning, TIGR was free to publish the complete genome in the spring of 1995, even though HGS had yet to see what the sequences looked like when they were put together into one complete, ordered series of letters with

Sutton's Assembler program. Haseltine's lawyers cried foul, claiming that HGS needed six months to peruse the assembled code before anyone else got a look.

Just when this question was on the point of being resolved, HGS abruptly rendered it irrelevant. The contract between the two enterprises allowed HGS to extend its proprietary rights to TIGR's data for an additional twelve months when it deemed the discoveries to have medical importance. In February 1995, when the H flu genome was nearly finished, Haseltine's lawyers sent a letter to Venter formally invoking the twelve-month extension period on the H flu genome, warning Venter not to publish the sequence or talk about it in public for another year. Venter refused.

"Right at the end, they fucked us," Mark Adams remembered. "But we said, 'Go ahead, invoke the extension and try to sue us. We're going to publish anyway.'" But from Haseltine's point of view, HGS was the one being fucked. "They had the data, but they wouldn't give it to us!" he said. "We were damned if we were going to pay ten million a year to somebody and not get to file a patent."

But Venter was adamant: he wasn't going to hold up release of the genome, anxiously awaited now by the academic community, for another year while HGS's lawyers scoured it for commercial value. Haseltine had two options: file an injunction in court to prevent TIGR from publishing, or file a patent right away, in effect staking a claim on any intellectual property burrowed away in H flu's genetic code before TIGR could spill it out on the Internet and into the public domain. But nobody had ever tried to patent a whole genome before. How could a convincing application be thrown together in a matter of weeks, especially with Venter holding back the final assembly?

Haseltine contacted patent attorney Jorge Goldstein, who was about to leave for vacation. Just before getting on the plane, Goldstein called a young patent agent in his firm named Robert Millman. It was a risk: at the time Millman had yet to even be admitted to the bar. But he had a quick legal mind, an almost scholarly devotion to the history of patent law, and a solid background in molecular biology. "I'm going to Florida, but I need an opinion," Goldstein told Millman over the phone. "Can somebody patent a genome?"

The short answer was "No." A patent presumes that someone has created something novel—an invention or new technique, or, as the

Patent Office calls it, a "composition of matter." The genetic code of a bacterium cannot by definition be invented, because it already exists in nature; one might as well try to patent a maple tree or a river. However, there *was* precedent for patenting potentially useful pieces of genetic code, such as genes themselves, that had been purified out of their natural state and therefore showed "the hand of man." For instance, Chiron Corporation, a biotech firm, had applied for a patent on the genetic sequence of the virus hepatitis C. Chiron's patent did not apply to the naturally existing virus but to the man-made version of it stored in the company's refrigerators, which after all was the form useful for developing a vaccine.

For starters, then, HGS—or, to be specific, HGS's benefactor SmithKline Beecham, which owned the intellectual property on TIGR's output—could indeed file a patent on useful bits of the H flu genome, such as genes shared with human beings that might have some medical implication. Defending a patent on the *complete* genome itself, however, was trickier. It was not a physical entity sitting in a refrigerator or in a test tube but a reconstructed sequence of DNA letters assembled by Sutton's software program and stored in the form of data on TIGR's mainframe computer. But this apparent stumbling block gave Millman an idea. Up to this point, computer software—with the exception of operating systems—had not been deemed patentable. In a recent case, however, the Patent Office had decided that an applicant could claim the *disk* containing the software as a composition of matter. If so, then what was the distinction between data comprising software code and data comprising the DNA letters of a living genome? Both had to have computers to make them useful—and both should thus be patentable on the same grounds.

It was a novel approach, but it was worth a try. To make either of his arguments stick, however, Millman needed to have the final assembled genome from TIGR in hand.

"Get over there and get the sequence," Haseltine said.

A few days later, Venter, convinced that he had successfully called Haseltine's bluff, sent the assembled sequence to *Science* for review, at the same time finally delivering it to HGS. Working around the clock, Millman wrote up a patent application and formally filed it on April 21, 1995—including twelve hundred pages containing the nearly 2 million letters of the bacterium's genome.

Publication of the first complete genome of a living organism was treated, rightly, as a watershed event. It wasn't just microbiologists and geneticists who took note. Anyone who had ever contemplated the fundamental meaning of life on a scientific level could find some new revelation in what was contained in that one organism's code. Even James Watson had to concede that the publication marked "a great moment in the history of science."

"I sat there looking at this piece of paper in my hands and thinking, That's *it*!" another scientist remembered. "There it is, the road map. This is what this organism really *is.*"

H flu's genetic script was contained in a single circle of DNA pictured on the journal's cover: 1,749 genes, amid very little "junk." A glossy foldout, like the centerfold of a men's magazine, displayed the genes stretched in an ordered linear array of little colored bars. There was an understated magnificence in that image that even a lay person could sense. Smith, Venter, and their colleagues had not simply ordered the genes and their base pairs into a complete sequence. They had also "annotated" the genome, checking each newly discovered *H. influenzae* gene against databases containing thousands of previously discovered genes in other organisms. Since the discoverers of these genes had often also figured out what the genes did, Venter's group could now assign similar roles to over half of the genes found in *H. influenzae.* Those designated with green bars, for instance, were involved in energy metabolism, while the yellow ones denoted genes for copying and repairing DNA, and the purple bars denoted those for the metabolism of fatty acids. Fully 40 percent of the bars, however, had no color. These were terra incognita, suggesting that there was a lot more to learn about the genetics of this one tiny organism, let alone the rest of life.

The paper in *Science* would become the most cited article in all of biology for the following year. In the meantime, while simultaneously sequencing human DNA with funds from the Human Genome Project, Venter committed most of his energy and staff to sequencing other bacterial genomes. Three months later, in October 1995, a team led by Claire Fraser published the genome of *Mycoplasma genitaliam,* a parasite dwelling in the genital tracts of various animals, including humans. *Mycoplasma* manages to live and reproduce with a total of only 470 genes, making it the smallest known genome on the planet, and hence a vital clue to understanding what separates life from nonlife. After the Lilliputian genome of *Mycoplasma* was published, TIGR scientist Scott Peterson and Clyde Hutchinson of the University of North Carolina began to dismantle it piece by piece, inserting bits of nonsense code that disabled one gene at a time. Their ultimate goal was to find the smallest number of genes needed to sustain and reproduce life. Their work immediately suggested another approach to the same question: why not build an organism from the bottom up, splicing one critical gene onto another until this artificial genome had the equipment to survive on its own? "Life from scratch!" Venter enthused. The closer the team came to being able to actually do this experiment, however, the more its ethical implications loomed. Later he put the project on hold and convened a panel of bioethicists and theologians to ponder the repercussions.

By the middle of 1996, Venter's group had finished their third microbial genome. From an evolutionary standpoint, this one was the most intriguing of all. *Methanococcus jannaschii* was a microbe found only in deep-sea hot vents. It was a representative of the archaea, a category of living things postulated by Indiana University evolutionary biologist Carl Woese to be a third "superkingdom" of life separate both from the bacteria and the eukaryotes (plants and animals) that make up the rest of life on Earth. Woese had argued for twenty years that the traditional lumping of bacteria and archaea into the single kingdom called prokaryotes was an egregious taxonomic error, all the more damaging to understanding evolution because the flaw was committed at the very base of the tree of life. The *M. jannaschii* genome spelled out the truth of Woese's idea in black and white: it shared only 11 percent of its genes with the bacterium *H. influenzae* and much more with the eukaryotes. Fully half of the genes, however, had never been seen before at all.

Venter soon began investigating several other genomes with more immediate commercial value, including pathogens like the microbes responsible for Lyme disease, stomach ulcers, syphilis, and, in the longer term, malaria and cholera. The codes of these creatures had enormous potential value to medicine. Infectious microbes rely on genetic subterfuges they have evolved to evade their host's natural antibodies as well as man-made antibiotics, and getting their genetic sequence was like stealing their plan of attack. Other projects got started on microbes that might have industrial value, such as making new fertilizers or cleaning up oil spills and other environmental messes. The most interesting of these was a remarkable microbe called *Deinococcus radiodurans,* which was known to be able to withstand 1.5 million rads of radiation—three thousand times the amount that would kill a human being. A dose of radiation blasts the genome apart, but after a few hours it stitches itself back together again exactly as it was. Insert the genes responsible for this kind of genetic repair into the genome of another bacterium that naturally gobbles up heavy metals, and the newly invented life-form could be put to work cleaning up nuclear-waste sites. "This thing can take more radiation than the Incredible Hulk," said Owen White, who led the project at TIGR.

In July 1997, Human Genome Sciences and TIGR issued simultaneous press releases: the two enterprises had agreed to dissolve their partnership. The divorce was hardly unexpected—Venter was completely immersed in what his institute was revealing about evolution, and he intended to continue focusing on simple organisms, comparing their genomes for clues to how life originated and diversified. Such basic research was of little use to William Haseltine at HGS, which had long since begun pursuing its own commercially driven course. The agreement freed HGS from its remaining obligation of support to TIGR, totaling some $38 million. Venter in turn was free to publish exactly what he wished, when he wished to—a freedom he flaunted openly in a second press release from TIGR issued on the same day. It was a simple statement announcing that the institute was posting on the Internet a massive load of new gene sequences, including the nearly completed genome of *Helicobacter pylori,* the ulcer-causing bacterium dwelling in the stomachs of over half the people on Earth. The gesture went a long way toward mending Venter's beleaguered reputation among academic scientists. "A lot of people viewed Craig as someone who was trying to take everything that there was to be taken," says Mitchell Sogin, a molecular evolutionist at the Marine Biological Laboratory in Woods Hole, Massachusetts. "But in the end he did the largest release of data that has ever been done by anybody. He has to be given credit for severing his ties and risking the well-being of his institute."

Venter was now almost fifty years old. With the development of the EST technique and the sequencing of the first living organism, he could claim to have fulfilled the promise he had made thirty years before to his brother Keith to do research that would "change the world." His battles with James Watson and William Haseltine had cost him some emotional pain and a stretch of his intestines, but he had mended his fences with the academic community and had no trouble digesting his meals. He had brought a brilliant Nobelist back to life and counted him among his best friends. He was working the way he wanted to, "doing academic science, with someone else's money."

This could be the end of a story, if it weren't the beginning of a bigger one. Late in 1997, Venter began getting calls from Michael Hunkapiller at Applied Biosystems (ABI) and Tony White, the head of ABI's parent company, Perkin Elmer. They wanted him to come out and take a look at the capabilities of a new, greatly improved sequencing machine that Hunkapiller's team had on the drawing board. What they were suggesting they could do with the machine was preposterous, however, and Venter paid them little attention—until he finally got around to visiting Hunkapiller, in February 1998.

ABI's headquarters is in Foster City, on the west shore of San Francisco Bay, within sight of Coyote Point, where Venter had raced around in his hydroplane as a boy. He only spent one day there, but he came back home to Potomac twitching with excitement. To his wife, he was sounding unbalanced again, like the old Craig who couldn't tell the difference between a sound idea and a stupid one. "Mike's machine is unbelievable, Claire!" he exclaimed. "They want to start a new company, with me as its head. With enough of the new machines, I think we could do it. We could get the entire human genome!"

Claire Fraser kept quiet, as she had long ago decided to do in these situations: just let him rave for a while and he'd come to his senses on his own. But he kept pressing her, following her around the house. She sat down in the sunken living room and turned on the television. He grabbed the remote and flicked it off. "So what do you think?" he said. "I really want to know."

"Are you insane?" she answered. "It's been less than a year that you got free of Haseltine. Do you want to give up your science, and have some company trying to control you again?"

"It's not like that. For godssake, it's *Mike.* We've worked together for years. We understand each other. We have the same vision."

"Yeah?" she said. "Well, what about this Tony White? Who's he?"

"Tony White is not important," Venter replied. "Trust me. He's not going to be a problem."

PART TWO

CHAPTER 9
A HUNDRED MILLION CUSTOMERS

In hindsight, it is hard to imagine that there would have been a race to sequence the human genome without Craig Venter, or for that matter, a Craig Venter without a race for the human genome to help him define who he was. But Tony White and his associates considered others for the job, and though it exhibited far more success sequencing genomes than anyone else, Venter's résumé was hardly without blemish. White liked team players, and nothing in Venter's past suggested he was one. His tendency to utter whatever came into his head might make him popular with the press, but it did not play well on Wall Street. Moreover, he had no practical experience in the private sector.

"We've got to get Craig to understand that this isn't some giant science project," Michael Hunkapiller told White. "It's a business."

"It could get dicey along the way," White agreed. "But I can see this guy is a winner. We need him."

In spite of his initial enthusiasm, Venter had some reservations about taking the job. He did not want a repetition of the Haseltine experience any more than his wife did. One thing he was adamant about: the genome that the company produced must be free and publicly available. He was not signing up for another genetic land grab. Hunkapiller and White agreed: at most, the company would seek patents on only a few hundred medically important genes. Venter also wanted his beloved TIGR to be taken care of. After some negotiations, the parties closed a deal that would give Venter a 5 percent stake in the company to be formed, with TIGR getting another 5 percent. Claire Fraser had reluctantly agreed to take over leadership of TIGR. In May, when Venter assembled the staff to tell them he was leaving, he broke down in tears.

The new company needed a name. Venter and his colleagues hired a firm specializing in naming new enterprises and explained what they intended to do. The consultants came back with a list of choices half an inch thick, indicating, perhaps, a certain degree of latitude in how to define a company that was moving into largely uncharted territory with a business plan that included giving away its principal product. The group narrowed the list down to thirty, including "Biotrek" and "Sxigen." Venter instead chose from the list the name "Celera," from the Latin root for "speed." With the accent on the second syllable, it had a sleek, sinuous, slightly romantic sound. You could imagine naming a racing yacht *Celera.* It was harder to imagine calling your boat *Biotrek* or *Sxigen.*

A building had been leased not far from TIGR, just off Rockville Pike, a six-lane suburban artery known simply as "the Pike" to the grim commuters and shoppers dependent upon it. The property was one of two twin four-story buildings joined by a glass walkway. Built for a defense contracting business that had moved on, the structures had been vacant for years and looked it. Their façades, once white, had absorbed so much exhaust from the Pike that they had assumed the color of unbrushed teeth, and the banks of one-way glass on each floor returned murky reflections of the overgrown landscaping.

One morning in August 1998, Venter took a visitor on a walk-through of the renovation under way. Shadow, the alpha male of Venter's three standard poodles, trotted along at an aloof distance. Contractors were bustling in and out of the front entrance. The surrounding parking lot had recently been resurfaced, and the sharp oily smell of newly laid asphalt hung in the air. Out on its far edge, a solitary workman with a long-handled roller was painting on parking spaces. From the rear of the building came a continuous crash of rubble disgorging through a chute propped against a giant hole cut into the wall of the third floor.

"Ready to get a look at the next century?" Venter asked.

The lobby was empty except for a sea of faded blue carpeting and a couple of contract security guards. In the hallways workmen were knocking down ceilings and walls, their faces covered with plaster dust. Ducts, metal tubing, and wires tumbled down in a disordered abundance, like jungle vegetation. Venter moved unhurriedly through the first two floors, pausing occasionally to stand in a musty empty room, as if he were watching things yet to happen in it. The whole third floor of the building had been gutted to a single, debris-ridden expanse the size of a football field. Here was where they planned to install the company's supercomputer. Since nothing like the sort of number crunching needed to assemble the human genome all at once had been attempted, Venter wasn't sure how powerful a computer would be required. To be on the safe side, he was estimating they would need the most powerful computer in the world. Bids had been solicited from several companies, and the competition had come down to IBM and Digital Equipment Company, which had recently been bought by Compaq. "Whichever one we choose gets bragging rights to the human genome," Venter said. "That puts us in a really good position. Either way, we'll get eighty million dollars' worth of equipment for half the price."

The fourth story had been gutted, too, including most of the floor, exposing the metal joists beneath. This was ground zero. If all went well, the room would soon be transformed into the world's largest DNA sequencing operation, row after row of the new ABI capillary machines running day and night. Work on the *Drosophila* genome was scheduled to begin on January 1, 1999, and be finished by early April. After that, *Homo sapiens.* Venter stood for quite a while in the middle of this second immense empty space, his hands on his hips, looking around and listening, mentally conjuring the machines into being, his face already aglow from their powerful heat. "GenBank now has about two billion base pairs in it," Venter said. "Once this facility is operational, we'll be doing that much every month."

The price for all this was estimated to be around $300 million. Perkin Elmer was not investing that amount of money in Venter without expecting something in return. But Venter was promising that the genome would be given away. A lot of people in both academia and business were wondering how he could make good on that promise without losing a lot of money. The academics were afraid Venter was hiding something. The market analysts were *hoping* he was.

"The basic code of the genome is just the beginning," Venter explained, back in his barren temporary office on the first floor. There was a magazine lying on his desk, open to a full-page advertisement for the company LexisNexis, which provides lawyers, businesspeople, journalists, and teachers with searchable access to billions of documents collected from sources around the world. He tapped it with a finger. "*That's* the business model," he said. "We're not a biotech. We're an information company, like LexisNexis. If you had the time, you could find the same information they have on your own. So why do they have two million subscribers? Because they've already done the legwork for you, so you can find what you want in seconds rather than hours. We're going to do the same for genomic information, on a global scale."

As an information company, Celera's flagship product would be a massive database of DNA with the human genome sequence as its heart. Any scientist could access the basic human sequence for free. But this peek behind the curtain would merely whet the appetite for the complete database, including volumes of data on genetic variability in humans, and the genomes of animals critical to biomedical research, like the laboratory mouse. Anyone, from a pharmaceutical company to an individual academic researcher, could mine this information for a sliding fee—millions for the pharmaceutical companies, a few thousand in grant money for an academic. By making the database quickly and widely available, Celera would become the definitive source of genomic information for the world, in much the same way that Microsoft had early on made its DOS operating system the standard for personal computers.

"Contrary to what you might be hearing, I am not the Bill Gates of the genome," Venter said. Then he added, with an eyebrow-arching grin, "At least not yet." There was no reason why the market should be limited just to scientists, either. Doctors, clinics, eventually even ordinary folk seeking information about their own genetic makeup could log on to Celera and take control of their health and their future. "This is the difference between us and Incyte that people don't seem to get," he said. "Incyte has twenty-two pharmaceutical customers. We don't want twenty-two customers. We want a hundred million."

A critical part of the Celera offering would be knowledge of how one person's genome differed from another's—especially differences consisting of a single base pair. Such pinpoint variations, called SNPs, short for "single nucleotide polymorphisms," were estimated to occur only once every thousand base pairs or so. But the genome itself is so exceedingly huge that the total number of SNPs (pronounced "snips") was believed to be as high as 3 million. Potentially, these variations might have as much commercial value as the genes themselves. Rarely, an alteration of a single base pair within a gene can lead to a disease. Sickle-cell anemia, for instance, occurs when someone has inherited two copies of a particular gene where at one point a single letter "A" takes the place of a "T." Just as easily, a SNP might bestow some benefit; indeed, if an individual inherits only one copy of the misspelled sickle-cell gene, he or she will be much less susceptible to malaria than the general population. The vast majority of SNPs, however, don't make people weaker or stronger, more evolutionarily fit or less. But they could still guide researchers to the genes that might. If a SNP showed up in people with a particular form of cancer, for instance, while healthy people had a different base pair at the same spot, then the SNP could function as a tiny bright beacon on the genome, broadcasting the location of an errant gene nearby. Using diagnostic markers to find genes in this way was nothing new; in fact, it was the essence of medical genetics. A huge catalogue of SNPs, however, would make the search much faster and more precise.

Even more exciting—and potentially profitable—was the prospect of using SNPs as the foundation for "individualized medicine": designing drugs tailored specifically to a person's genetic profile. Let's say, for instance, that a group of people with a particular pattern of SNPs responded well to a new cancer drug, but another group with a different profile did not. If a new patient turned out to belong to the first group, the doctor could prescribe a treatment likely to be beneficial. If a patient showed the other pattern, the doctor could save the patient money and precious time by rejecting that drug as a treatment, possibly avoiding serious side effects as well. The worst possible side effect of a drug—and an alarmingly common one—is death. An estimated 106,000 people die from drug side effects a year; it is the fourth largest cause of death in the United States. In the late 1990s, some of those deaths included a small number of patients who took the obesity drug combination Fen-Phen. Naturally, the product was taken off the market. But what if a specific genetic pattern unique to those ill-fated people had been known in advance? They could simply have been warned not to take Fen-Phen in the first place. The drug might still be on the pharmacy shelves, available to the vast majority of patients for whom it was perfectly safe, and making huge profits—instead of incurring huge lawsuits—for its manufacturer. That was the sort of information that Tony White was betting

$300 million that the pharmaceutical companies would want to have. And he was counting on Venter to get it.

On the way out, Venter took his visitor on a quick detour into the basement. Down there workmen were readying space for the backup power system, which would prevent the disaster of an outage during the continuous sequencing process. Were the flow of DNA letters from the machines to stop in midsentence, the company could lose millions. Another room was given over to the backup to the backup. All the infrastructure of the operation—the air-conditioning, the miles of fiber-optic cable, the elaborate security measures to prevent data theft and sabotage—had to be inflated to the grotesque dimensions of the sequencing and computer operations. Celera's electric bill promised to top $1 million a year; the city of Rockville, Venter boasted, had been forced to redesign its power grid to accommodate its newest, largest consumer of electricity. But pumping up lab science to these proportions wasn't simply a matter of multiplication. Entirely new problems emerged. For instance, standard, disposable plastic pipette tips, ubiquitous in bench science, cost about a nickel apiece—hardly a significant budget item in most labs. At a meeting to go over Celera's prospective budget, somebody pointed out that to keep the sequencing machines fed, the bill for pipette tips alone would run to $14,000 a day. Perkin Elmer could not afford that many nickels, so a robotic system for loading samples of DNA had been engineered that did not require disposable tips.

"People really don't get the scale of this," Venter said, stepping over a pile of twisted metal blocking his path. "Francis doesn't have a clue."

CHAPTER 10
THE GENE HUNTER

Craig Venter was wrong about Francis Collins: he really did have a clue. One might be fooled by the thick glasses, the home-cut hair, the Ned Flanders mustache and hi-diddly bonhomie. The way his voice bounded onto syllables for emphasis, sometimes with a wag of chin to drive the point home, wasn't the most stylish manner of speech either. But Collins had not become the leader of the largest organized effort in the history of medical science by being clueless.

Before the meeting in the Red Carpet Club lounge at Dulles Airport the previous May, Collins had been no more aware of what Perkin Elmer was planning than anyone else. But by coincidence, he and Michael Hunkapiller were booked on the same flight to California immediately following the meeting. They had a cordial conversation on the airplane; after all, the public genome program was ABI's biggest customer. By the end of the flight, Collins had Hunkapiller's assurance that the Human Genome Project's sequencing centers would enjoy the same access to the new capillary machines that Celera would receive. Even with an adequate supply of sequencers, however, Collins knew he could not compete with Venter's announced delivery date of 2001 and still keep to the lofty goal of a highly accurate, 99.99 percent complete genome. It wasn't humanly possible. Abandoning that mission was unthinkable; yet it was equally unthinkable to allow Venter to triumph. The Human Genome Project, Collins told himself, was about community, about rules that applied to all, about the sacrifice of individual motives for the collective good. It was even a bit about God. Craig Venter was about Craig Venter. Collins hadn't forgotten that Venter had been one of the few scientists at the Bermuda meeting to oppose the requirement that newly sequenced DNA be revealed immediately, in order to prevent anyone from profiting from the genomic information. Collins had warned later that anyone who refused to abide by the Bermuda rules would have their funding terminated, including Venter's group at TIGR. But Venter had found a way to elude his control. He was a renegade now, and he was dangerous. Collins did not believe for a moment that Venter was going to give away his version of the genome for free—even if Venter himself believed it. Sooner or later, Collins thought, Craig would find himself shackled by the interests of his shareholders, just as he'd been captive to William Haseltine's ambitions in the EST affair. If something wasn't done, the human code of life would be subject to the control of a single corporation. God's language, in the mouth of Mammon.

This was how "clueless" Collins was: the evening after Venter made his May 1998 presentation at Cold Spring Harbor, Collins invited Philip Green, the computational biologist from the University of Washington, and the heads of the three largest HGP genome centers—Eric Lander of the Whitehead Institute, Robert Waterston of Washington University, and Richard Gibbs of the Baylor College of Medicine—out to dinner at an Italian restaurant in a neighboring town, away from the cloud of apprehension that Venter's visit had generated. During the meal the five men discussed their options. Among these was a new approach to going after the genome sequence, one that would have been considered heretical just a few days earlier. It was Lander, of the red hair and the brain perpetually on fire, who pushed the new plan the hardest. If Venter was going to make a play for the complete human sequence by 2001, then the public program had to get a "draft" version out by then as well, even if it was full of gaps. The gaps could be filled in later. But if something wasn't done to counter the threat from Venter's proposed company, Congress would cut off funds and there wouldn't be a public genome program at all.

The idea of a draft was sure to draw bellows of protest from the purists in the Human Genome Project, and none was purer than Maynard Olson, Philip Green's colleague at the University of Washington. For years Olson had functioned as a sort of moral rudder for the HGP, keeping it on course toward a finished genome that would stand science in good stead for ages to come. Since taking over the reins of the program, Collins had shown himself a purist, too, siding with Olson against any proposal to take a shortcut that would harvest *most* of the genome quickly, leaving the rest to finish as time and money would allow. Over dinner, Collins listened to Eric Lander make his pitch, and he began to wonder whether he and Maynard would have to concede that the landscape had changed.

The group met again when John Sulston of the Sanger Centre arrived at Cold Spring Harbor later in the week, along with his benefactor, Michael Morgan of the Wellcome Trust. The British representatives refused point-blank to go along with the notion of a draft. Collins didn't force the issue. He needed more time to think about it himself. In the meantime, it was critical to manage how the government program was now being perceived—in Congress, with the public, and within the HGP community itself. Collins bristled inside at the idea of cooperating with Venter, and he certainly wasn't going to take his advice and switch to the mouse genome. But as long as Venter was promising to make his data public and freely available, Collins had to consider the possibility of a collaboration, or at least leave the impression that the door was open. Congress would hardly look favorably on a government program that rejected an alliance with the private sector, one that might potentially save the taxpayers hundreds of millions of dollars, when there was no good reason not to go along.

In June, a congressional subcommittee hearing was called to review the federal program's future in light of the Perkin Elmer announcement. Collins arrived to find Venter wearing a sports jacket, tie, and trousers nearly identical to his own. He quickly seized his opportunity. "Let me assure you, we will work together . . . ," Collins told the subcommittee. "If you doubt that, notice that Dr. Venter and I seem to have worn the same clothes today without intending to. We are intending to be partners in every possible way, so let this be a symbol thereof."

But while he had to assure Congress that he was open to a collaboration, Collins knew he should not appear too eager, lest he crush morale and alienate the more rabid supporters of the public effort. None was more rabid than the Wellcome Trust. The trust's Michael Morgan had made it clear that if the American program collaborated with Venter in any manner that compromised the principles of the Bermuda Accord, the trust and its millions would go their own way. Collins did not always care for Morgan's scorched-earth tactics, but Morgan was powerful and passionate in the cause and had to be dealt with carefully. Smaller genome centers in France, Germany, and Japan that were contributing to the public project also had to be treated with respect. The Human Genome Project was not simply a scientific enterprise; it was the embodiment of an ideology holding that the human code of life belonged collectively to all humanity. Any lab in any country capable of contributing had been invited to do so. But if the public program was to compete successfully with Venter, it would no longer be feasible to have so many separate entities involved in the decision making. The project would have to become more centralized and streamlined. The people who had proved that they could deliver efficiently would have to get a bigger slice of the pie, which meant others would go hungry.

Meanwhile, public perception had to be adroitly handled. In his articles in *the New York Times,* Nicholas Wade was portraying the Human Genome Project as a lurching dinosaur. Collins found that gallingly unfair. But he wasn't clueless about dealing with the media, either. Your message had to be crystal clear, and you had to drum it in, again and again. The Human Genome Project, he told reporters through the summer of 1998, was a rousing success that had consistently met its milestones *ahead of schedule and on budget* due to the effort of *all the hardworking folks* in the United States and around the world who were dedicated to finishing a version of *the book of life* that would *stand the test of time.* Above all, Collins pressed home the message that there *was* no race between his program and Venter's new company, because the two initiatives were after very different prizes, with different finish lines. Instead, the public program was "racing itself." Its goal was perfection, because the code of human life demanded it. Celera's goal was—well, something less, tied to making money.

During an interview with Tim Friend, a reporter for *USA Today,* Collins referred to Venter's proposed assembly as the *"Reader's Digest"* version of the human genome. Half an hour later, Friend's cell phone rang. It was Collins calling him back. "Come to think of it, I think even calling it a *Reader's Digest* version is optimistic," he told the reporter. "More likely theirs will be the *Mad Magazine* version of the genome." "*Mad Magazine*?" Friend asked. "Are you sure you want to say that?"

Yes, Collins replied. He was sure he wanted to say that.

In contrast to Venter's troubled childhood, Francis Collins's upbringing had been spectacularly wholesome. He was raised on a farm in Virginia, where he was home-schooled through sixth grade by his mother, a playwright. His father was a medieval scholar and folk-song collector who had worked with Eleanor Roosevelt during the Depression at Arthurdale, West Virginia, the experimental community for impoverished coal miners. After the war, Collins's father had moved his family from Long Island to Virginia to live off the land, eschewing modern machinery and working his farm with animal-drawn plows. When it proved too hard to make ends meet, he took a job teaching drama at Mary Baldwin College for Women nearby. He made their home a nexus for actors, activists, and folk musicians, like Mike Seeger, Alan Lomax, and the Greenbriar Boys, and in the summer he produced plays. For Francis a day in summer might begin with milking the cows or other chores. Then he'd practice piano or read, and in the afternoons there were rehearsals for the theater's latest production. He would help build sets and run the lights, and when a boy was needed he'd join the cast. In the evening, whoever was around would gather on the porch with a guitar, banjo, or fiddle and pick out some tunes, sometimes till two or three in the morning. If you wanted to talk about art, politics, or life, there was always somebody else around in some part of the house. If it takes a village to raise a child, Francis Collins had one in his own backyard.

He entered the local high school two years early and graduated at the age of sixteen, valedictorian. At nineteen, the age Craig Venter had been when he was surfing in Newport Beach, Collins had a graduate fellowship at Yale, where he worked for his dissertation on a theoretical problem in quantum mechanics, exploring the transition of two colliding molecules from a translational state to a vibrational state. He and his girlfriend had conceived a child, and since abortion was out of the question for them he now had a wife and a baby on the way. It was 1969. He had always gravitated toward the hard pure sciences of chemistry and physics because they thrilled his mind, but now he took stock of where he was going, wondering if mind-thrill was something worth devoting a life to, and noticing, too, that those who had gone before him in physical chemistry didn't look like they were all that excited anymore. Meanwhile he'd made friends with a graduate student in molecular biology who was trying to clone DNA to study its function in bacteria, and he did seem excited. Since high school Collins had avoided biology as a tedious, descriptive science. But now he took to reading library books from Cold Spring Harbor Laboratory, full of the ideas of Delbrück, Watson, Jacques Monod, Barbara McClintock, and the others who had invented molecular biology and whose thrilled, confident faces shone in the photos. He imagined what it must be like to be a part of such a community.

After getting his Ph.D., Collins entered medical school at the University of North Carolina, earning the highest grades in his class. A year later, he took a six-week course in medical genetics. The concept that a single change in a base pair could wreak havoc through multiple organs of the human body seemed an inestimably profound statement about human life. While terrifyingly fragile, it still ran according to a *set of instructions,* a logical system, and was therefore capable of being understood, and—someday—repaired. Disease was cruel but not random. If your mean uncle smoked all his long life, while your sweet uncle who'd never touched a cigarette died of lung cancer, it wasn't luck, fate, or irony. It was a difference in code, or a difference in code curled around and through the effects of environment.

With this revelation, two parts of Francis Collins—the logician and the healer—bonded into a synthesis he would never have thought possible. This was in 1973. Collins was twenty-two. At that age, just a few years earlier, Craig Venter had been bending over the young, still warm corpses of his fellow draftees, learning how little it took to kill a man and wondering what he could do about it. Human frailty, it seemed, had sent both men on their careers. Now their paths were converging.

In another sense, however, they were headed in opposite directions. Venter's family was Mormon in heritage, but his father had been excommunicated from the church for, as Venter put it, "drinking, smoking, and getting an education." Collins had grown up with the values instilled in him by the intellectually rich secular humanism of his parents. Religion was acknowledged as an important part of Western tradition, but you went to church to experience not so much the glory of God as the beauty of seventeenth-century choral music. As a young intern watching his patients die, however, he was struck by the power that faith granted the ones who believed in God—a power that seemed to far exceed the strength of his own palliatives. Ever the rationalist, he took stock of the situation and conceded that he hadn't really "looked at the data" on the question of the existence or nonexistence of God. He spent a year correcting that oversight, fully expecting his investigation to shore up his atheism.

Just the opposite happened. His readings, especially of the Gospel of John and *Mere Christianity* by the Oxford don C. S. Lewis, convinced Collins that logic and faith presented not an either/or choice but a perfect fusion. Science was a means of revealing fraction by fraction the omniscience of God, a light teasing a sparkle from one tiny facet on the infinite jewel of His mind. Perfect and holy Himself, He had created human beings who, while sinful and imperfect, were endowed with an understanding of right and wrong, a universal law of morality. And how better to bring such a creature into being than through evolution? Far from emptying the universe of the necessity of God, Charles Darwin had simply revealed another tiny sampling of His infinite ingenuity. Collins believed that he too had been given the gift of unusual intelligence in order to understand the basis of life, which was understanding God. At the age of twenty-seven, the same year he was named "intern of the year" at UNC's hospital, he was born again.

He finished his medical training, returned to Yale for a fellowship in human genetics, and in 1984 joined the faculty at the University of Michigan. He wanted to look for genes that cause disease. At that time, only a handful of disease genes had been identified. With so little known about the landscape of the genome, gene hunters had to approach their quarry from the disease on down: first figure out the physiological system gone awry, then try to identify a protein whose absence or malfunction had caused the failure in the first place. In the rare event that you managed to isolate such a protein, you could read its sequence of amino acids and reverse translate them into the coding sequence of base pairs in the errant gene. Even if you got this far, however, you would still know nothing about the location of the gene on its chromosome and hence nothing about the surrounding genomic territory—including messages in the DNA that told the gene when to turn on or off and what other genes nearby might be performing in tandem with your target gene. You would have isolated the gene but discovered little about how it worked, because genes don't work in isolation. It would be like trying to figure out how internal combustion works by examining a carburetor or a fuel pump.

Collins, beginning his career, was excited by an alternative approach just then becoming possible. Instead of working from the top down, he could start by locating a gene on a chromosome without knowing anything about its sequence or function, and work up from there. Once he knew the gene's exact location, he could read its code and from it reconstruct the amino acid sequence of the malfunctioning protein. Both sequences could then provide clues to the protein's normal role, illuminating the cause of the disease and perhaps even a path to a cure.

But how do you locate something when you don't know what it is you are looking for? The answer—and this was the new approach—was to look for something nearby that you *could* identify, and feel your way from there. The human genome is sprinkled with random spots where one person's sequence differs from another's in some small way. For instance, at a given spot in your brother's genetic code the sequence CCAA might show up twice in succession—CCAACCAA—while at the same spot your sister may have an extra stutter inserted: CAACAACAA. While usually harmless in themselves, these individual differences, known as polymorphisms, might by chance be found in the general proximity of a gene of interest, and thus would tend to stick with the gene through the recombinations occurring from parent to child through generations, like a card sticking to another each time a deck is shuffled.

In the early 1980s, a new technique in molecular biology made it possible to identify and mark the location of these variable regions. While the gene itself might be as elusive as a buried ace of spades, if you could turn up a marker region where sick people had one version of the polymorphism and their healthy relatives another, there was a good chance that the gene was lurking in the general vicinity. Depending on how frequent the correlation was, you could trace the affected gene to its chromosome, then to a particular arm of that chromosome, and so on through ever-narrowing circles, until you'd tracked down a marker so prevalent that you knew the culprit gene was in the immediate neighborhood. It was in the last stages of such a dragnet that Collins made his brilliance known.

"Woe to the child which when kissed on the forehead tastes salty. He is bewitched and soon must die." This folklore adage comes from northern Europe, where history has seen more than its share of such infants. Their bewitchment is now called cystic fibrosis. It is caused by a recessive Mendelian mutation, which means that a person must inherit two defective versions of a particular gene, one from each parent, in order to be born with the disease. When this happens, something goes awry that alters the amount of sodium and chloride ions passing through the surface of some cells, including skin cells—hence the salty forehead. In lung cells the same faulty transport mechanism produces an abundance of sticky mucus, leaving the child vulnerable to repeated infections that erode the lungs' tissues and eventually the ability to breathe. The mucus accumulates in the pancreas, too, clogging the flow to the intestines of enzymes needed for digestion. Cystic fibrosis is the most common hereditary disease among Caucasians, affecting thirty thousand adults and children in the United States alone. With modern antibiotics and other medical interventions, they no longer all die as soon as they once did. But still half never reach the age of thirty.

For convenience, people refer to the affected gene as "the cystic fibrosis gene." But it must be kept in mind that *genes do not cause diseases, they only "cause" proteins.* One in twenty Americans harbors a single defective copy of the gene, but since those people also have a normal, dominant version inherited from the other parent, they exhibit no symptoms. Using words like "defective" and "normal" to describe a gene is useful shorthand, too, but not always accurate. The same variation in a gene can be good, bad, or indifferent, depending on its context. The gene variant that predisposes people to developing Alzheimer's disease may help protect them from kidney damage. Or consider sickle-cell anemia. When a recessive form of the gene for the production of hemoglobin finds itself paired with another of the same form, for instance, it causes the ordinarily doughnut-shaped red blood cells to collapse into sticky crescents that clog the blood, causing pain, strokes, kidney damage, jaundice, and a host of other ailments. Sickle-cell anemia is much more common in African Americans than in European Americans, and for sound evolutionary reasons. When only one recessive form of the gene is present, the individual carries a markedly higher resistance to malaria. In West Africa, where malaria is rampant and from whence most African Americans derive, this is a very good gene to have. But in the United States, where malaria is all but nonexistent, the bad side of the gene comes to the fore when it is passed on in a double dose.

Cystic fibrosis appears to be governed by a gene with a dual nature, too. In cooler climates, carrying a single copy of the recessive form protects the individual against typhoid fever, which plagued Europe for centuries until antibiotics were discovered. This selective advantage explains why cystic fibrosis is more common among Europeans than in people adapted to the tropics, where typhoid fever was less a threat. The search for the responsible gene, an epic in the history of human genetics, began in 1981 when Lap-Chee Tsui, a postdoc at the Hospital for Sick Children in Toronto, started examining the clinical records of cystic fibrosis patients and their families in Ontario. He was looking for a polymorphism marker that had traveled through the generations along with the affected gene. Soon afterward, Ray White at the University of Utah and Robert Williamson at St. Mary's Hospital in London, among others, began to compile and analyze their own pedigree databases.

By 1984, a private company near Boston called Collaborative Research had entered the race. It was a great surprise to Tsui when Collaborative Research scientists came up to Toronto and offered him two hundred probes to ferret out the location of markers throughout the genome, any one of which might turn out to correlate with the cystic fibrosis gene. Of course, Collaborative wanted a stake in the intellectual property if Tsui discovered the gene. In the United States alone, 20 million people harbor the defective variation. Were any one of them to conceive a child with another member of that not-so-select group, there was one chance in four that the baby would be born with cystic fibrosis. The market for an accurate genetic test to be used in family planning and prenatal testing was thus thought to be worth billions.

Using the Collaborative probes, in 1985 Tsui was able to trace the cystic fibrosis gene to chromosome 7. He and Collaborative tried to keep the discovery a secret, but inevitably it leaked out to the competing labs, who turned their attention to chromosome 7 as well, looking for markers that were closer and closer to the culprit gene. After two more years of crawling around in the code, Tsui found two additional markers that surrounded the gene like bookends, greatly narrowing the search. But there were still over a million and a half base pairs between them—a lot of territory where the gene could be hiding. While the competing labs tried other methods to get closer to the gene, Tsui used a technique he called "saturation mapping," essentially bombarding the stretch of DNA between the two identified "bookends" with hundreds of probes for more proximate markers.

Meanwhile, Francis Collins and his lab in Ann Arbor had joined the hunt, using a technique he had invented back at Yale. The other labs were hampered by the fact that human DNA at the time could be cloned—put into a virus and copied—only in very small pieces, just 40,000 base pairs or so at a time. To feel one's way from one marker to another, and ultimately to the gene itself, one had to "walk" from one of these little clones to another: sequence a clone, then find another one with a sequence overlapping the first clone on one end, look for a third one overlapping the second one on its opposite end, and so on. Since not all regions of human DNA were amenable to being cloned, inevitably the researcher would reach a barrier, like a hiker coming upon an impassable stream. In effect, Collins had figured out a way to hop over the stream. Using restriction enzymes, he cut DNA into fragments 100,000 base pairs long, then joined the fragments top to bottom in a circle, like a snake eating its own tail. He then cloned and sequenced just the places where the circles joined, and used these to skip ahead 100,000 letters at a time. He called the technique "chromosome jumping."

Tsui could see that Collins's method was complementary to his own, and the two young researchers decided to collaborate. But they were just getting under way when Williamson's team in London declared that they had found a prime candidate for the cystic fibrosis gene, speaking about it with such enthusiastic conviction that Ray White in Utah and most of the other teams conceded the race and wistfully turned to other projects. On the suspicion that Williamson might be mistaken, however, Tsui and Collins pressed on. To prove its case, Williamson's team needed only to sequence two versions of the gene they had found, one from a normal cell and one from a cystic fibrosis cell. The two versions would have to be different, of course, the variation between them representing the mutation responsible for the disease. But when they finished, they were horrified to find the sequences identical. The heady effusion of a few months before had been a classic case of premature eureka. Williamson had found a gene, but it wasn't the right one.

Williamson's gene did, however, serve as another marker to narrow the hunt. Labs began jumping back into the game, but by this time Williamson and the Tsui-Collins collaboration were well out in front. Both were concentrating their efforts in a region spanning about 200,000 base pairs, and getting closer to the gene by the week. Now the hunt turned to stretches of DNA that looked "gene-ish," exhibiting structural similarities to genes known from other organisms, especially those coding for membrane proteins. Williamson's group grabbed on to the tail end of what looked like the gene, but their path to the rest was blocked by an unclonable region that could not be "walked" across. Tsui and Collins, meanwhile, had latched on to the front end of the same gene. In the summer of 1989, with Williamson breathing down their necks, they announced that they had found the correct gene, a large stretch of DNA coding for almost fifteen hundred amino acids that bore all the earmarks of a membrane-channel protein. Seventy percent of the people afflicted with cystic fibrosis carried a mutation that deprived the protein of a single amino acid. It was this tiny flaw that for centuries had caused so many children to die.

The news that the cystic fibrosis gene had been found broke one week before papers by Tsui and Collins and their colleagues were due to be published in *Science.* The Howard Hughes Medical Institute, which had funded both labs, provided a private plane to whisk the two researchers and their colleagues between same-day press conferences in Washington and Toronto. The success brought Collins fame, grant money, and a share of the patent on the gene.

It also brought him closer to God. "If you drew a circle around what God knows, it would be unimaginably huge," he later said. "What I know is a teeny, teeny dot within the circle. But every once in a while we humans get to sneak out of the little dot and find something that wasn't known before. That's the way it was with the cystic fibrosis gene. I felt I was getting a tiny glimpse into God's mind."

Over the next several years Collins was granted several more forays out of the teeny dot. He played a key role in the discovery of more genes, including the one malfunctioning in Huntington's disease. He raced Ray White at Utah again to capture the gene for neurofibromatosis, a neurological condition that afflicts its victims with benign but disfiguring tumors, "café au lait" spots, and sometimes malignant tumors and learning disabilities. Perhaps even more than the scramble for the cystic fibrosis gene, the contest with White demonstrated an inveterate urge in Francis Collins to get to the finish line first.

After starting out as collaborators in the hunt for the neurofibromatosis gene, Collins and White parted ways over a difference in understanding their respective roles. They then found themselves bumping elbows in a frantic search through the same stretch of DNA on chromosome 17. White protested that Collins's effort benefited greatly from clones White had sent to the University of Michigan when the two labs were still collaborating, which were then used for an entirely different purpose, enabling Collins to hunt for the gene on his own. Collins maintained that he was free to use the clones any way he wished. The two teams found the gene nearly simultaneously, but Collins sent his paper to *Science* first. When White found out that he was about to be scooped, he quickly wrote up his own discovery for the journal *Cell,* which pushed it into print less than three weeks after receiving it. Meanwhile Collins and *Science* had gotten word of White's imminent publication, so *Science* pulled a last-minute switch and pushed Collins's publication up a week. Both papers ended up being published on the same day, accompanied by a joint press conference at the Howard Hughes Medical Institute, which had funded both labs. White spoke for five minutes, then Collins got up and spoke for twenty, leaving a lot of people with the impression that he was the primary discoverer. Ray White, forgotten in Utah, remained bitter even ten years later.

The ambition and political savvy Collins evinced in these gene hunts should have been enough to convince anyone that this was a dangerous competitor to dismiss, especially if the goal was not a single gene but the code of human life itself. In September 1998, Collins went public with a change in plans for his program. The Human Genome Project was now confident that it could finish the human genome in 2003, two years ahead of the original schedule. What's more, they planned to have a "working draft" covering 90 percent of the code completed by 2001—the same year that Venter was promising to be finished with his own *"Mad Magazine"* version.

CHAPTER 11
ALL HANDS

By September 1998, the Celera building off Rockville Pike no longer smelled of mildew and mouse droppings. But there was no sign yet of the promised sequencers from ABI, and the constant drilling and hammering indicated that the space to put them in wasn't ready anyway. Most of the building still belonged to the construction workers, who strode around with their walkie-talkies and pendulous tool belts like soldiers of an occupying army. Venter and the twenty-odd other scientists and employees were camped out in temporary quarters on the first and second floors. Sam Broder, a former director of the National Cancer Institute, had joined the company as its chief medical officer. Peter Barrett, who had helped orchestrate Tony White's overhaul of the parent company, would serve as chief business officer. To oversee policy issues, Venter had hired veteran science administrator Paul Gilman. He had brought Hamilton Smith, Mark Adams, and a dozen others over with him from TIGR. Software engineer Anne Deslattes Mays was given the monumental task of developing the software systems that would run the pipeline, from feeding the machines, to making sense of the code coming out the other end, to delivering a huge, stable, user-friendly database to customers. Fortunately, the company wasn't expecting customers for months or even years, which would give her group time to focus on developing the up-front component.

Meanwhile Granger Sutton, the computer scientist who had written the computer algorithm to assemble the first bacterial genome, would try to repeat that triumph on a much grander scale. Venter called Sutton his "quiet warrior." Tall, with the full wavy hair of a Kennedy, he might have attracted attention were he not constitutionally inclined to duck beneath it. At TIGR, he had been asked to write a program on his own that could put microbial genomes together, so that's what he had done. At Celera, he was being asked to put together a genome a thousand times larger and infinitely more complex, and he was prepared to go off in a corner and do that, too. Worrying seemed beside the point. But Sutton did think he could use some help. Soon after Venter's announcement the previous May, Sutton had gotten a call from Gene Myers at the University of Arizona, who with Jim Weber at the Marshfield Medical Research Foundation had cowritten the theoretical paper on assembling the human code using whole-genome shotgun. Celera was attempting to put into practice what they had only speculated might be possible, and now Myers was asking "Can I play, too?"

Sutton mentioned the matter to Venter. At first Venter believed hiring Myers would be too risky. Myers's work was more theoretical than practical. His paper on assembling the human genome with the shotgun technique had been badly received and forgotten; Venter himself had not even read it. Myers also had a reputation for being moody and oversensitive, with a tendency to go off the deep end in conflicts with colleagues. On Sutton's recommendation, however, Venter decided to take him on anyway. "If he implodes," he told Sutton, "you clean up the mess."

Venter liked to say that his greatest talent was hiring people who were smarter than he was. A lot of successful people affect this modesty, but he did seem to have a radar for detecting dormant talents in others that he himself lacked but were critical to realizing what he wanted to accomplish. When it came to bench science, Venter knew he hadn't the patience or the genius—but Hamilton Smith did, so Venter raised him from the dead. Likewise, he had only a vague conception of the computations needed to assemble the human genome by the shotgun technique. But without knowing it, his ambition had awakened in Gene Myers an angry need to vindicate his trampled theory, and now that anger was a sharp glint in Venter's overall scheme.

He had picked Paul Gilman, Celera's new director of policy and planning, to fill another lacuna. Venter was a master at making great leaps into unknown territory, Gilman a patient craftsman of the bridges needed to bring everybody else across. He had worked behind the scenes in government for two decades, most of it spent in federal science policy. He'd also served six years as Senator Pete Domenici's chief of staff; it was Gilman, in fact, who in 1986 had alerted Domenici to the virtues of the budding Human Genome Project. Until a few weeks before, Gilman had been employed, and reasonably content, as a senior administrator in the National Research Council at the National Academy of Sciences. But there was a long-buried restlessness in him, just as there was in Myers and Smith. Years ago, when he was still an undergraduate at Johns Hopkins, he was assigned to read Milton Friedman's *Capitalism and Freedom.* Friedman argued the case that certain massive governmental programs such as the draft could be more efficiently accomplished by the private sector. The book had made a deep impression on Gilman back then, but he'd forgotten about it until Venter and Hunkapiller had announced their intention to sequence the human genome. Then out of the blue, Venter called him at home. "A little bird tells me you'd like to work in a place like this," he said. Gilman was astonished. He hadn't told anyone about his personal interest in Celera except his wife. He drove up to Rockville and got the grand tour. "We're going to establish a new paradigm," Venter told him. "We're going to prove you can do open research and make money at the same time." The next day, Gilman gave notice at the National Academy.

There was another key position to fill. Venter loved to talk about the gigantically powerful supercomputer that Celera was going to need to sequence the genome, but he had no idea how to put together such a system. No one else did, either, since a computing problem the size of the entire human genome had never been attempted before. Compaq was wooing Celera, and Marshall Peterson was a consultant working for them on the project. Venter needed someone who could appreciate the scale of the project, and knew how to work under pressure. Peterson seemed to fit the bill. He had done three tours of duty as an army helicopter pilot in Vietnam, where he was shot down four times. During the war Peterson's buddies had called him Mad Dog, but he didn't look ferocious, at least not anymore. In fact, he seemed to exude a calm, slightly melancholy attentiveness, as if he'd made peace with his demons and was hoping it wouldn't cost you as much to do the same. His body was compact and slope-shouldered, like a full-size Yoda; the power of his handshake came as a shock.

After the war Peterson served as an army test pilot, which isn't the most carefree occupation, either. In the early 1970s, complex computing systems were just beginning to become part of aircraft design. Computer-assisted fighter jets were far more maneuverable but required the computer to be functioning for the plane to fly at all. Peterson took a keen interest in the instruments that were now holding him up in the air. He got a degree in aeronautical engineering and began designing computer-assisted weapons systems. After leaving the military, he took a job at Digital Equipment Corporation, designing ultrareliable systems for Dow Chemical and the World Bank. Digital was bought by Compaq while Peterson was installing systems for the Erickson cell phone company in Sweden. Then in 1998 he was called back to consult on Compaq's bid for the Celera contract. He got along well with Venter, also a Vietnam vet, and was hitting it off particularly well with Gene Myers, who would be by far the supercomputer's most demanding user. Venter offered him a job. "I thought about how I could keep on helping Erickson make better phones, or I could come here and help cure cancer," Peterson said, soon after he arrived in Rockville. "Which would you choose? This is the most exciting project on Planet Earth."

By mid-September, Celera had fifty employees and offers out to a hundred more. Venter called the nascent staff together in the basement cafeteria for an "all-hands meeting." Bob Thompson, the plant facilities manager, had made sure the cafeteria was up and running in time for the meeting. Thompson was the point man for virtually everything having physical substance, from crafting the complex infrastructure for the temperamental sequencers and supercomputer to installing the plumbing, electrical circuits, data networks, cold rooms, telephone systems, security systems, carpeting, and toilet fixtures. He had set up operations for Perkin Elmer before, and he knew it wasn't just about buildings. At the lunch break, food service staff in crisp starched shirts and ties handed out free plates of chicken marsala and trembling slabs of tiramisu. "I wanted the cafeteria open today because we've got a lot of people here who don't know each other," Thompson said. "Eating together makes people instantly less strange to each other."

After lunch, some folding chairs and a slide screen were set up in an open space next to the cafeteria. A collective buzz of anticipation energized the room, as if the group were meeting on a dock to embark on some great expedition. The staff was still small enough so that people could stand up one at a time and introduce themselves. "My job is to keep one step ahead of everybody," Thompson said when it was his turn. He stood square and compact, wearing jeans and a gray photographer's vest over a faded paisley shirt. "Today the goal was to have meal service. The goal by the end of October is to have some science in this building. Someday I'm going to tell my grandchildren I was part of this."

The meeting got down to business. In a few days, Venter and some other senior people were flying down to Miami for the annual Genomic Sequencing and Analysis Conference. Started as a small academic gathering by Venter when he was still at NIH, GSAC—pronounced "Gee-sack" by those who attend it—had grown in size and influence through the nineties until it rivaled Cold Spring Harbor's spring meeting as the most important yearly gathering of the genomics community. The public Human Genome Project had always been well represented. But suddenly genomics had become the hot new thing in drug development, and this year registrations were flooding in from the pharmaceutical and biotech industries, including competing firms like Incyte and Millennium Pharmaceuticals. "Everybody who's going down to Miami should be aware that there will be people aggressively trying to recruit you," Peter Barrett warned the group. "Instead, plan on coming back with two new prospects. Steal. Don't get stolen."

GSAC would be Celera's public debut, and Venter had scheduled the all-hands partly to provide a captive audience for a rehearsal of the talks. First Hamilton Smith shuffled up to the front of the room, like a bashful teen asked to present his science project, and described the intricately crafted DNA libraries he was constructing to keep Celera's code factory fed with raw material. Mark Adams followed with an overview of the subsequent steps in the pipeline, from reading out the millions of DNA fragments from Smith's libraries, to reassembling them into a coherent genome, to the critical step of "annotation"—developing and using special pattern-finding software tools to look for hot genes and other meaningful information within the code. Venter sat in the front row, his arm draped over the back of his chair like a theater director watching a dress rehearsal, interrupting occasionally with polite but pointed critiques. He would not be giving a talk in Miami. But there was no doubt who was running the show.

Sam Broder was next up. He was in his mid-fifties, slight and dark-haired. His team, which so far consisted of himself, would represent the ultimate step in the Celera pipeline: devising ways to increase the usefulness of the genome as a tool to cure disease. Broder had spent twenty-two years as a medical researcher at NIH, where he had been instrumental in developing AZT, the first anti-AIDS drug. He had risen through the ranks to become the director of the National Cancer Institute, the largest single agency in the National Institutes of Health. Throughout his career, Broder had often given lectures about cancer to the public. After every one of his talks, someone from the audience would take him aside and ask, Can you help my mother? Can you cure my son? Can you save my wife? He had had to learn how to prepare himself emotionally before a lecture in order to bear the questions that would inevitably follow. It made him angry that he could do nothing to help—indeed, that a huge government agency full of the best and brightest scientists available was still so powerless against the disease. In his office at NCI he had hung a wooden sign on the wall with the words IT'S THE DISEASE, STUPID! burned into it. When he left in 1995 to take a job at a small drug development company, he brought the sign along. Now it was on his bookshelf at Celera.

Broder's talk in Miami would be about the value to the medical community of finding single-base-pair variations—the so-called SNPs sprinkled throughout the human genome. SNPs were the hottest topic in medical genetics, and Celera was positioned to find them en masse. This was a key message to the pharmaceutical companies that the company hoped to attract to its database, and in Miami those companies would be represented in force. "The official party line from NIH is that there may be something on the order of three million SNPs in the human population," Broder was saying. "Look long enough and you'll eventually find—"

"So let's say we find all three million first," Venter interrupted. "What will you say if you're asked in Miami if we plan to patent them all?"

"What I'd like to say," Broder replied, "is that if there is an important nonobvious contribution to science or medical practice, we will not give up our right to file a patent."

"You really don't want to get into this down there," Venter pressed. "Just say that we're thinking about it and we'll let you know after we decide."

It was Gene Myers's turn to speak next. He was standing in the back of the room, apart from the others. Handsome in a craggy, Charles Bronson kind of way, he had straight, glossy black hair, and a face roughened by old bouts of acne and long exposure to the sun. He was wearing jeans and a paisley silk shirt, and there was a diamond stud in his left ear. He was paying such forceful attention to Broder's talk on SNPs that his forehead was bunching up in furrows. He liked the science, but the bantering about patents made him uncomfortable. It made him wonder, as he had a hundred times already, whether he'd made the right move in leaving Arizona. Among his academic friends and colleagues, abandoning a tenured professorship for a corporate job was synonymous with selling out and considered pretty much a one-way street. He missed the desert and hadn't gotten used to living in a Washington suburb. He didn't know if he could trust Craig Venter or these other people talking about gene patents and stealing people from other companies. Then he remembered that he was being given the chance to assemble the human genome by shotgun.

Still frowning, Myers gathered his laptop and strode to the front of the room. He knew there was nothing he could say in Miami that would convince anyone that a computer algorithm could be devised capable of assembling the whole genome of a human being, or even the much smaller genome of a fruit fly, in one fell swoop. He might as well try to persuade them that he could write a program that could teleport him across the room: it was the kind of thing that no one would believe until they saw it. But he might be able to address at least one point of contention. The theoretical paper he had written with James Weber had been trashed in part because the human genome simply covered too large a landscape to make sense of in a single mathematical problem. A shotgun assembly of it would be riddled with holes—a "Swiss cheese genome," one critic had called it. This was one reason the government program had committed to the more cautious, "hierarchical" approach of assembling tiny bits of the genome at a time, then piecing together those assembled bits—like doing thousands of small jigsaw puzzles rather than facing the daunting task of assembling a single gargantuan one.

With some PowerPoint graphs and figures, Myers now demonstrated that at least in one sense, size did not matter. As long as you had enough computing power, whether you were trying to put together one tiny piece of the genome at a time or doing the whole thing at once, the number of gaps in the resulting sequence would come out to be the same. Though this was only a rehearsal, he bulled through his proof, as if defying anyone to doubt his calculations. "In terms of coverage, if the public-program people argue that our assembler won't work, then they have to admit that theirs won't either," he concluded.

Venter was impressed. He had interrupted the other speakers frequently, but with Myers he confined himself to a single comment at the end. "The red in that last slide won't show up very well in the back of the room," he said. "You should change the color."

A few years earlier, Gene Myers had had a strangely vivid dream. In the dream he was dressed in a tuxedo, waltzing round and round a ballroom in Washington, D.C., with his beautiful blond wife in his arms. He woke up that morning spinning and perplexed. He didn't have a beautiful blond wife—or any wife, for that matter. He lived alone in Tucson, a mathematician, a computational biologist, a recovering alcoholic and drug abuser in his mid-thirties who had never owned a suit, much less a tuxedo, and didn't know how to dance. Just a few weeks after the dream, however, Myers met M'Liz Robinson, a business student at the university who was pretty, blond, and liked to dance. She persuaded him to take ballroom lessons with her. They became so adept as a pair that other couples were asking them for lessons. A few months later, they were married. Now they had moved to a Washington suburb and M'Liz had a job at Celera, too, in business development. Myers figured it was only a matter of time before the dream came true.

After the all-hands meeting, Myers retreated to the second floor. The workmen had made a corner of it habitable, and he and Granger Sutton had adjoining cubicles up there, overlooking Rockville Pike through a thin scrim of trees. The office was bare except for a desk, a chair, and a bookshelf. He swiveled his chair around to face the window and stared out at the streetscape. Huge bright blue letters spelling VOLVO ornamented the car dealership across the Pike. Otherwise it looked pretty gray.

Myers had put some mementos out on the bookshelf, to make the cubicle feel a little more like home. Among them was a test tube containing a pus-stained Band-Aid—a tribute to Friedrich Miescher, the Swiss scientist who first discovered DNA in 1868 in the used bandages he had scrounged from a local surgical clinic. Some computer science friends back in Arizona had given it to Myers as a joke, when he first got involved in biological problems. Myers had no training at all in biology, apart from what he'd picked up on the job. He would like to learn more, but squirming, living biology did not enter into the equation he was going to solve.

When he was five years old, Myers decided to write down all the numbers from 1 to 1,000. At the time it just seemed like a fun thing to do, but by the end he'd introduced himself to one thousand lifelong friends. He was finishing college-level math courses before he reached puberty. His family, amazed by this strange talent among them, enjoyed tossing problems at him, which he would go at like a dog thrown a bone. "Once they gave me a puzzle of wooden blocks that could be arranged into a hundred and fifty different shapes," he remembered. "I wrote an algorithm to solve them all."

When Myers was in his twenties, somebody handed him a Rubik's Cube, the famous puzzle toy invented in Hungary in 1974. A Rubik's Cube has six colors and is composed of a set of interlocking panels that can be moved independently. The trick is to rotate these panels in your hands so that each side of the Rubik's Cube is composed of only one color. The puzzle has 43,252,003,274,489,856,000 different possible configurations. Only one of these possibilities solves the puzzle. If you toyed with it randomly and allowed one second for each turn, it would take you 1,400 million million years to find the correct solution. Myers put the cube down on his desk and frowned at it for an hour, scribbling numbers now and then on a pad. Then he picked up the cube and twisted it this way and that, according to the set of subroutines he had written down. The colors slipped into place, forming solid colors on all sides. He wasn't particularly proud of this accomplishment. If he hadn't been stoned at the time, he remarked later, he probably could have devised a solution in half the time. But a *real* genius could have figured it out in five minutes.

Myers wasn't about to congratulate himself for the little proof he'd presented at the all-hands meeting, either. Any decent computational biologist could have done the same thing. In the abstract, assembling a genome by the shotgun method was a classic example of an inverse problem. Rather than break something apart to see what it is made of, you sample bits of information about it until you can re-create what it is you've been sampling. A jigsaw puzzle is also an inverse problem, and up to a point it is a convenient analogy to putting together a genome by shotgun. The sequenced fragments of DNA coming out of the machines are the puzzle pieces, and the genome is the image they form when all the pieces are locked into place. But there are crucial differences. First, jigsaw puzzle pieces lock together by the shape of their edges, while each piece of a genome puzzle must *overlap* its neighbor by a handful of base pairs—around fifty, to be statistically sure—in order for them to be candidates for a match. Second, a jigsaw puzzle's pieces are all carved up from a single image, so each one is unique. The fragments of DNA to be assembled aren't unique at all. On the contrary, each one *must* contain bits of sequence that are also represented on other pieces, or else there would be no way to overlap them, and no way to put the picture together.

The upshot of this is that one needs a whole lot of pieces in order to put the picture together. The *Drosophila* genome, which would be Myers's first challenge at Celera, was made up of an estimated 120 million base pairs. Each sequenced fragment of DNA coming off the automated machines would be approximately 500 base pairs long. This meant that in theory the *Drosophila* genome puzzle consisted of 240,000 pieces, or 120 million divided by 500. But if Myers were to take the first 240,000 fragments from the sequencers and try to assemble them, he wouldn't be close to getting a representation of the entire *Drosophila* genome. The shotgun process was random, so some parts of the picture would be represented many times over, while others would not be represented at all. Myers would have to wait for another batch of 240,000 fragments, then another, and another, until he had enough pieces to ensure the whole territory of the genome had been sampled at least once. Mathematically, he could not be absolutely sure of that until he'd sampled the genome ten times over—or 10x, in the shorthand of genomics. His assembly program for *Drosophila,* therefore, would have to handle a total of 2.4 *million* puzzle pieces, or 10 x 240,000. In the initial step, the computer would have to compare each fragment with every other fragment to look for overlapping sequences, which meant 2.4 million times 2.4 million individual calculations—close to 6 trillion calculations. Not a problem at all, with enough processing power.

The human genome would be more challenging. A genome 3 billion base pairs long meant 6 million pieces to cover the genome just once; 10x coverage meant that 60 million fragments of DNA would need to be sequenced, each five hundred letters long. That was an order of magnitude more of sequenced DNA than currently existed in the entire world. Venter had said that Celera would have enough sequencing power to provide the raw data, but Myers couldn't help noticing that ABI had yet to deliver a single machine. As for computer power, 60 million times 60 million was . . . well, lots and lots. Whatever the sum, the sheer bulk of the computation would bring any existing computer to its knees. Venter said he had that matter covered, too, but so far the only computers in the building were desktop PCs. Unless the resources started showing up soon, Myers thought, there was no way he could get an assembly program up and running in time to deliver *Drosophila,* never mind *Homo sapiens,* on schedule.

He checked himself, remembering that these were other people's problems. He had enough to worry about on his own. It wasn't the mere size of the human genome that made a shotgun assembly so daunting. It was all those repeated sections of the DNA code. Using the gigantic jigsaw puzzle analogy, perhaps as much as 40 percent of the genome consisted of pieces that were identical to some other piece. Imagine trying to put together a jigsaw puzzle of a forest scene, with nothing to go on but matching the shape of a fragment of leaf to that of another bit of leaf, all the time knowing that the shape and size of 40 percent of the leaves are exactly the same. This was why the critics were so sure the attempt would end in catastrophe. Like Rubik's Cube, the human genome could be assembled in a nearly infinite number of ways, but only one was correct.

On a good day, that thought excited Myers in a way that nothing ever had before. People had been challenging him with puzzles all his life, and in computational biology—arguably in all computer science—this was the Mother of All Puzzles. He was dying to get his brain into it. Like others in his field, he had spent most of his professional life improving the efficiency of existing algorithms, and now he had been handed the chance to create something completely novel to solve a problem of historic proportion.

He and Sutton were working well together, too. Their minds matched well. Assembling the genome was a huge practical problem, and Myers was used to thinking hugely, while Sutton was skilled at solving practical problems. They spent hours together talking and thinking, keeping each other inspired with a couple of pencils and a shared pad of paper. In the process, they had developed a personal shorthand notation for visualizing components of an algorithm capable of handling the human genome's dizzying eccentricities. They had a breakthrough on a plane trip to California. In the theoretical paper with Weber, Myers had thought the solution would come from the top down: start with two known points on the genome and mathematically construct a series of bridges between them. On the flight, the yellow pad passing back and forth across their seat trays, Myers and Sutton suddenly saw that a cleaner solution rose from the bottom up. They visualized tiny islands of sequenced code like growing crystals—far-flung stars in a great dark sky gradually extending their perimeters toward their distant neighbors until, finally, they touched. When Myers and Sutton got back to Rockville, they wrote some simulation programs that could be run on a desktop PC. The algorithm chopped up some known microbial genomes into random bits and put them back together again in a fraction of the time that Sutton had originally needed with the TIGR Assembler.

The success of the simulations was gratifying and instructive, but a long way from the reality of what had to be accomplished. On his bad days, Myers thought about the problem of the repeats and felt a cold anxiety opening up in his stomach. He had taken personally the trashing his theoretical paper with Weber had received. But he knew it hadn't been meant that way. Philip Green, who had led the attack, was not a mean guy. He wasn't a fool either; indeed, he was among the most respected computational biologists in the world. What if Green's blistering rebuttal to his theory had not been the knee-jerk reaction that Myers had labeled it in his mind? What if Green and the others had insisted a whole-genome shotgun assembly of the human code couldn't work because in fact it couldn't?

It was much too early to tell. Metaphorically, whole-genome shotgun assembly was like spilling the pieces of a jigsaw puzzle out of a coffee can onto a table and having the puzzle come together all at once. But the algorithm would actually entail a series of discrete problems, each of them fanning out in an array of subprocesses, and those subprocesses would be further broken down into their own components, and so on, until one reached the level of the hundreds of thousands of individual lines of computer code that would have to be written to make the whole thing work. Ways of checking and refining each component would be incorporated into the program. But whether it worked or not would be revealed only in the last step, when the time came to upend the can and see what tumbled out. Myers imagined himself in both scenarios. If the pieces fit together—if the crystals had grown so that they all aligned and touched, like a diamond necklace stretching from the beginning of the first chromosome to the end of the last—it would do more than just vindicate his theory. It would make him a celebrity in his field, even a legend. He liked the sound of that. But if the result was the mush that most of his colleagues were predicting, he would be a sold-out scientist sitting in a cubicle, staring out the window at a Volvo dealership and defenseless against the agonizing thought that there was no way back to his old life.

Myers's gloom did not penetrate much into the general mood. The others took their cue from Venter, who in his confidence seemed to emit light, as if the sheen off his pate were not a reflection but a source of radiant energy of its own. He wasn't any surer the assembly would work than Myers was, but rather than let his doubts worry him, he turned them into an inspiration, as if the uncertainty of the enterprise was the very thing that gave it flavor and gusto. Why do something that you knew in advance was going to succeed? In early October, Robert Waterston and John Sulston, who were on the verge of completing the genome of the nematode worm *Caenorhabditis elegans* and were considered the best genome assemblers in the world, wrote in *Science* that Celera's whole-genome shotgun method would prove "woefully inadequate" for a puzzle the size and complexity of the human code. Venter waved the article around as if it were an endorsement. Waterston and Sulston had earlier doubted the success of ESTs and the *H. influenzae* genome, so it stood to reason that their dour predictions now were a hopeful sign. Whether this logic made sense was beside the point; what mattered was it inspired everyone else to think the same way.

"Nobody can say for sure at this point whether it can be done," said Hamilton Smith. "The difference is, we have the guts to try it." Then he laughed, as if he were surprised and a little pleased by what he'd just heard himself say.

By the end of the month the staff had doubled in size, and would double again in another month. Boldface employment ads in Sunday newspapers beckoned not only with good salary and benefits but with an invitation to "become a part of history." The building off Rockville Pike had proved too small to contain the company's hopes, so the adjoining, twin structure was purchased and summarily gutted. Hallways were stacked to the ceiling with cartons of new computers and monitors waiting to be opened. But there was a major piece missing. Out at Applied Biosystems in Foster City, California, the engineers were having trouble getting the new automated sequencers to work consistently. By November, not a single one had been delivered to Celera. It was beginning to seem impossible to keep on schedule.

On November 16, Venter called a meeting of his senior staff. The purpose was to review a list of the company's specific goals for fiscal year 1999, in preparation of a stock offering to occur in a few months. Peter Barrett stood in the front of the room next to a whiteboard, jotting down suggestions with a red marker. At the top of his list he had written "Develop scientific and business infrastructure," "Complete fly genome," and "Close deals worth 50+ million."

"Any one of these goals is mind-boggling," Venter said.

"And every one depends on the delivery of those machines," said Hamilton Smith. "We're making all these projections based on having them up and operating early next year."

"We can't write down that these goals depend on delivery," said Barrett. "That would be insane."

"The public promise was to have the *Drosophila* genome published in *calendar* year 1999," Venter said. "That gives us some leeway."

"If we can't get *Drosophila* sequenced by June," said Adams, "we're in trouble in any case."

"Even if we got the sequencers tomorrow, we still couldn't start sequencing, right?" asked Gene Myers. For weeks he'd been wearing a Polarfleece jacket and a long black-and-white wool scarf around his neck, as if he were perpetually chilled. "The third-floor facility isn't ready for them yet."

Bob Thompson, the facilities manager, stiffened in his seat.

"So Bob, when will it be ready?" Venter asked him.

"January."

"When in January?"

"January has thirty-one days."

"You weren't saying that last summer," Venter said.

"Things have slipped," Thompson replied.

Venter fidgeted with a soda can on the table. "I'm going to need some Advil for this," he said. Peter Barrett turned and wrote on the whiteboard, "Source Advil." Nobody spoke for a while.

"Well, we should break out some champagne when these sequencers come in," said Smith, trying to lighten the mood.

"No," said Barrett. "Somebody would spill it on the machines."

The first sequencer finally left ABI for Celera on December 6. Its wooden crate was too large to fit through an airplane cargo door, so the machine traveled by ground, handed off from one truck to another. On the morning of the eighth, word came that the last handoff had been made at Dulles Airport and the truck was due to make delivery any minute at TIGR, where the machine would receive some initial testing. Mark Adams and Bob Thompson rushed over from Celera. Some curious TIGR scientists had already gathered on the loading dock, along with some people from Perkin Elmer and ABI who had flown out to monitor the delivery and installation. The sky was a dull sheet of platinum and a steady, phlegmatic rain began to fall, slicking the cement floor of the dock on its outer fringe. Venter was not there but had given instructions that he should be called as soon as the truck arrived. Hamilton Smith drove up in his old Mercury and joined the others, his mop of white hair bobbing above them in spite of his attempt to blend in. He drifted over to a reporter standing in the crowd. "It's the end of one era and the beginning of another," he said solemnly. In his normal voice, he added, "I saw you taking notes, so I thought you might like a quote."

The minutes ticked by. A determined shipping clerk with a clipboard went about his business, weaving around the human obstructions as he sorted out the ordinary parcels. The atmosphere resembled that of a maternity waiting room, with all the people waiting for the same baby. Thompson was officially in charge of the delivery, but there was an unspoken feeling that Mark Adams, who would be overseeing the sequencing effort itself, should be the one to sign for it. He was containing any expression of excitement, except for a slight acceleration of his gestural tempo. "In the latest fad terminology, this is an inflection point," he allowed himself to say. "But I won't believe it until I see that truck."

An hour went by, and there was still no sign of it. Dulles Airport was only thirty minutes away. Thompson paced from one side of the open dock to the other. Every few minutes he dodged out into the rain to glare furiously up the street, as if that might make the truck arrive sooner. A Perkin Elmer marketing representative in a business suit was on her cell phone. "I don't *care* if he's a new driver," she was saying. "That's *their* problem!"

"Not good," Adams muttered. "Not good."

Then, just before noon, a truck turned into the parking lot. The driver did a double take when he saw all the people on the loading dock, the cigarette in his mouth swinging like a loose needle on a gauge.

"Carl, it's here," the marketing rep said into her phone. "Tell the team congratulations." The driver backed the truck up to the dock and warily climbed down from the cab. "Am I interrupting something?" he said, not sure whom to address.

"On the contrary," Adams replied, breaking into a big smile. "You're the main event."

CHAPTER 12

DEAD ON ARRIVAL

Paul Gilman, back at Celera from a visit to Capitol Hill, took off his suit jacket of light gray herringbone and hung it on the door of his windowless office. At forty-six, his wavy hair was the same shade of gray as his suit. His face was long and pale, but his gray-blue eyes were keenly bright. The jacket on the door looked like a skin that he'd just shed. He stood for a moment with his hands in his pockets, taking in the anonymous furnishings as if they were new acquaintances he was eager to get to know.

As policy director, Gilman's job was to convince people that Celera was a good thing—good for science, good for business, good for humankind. He was not at all surprised that Francis Collins and his colleagues in the public genome project were equally intent on portraying Celera as bad. In their position, he would be doing the same thing. But if the company were to reach beyond the pharmaceutical industry to attract 100,000 customers, much less the 100 million that Venter envisioned, the world would have to understand that Celera was an opportunity, not a threat. This was the message he was taking to his former colleagues on Capitol Hill: The human code was the portal to the future, and Celera was the instrument to open that gate years earlier than expected. But the effort to sequence the human genome shouldn't be viewed as a competition. In the best of all possible worlds, Gilman believed, Celera and the government should find a way to work together. Fortunately, there was an opportunity waiting for him to develop.

Back in September, Ari Patrinos, the head of the genome program at the Department of Energy, had dropped by Celera for a visit. Venter gave his visitor the accustomed tour of the empty spaces; then they sat down to talk. Patrinos had come to broach the possibility of a collaboration between DOE and Celera. In spite of the animosity that the DOE's partners in the public genome program felt toward Celera, collaboration made sense. The DOE had been funding Venter's operation at TIGR for years, and Patrinos had never thought of him as a threat. On the contrary, Venter's contempt for playing by NIH's rules made him an intriguing ally. The DOE had initiated the Human Genome Project but had long since become NIH's poor cousin, outmuscled financially and intellectually. As a Greek boy growing up in Egypt, Patrinos had learned that a slightly built member of an ethnic minority had to roll with the punches, literally and figuratively. His diplomatic skills had served him well in his career, but he had never completely overcome his resentment of bullies. He admired Francis Collins personally and counted him a friend. But he did not share the "All for One, One for All" embattled mentality that had taken possession of the Human Genome Project since Venter's big announcement the previous May—especially when the DOE, as usual, had so little influence over what direction the "All" would be taking.

Patrinos had no illusions about where the DOE fit into the public genome program politically; his agency did not enjoy the beneficence that Congress lavished on NIH, and money was power. If a mutually advantageous arrangement could be worked out between the DOE and a private-sector company like Celera, it could improve his leverage in Congress, not to mention bringing some long-deserved attention back to the DOE genome program. The ultimate goal for everyone, after all, was to get the human code finished and available to scientists as quickly as possible. If Venter could help with that, why *not* work with him? If such a collaboration appeared possible, of course Patrinos would have to bring Francis Collins into the loop. But in the meantime it wouldn't hurt to talk to Venter privately.

"We're open to anything you want to propose," Venter told him. "But it might be better to wait until the dust settles a bit." At that moment a ceiling panel crashed open from above and a workman's head and shoulders popped out, like an inverted jack-in-the-box. "Oh. 'Scuse," the workman said, giving a short blast with his drill before retreating.

By November, Venter figured that enough dust had cleared to explore the possibility further. He asked Paul Gilman to see if he could work out terms of a deal.

"Does Collins know about this?" Gilman asked.

"If he does, it's not because *I* told him," Venter replied. "If Francis doesn't pee all over this first, it could be a real boon for everybody involved."

Gilman got in touch with Marvin Frazier, the program manager in Patrinos's department immediately in charge of the agency's human genome effort. Frazier was seriously ill with stomach cancer, but told Gilman that he was eager to cooperate and did not think his health would impede him. They began exchanging ideas over the phone and sending drafts of a memorandum of understanding back and forth to be shared with their superiors. Scientifically, the terms of the agreement amounted to a fairly simple instance of mutual back-scratching. On its part, the DOE would alter the sequencing tactics it was using at its Joint Genome Institute in Walnut Creek, California, so that it could provide Celera with data that could help the company put its whole-genome shotgun version of the code together more reliably. In return, Celera would give DOE scientists access to the sequence information it was accumulating on chromosomes 5, 16, and 19—the territory that the Joint Genome Institute had been allotted by the NIH. In addition, both sides agreed to share expertise in the subsequent process of "annotation"—finding genes and other meaning in the raw stream of DNA letters. Finally, they agreed to jointly publish their results.

To Gilman, the memorandum of understanding was looking like a classic win-win situation. While the scientific exchange would benefit both sides, both would benefit politically, too. The DOE would get a boost for its struggling genome program and a little respect back after years in NIH's shadow. For Celera, the document would show that the company was, as Gilman liked to put it, "wearing the white hat." If a major public agency was willing to work with Celera, NIH would have a more difficult time demonizing it. To that end, Gilman had carefully modeled the wording of the memorandum on a similar document being prepared for the fruit fly collaboration between Celera and Gerry Rubin's lab at Berkeley—a project that had the full participation and blessing of Collins, whose institute funded Rubin's lab. This was a tactic that Gilman had learned on the Hill: lift the language for your bill amendment from a document your opponent has already agreed to; that way he can't reject it without appearing hypocritical.

On the first weekend in November, Gilman and his wife entertained some friends for dinner, including Martha Krebs and her husband. Krebs was the director of the Office of Science at the DOE, and thus Ari Patrinos's boss. Gilman complimented her on how smoothly things were going between the DOE and Celera. "This is the way the world should work," he said. "I just wish NIH would do the same."

"Well, I wouldn't be surprised if they get something going with the private sector soon, too," Krebs replied. "If not with you, then maybe with Incyte."

A little alarm bell rang in Gilman's head. Martha Krebs was a physicist. How would she be able to pull the name of a small biotech company like Incyte out of the air like that? He mentioned his suspicions to Venter on Monday morning. But Venter already knew what was going on. "Ari told me that Maynard Olson is contracting with Incyte to do ten million reads of human DNA sequence," he said. "That nearly doubles the public program's capacity. And Maynard gets his money straight from Francis."

"So," said Gilman. "The plot thickens."

In an extraordinary display of unfortunate timing, Collins himself was scheduled to pay a courtesy call on Celera the following day. Gilman met him in the makeshift lobby and escorted him back to Venter's office, where Hamilton Smith, Mark Adams, and Sam Broder had gathered. Though Collins could hardly have been looking forward to the visit, there was a resolute smile on his face.

"So is it true that you're giving a major grant to our competitor?" Venter asked, as soon as everybody had sat down. Collins's smile vanished, but only for a second. "Gosh, Craig," he said. "You know as well as I do that I don't control what extramural grantees do with their grant money."

"That's not true," Broder said. "Nobody can get a grant from you unless they abide by the Bermuda rules, which are unique to your institute."

"That's a very different issue," Collins responded. "But since you mention it, any data that Incyte would produce on NHGRI funds would of course have to be released nightly, as the Bermuda Accord requires."

"Come off it, Francis," said Venter. "You're using Maynard as a conduit to channel funds to Incyte, and cutting into our market value in the process. If you ask somebody to kill someone for you, you're still guilty of murder."

Gilman attempted to turn the conversation to less flammable topics, but it was hard to find one.

"I went there with the best of intentions," Collins later insisted, "but went away feeling that it had been a pretty difficult interaction."

Not everybody at Celera was thrilled by the prospect of a deal with the DOE. While Paul Gilman was putting the finishing touches on a draft of the memorandum of understanding, Peter Barrett and his four-person team in business development were working equally hard to sign up Celera's first paying customers. He was surprised by how much interest there was so soon. Sequencing of the human genome would not begin until after the fruit fly was finished, and obviously the fruit fly couldn't begin until there were some sequencers installed and running. But customers were knocking on the door nevertheless, and Barrett wasn't about to tell them to go away. He had put together an "early access" program that would allow well-heeled drug companies and biotechs a privileged peek at the DNA sequence pouring fresh out of the Celera pipeline before it was released publicly—data on both the fruit fly and, when the time came, the human code. In an article in *Science* in June 1998, Venter and his colleagues had promised that raw DNA sequence would be released every three months. If there was going to be a gold rush to find genes, however, three months wasn't a bad head start. Barrett had deals pending with the giant Swiss pharmaceutical firm Novartis and with Amgen, whose billion-dollar anemia drug Epogen had endowed it with plenty of cash.

The likes of Novartis and Amgen weren't going to sign up, however, just to peek through a window that was going to be flung open for the world at large three months later. The real value in paying millions to become an early-access partner was to get more information on the genome than an academic lab was going to get for a few thousand, and certainly more than an individual researcher cruising through the data would have for free. But these "levels of access" had never been clearly defined. In Barrett's view, it didn't take a genius to see that the more Celera was going to give away, the less it would have to sell. Worse, if Celera made its human and fly DNA sequence information available in a way that allowed other researchers to download it into their own databases—for instance, by trading it away for help from the DOE—what was to stop companies like Incyte from scooping up the data, adding their own software bells and whistles, and reselling it cheap to the very companies that Celera hoped to attract?

"If we can get this MOU done properly, it's a tremendous coup for Celera," Venter said at a senior staff meeting in late November. "It's a legitimization of what we are. And it puts pressure on the top dog to set up a similar thing with us." But the collaboration with the DOE wasn't simply calling NIH's hand; it was also forcing Celera to lay down its own cards and reveal whether Venter's vision of "open research" was a new paradigm or a giant bluff.

"Our customers are going to wonder why they should pay us if they can just wait a few months and get the data for nothing," Barrett objected.

"On the other hand, if we collaborate with the DOE, we show we aren't wearing the black hats," said Gilman. "Academics will buy subscriptions, and we make money."

"Paul, how many universities are you going to need to sign up to make up for losing a single pharma?" said Barrett. "A hundred? Two hundred?"

"Is your subtext that you don't like this whole MOU?" Sam Broder asked.

"As far as intellectual property goes, yeah," Barrett replied. "The whole point is not to let the data out in the same manner that we are asking customers to pay for. It can't be downloadable."

"But it *has* to be downloadable," Mark Adams protested. "Just not in a way that Incyte or some other company can grab it and resell it for themselves."

"That's easy," said Gilman. "We have people click on a button on the web page that says 'I agree not to resell this data.' That's all there is to it."

"If we make the data downloadable at all, that will kill the Novartis deal right there," said Barrett.

"We can't be held up by one deal," said Venter.

"It will kill *all* the deals," Barrett replied. "It's going to look like we can't decide what we're doing, and therefore can't be trusted."

"This MOU isn't new information to you," Venter said impatiently. "This is what happens when you push contracts too fast."

Barrett reddened. "What happens when you push contracts is called *revenue,*" he said. "What's the huge downside of saying the data is searchable but won't be downloadable?"

"For one thing, people will say I lied to Congress."

"We *said* we'd make it completely available!" Adams added. He pulled out a copy of the *Science* article they'd published the previous June and read from it, articulating each word with exaggerated emphasis. *"An essential feature of the business plan is that it relies on complete public availability of the sequence data."*

"Come on, Mark," said Barrett. "What does that mean, really?"

Viewed from the air, the headquarters of the Department of Energy in Germantown, Maryland, resembles the kind of maze you run rats through when you want to test how fast they can find a piece of cheese. Inside, you see how it looks from the rat's point of view. The corridors are painted in two tones of beige. They are studded with beige doors that open into beige offices, and some corridors are so long that they tunnel to a vanishing point in your vision. Fluorescent bulbs hum overhead, and there is a scent of Xerox ink and floor polish. At measured intervals down each hallway is a clock bearing on its face the spread-eagle symbol of the agency. If you placed all the halls in the four-story structure end to end, they would run for seven miles. New visitors are given a map to keep them from getting lost.

Marvin Frazier occupied Room G-162, near a bend in a corridor on the first floor of Wing G, about a quarter mile from the entrance. He was fifty-seven years old and a microbiologist by training, but thought of himself as a bureaucrat, which to him meant someone who is supposed to get a job done. In his spare time, he liked to read, spend time with his grandchildren, and fly fish. He had smooth, capable-looking hands and warm, kind eyes. Though he plays only a small role in this story, in a sense Frazier is at the very heart of it. In January 1998, after complaining of stomach cramps, he was diagnosed with a deadly cancer called gastrointestinal stromal cell tumor, or GIST for short.

GIST is believed to be caused by a mutated gene in a signal transduction pathway—a cascade of protein interactions that transmits a signal from outside a cell to its nucleus. Along the pathway instructing certain cells in the lining of the digestive tract to divide, a gene called c-Kit gets stuck in the "on" position, like a broken accelerator pedal, telling the cell to divide and divide, faster and faster. To fix the pedal, it would be a great advantage to know the precise difference between the c-Kit gene in a healthy person and the c-Kit gene in an afflicted person, like Marvin Frazier. Furthermore, whether c-Kit is acting alone in the breakdown or (more probably) has allies cannot be determined without understanding all the protein interactions involved in the pathway, which cannot be determined without knowing all the genes involved in the pathway, which cannot be determined without knowing all the genes in the human genome. Thus Marvin Frazier not only worked for the Human Genome Project, he was a living expression of its need to hurry toward its goal.

Unfortunately, though GIST afflicted only some two thousand people a year in the United States, most cases were fatal. The only treatment was quick, radical surgery. A week after being diagnosed, Frazier had fifteen pounds of his innards excised, including his entire stomach and part of his liver. But GIST tumors have a surface like coffee cake; they tend to crumble under the scalpel, making it extremely difficult for even the most skilled surgeon to remove all the affected cells. Divide, divide, faster faster faster. GIST tumors also develop resistance to drugs very quickly, making all known drugs at the time virtually useless. Frazier knew this, but after the surgery he nevertheless underwent four grueling rounds of chemotherapy, interspersed with blood transfusions to counter the infections he'd developed as side effects, because he was concerned that if he didn't his family would think he had given up hope. In hindsight, he would not have made the same decision. The chemotherapy was the worst experience of his life, much worse than having to learn to eat without a stomach. What GIST patients like him needed was a drug designed to target and correct the errant message in the signal transduction pathway itself. Unbeknownst to Frazier, while he was negotiating with Gilman on the memorandum of understanding, scientists at Novartis and the University of Oregon were sifting through hundreds of likely compounds tailored for just that purpose. Their search, too, would be greatly accelerated if they had access to the whole genome.

Like Gilman, Frazier saw the developing agreement between Celera and the DOE as a win-win. The exchange of data would speed things up on both sides and save millions in sequencing costs. But even more important, he thought, was the exchange of expertise. The DOE had people who were intimately familiar with the genetic landscape of chromosomes 5, 16, and 19 and could serve as guides to the territory for the Celera scientists. In return, the DOE would profit not just from Celera's sheer sequencing muscle but from the unequaled proficiency that Venter's team would bring to the annotation process as a result of their experience with microbial genomes at TIGR. Of course, the proposal would need the blessing of the DOE's partners in the Human Genome Project—or better yet, their active participation in the collaboration. It was the obvious way to go. As long as Celera was going to put its data in the public domain, why should the two efforts proceed on separate tracks? It made a lot more sense to join forces and get as good a product as they could, as quickly as possible. Frazier wasn't concerned about who got the credit for delivering the code of life. There was plenty of glory to be shared, and he had never figured on getting any glory anyway. "This is a beautiful project," he said. "I'm just honored to be associated with it."

As soon as his colleagues at the DOE and the team at Celera had agreed on the language of the memorandum, Frazier sent a draft by e-mail to Collins's National Human Genome Research Institute, asking for input. From the little feedback he got, he could tell they weren't happy about the initiative. But Collins did not voice any major objections. One criticism from NHGRI was that under the draft terms, Celera would share its sequence data with the DOE as soon as it was processed through the company's pipeline, which meant that DOE scientists could peruse information about the location of genes that other scientists in the public program would not have until three months later. That was unacceptable. Frazier edited the proposal to make it explicit that no data would be exchanged between the DOE and the company that was not available to the public. He sent the revision to Celera, and Venter and Gilman OK'd the change. Ari Patrinos sent a copy of the draft to Martha Krebs, who gave it her full support. Rita Colwell, the director of the National Science Foundation, also gave the draft her approval.

More important, of course, would be the reaction within the Human Genome Project. Patrinos and Frazier were fairly confident that Collins would not try to block the initiative. After all, he had had plenty of opportunities to object to it but had sent back drafts with only minor suggested changes. The DOE officials were more concerned about the reaction of the HGP genome center directors—Eric Lander, Robert Waterston, and Richard Gibbs. But surely if they had any major objections to the collaboration they would have voiced them by now. The first sign that there was trouble brewing came when Rita Colwell telephoned Venter in early December. The previous evening, she had attended a reception at the Canadian embassy. "I ran into Sir Robert May, Tony Blair's science advisor," Colwell said. "He asked me if it was true that an agency of the U.S. government was teaming up with a private company to patent the human genome."

The proposed collaboration was the first item on the agenda of a routine meeting of the American genome center leaders scheduled for the evening of December 3. Just before the meeting, Francis Collins called Ari Patrinos to let him know that he did not need to attend. "In hindsight," Patrinos said later, "I realized it was to let me save face." Marvin Frazier, meanwhile, had no idea that the agreement with Celera had come to the attention of people in the British prime minister's office, or that he was walking into a trap. But he must have had an inkling as soon as he and Elbert Branscomb, director of the DOE's Joint Genome Institute, entered the conference room at the Bethesda Holiday Inn on the evening of December 3. In addition to Collins, Lander, Waterston, and Gibbs, there were two other men at the table: the Sanger Centre's John Sulston and Michael Morgan of the Wellcome Trust. Sulston seldom attended routine administrative meetings of the American program. The presence of Morgan, his benefactor, was an even bigger surprise. Everyone was looking grim. Before anyone said a word, Frazier was suspecting that he may have miscalculated the reaction to the DOE's initiative.

If he had any doubt, it disappeared when Harold Varmus, the director of NIH itself, walked into the room—"kind of accidentally," Frazier remembered, as if Varmus had just happened to be in the neighborhood that evening and thought he'd stop by. Instead of presenting the proposal to his colleagues, Frazier instead found himself being furiously reprimanded by a man who had won the Nobel Prize and headed the most powerful health agency on Earth. Even months later, talking about the affair seemed to cause him pain. "Suffice it to say that everybody was very upset," he recalled. "They felt I'd put them in a terrible spot. NIH was the leader of the genome program, and any overture to Celera should have come from them, not us. They took turns on me. Varmus, Morgan, Francis, and Lander were the harshest critics. Sulston, too. They told me that such a decision as this was way above my pay grade—in some ways above Ari's, too. They said if you go ahead with this, the DOE will be a pariah in the human genome program. Varmus let me know the ramifications were huge, beyond the genome program. We also have a program in structural biology with NIH, and he intimated that this would destroy the interaction of the two agencies on that, too."

From the perspective of the others at the meeting, the anger was fully justified, though perhaps much of the wrath directed at Frazier was meant for someone else. In their view, the DOE was being used as a pawn by Venter. They saw the collaboration attempt as nothing more than a divide-and-conquer move designed to embarrass NIH into a similar agreement. Frazier insisted that the proposal had been DOE's initiative, not Celera's, but that only brought more anger down upon him. As the junior partner in the genome program, the DOE should never have gone as far as to put the terms of an agreement in writing without bringing NIH and the Wellcome Trust into the loop. Of course, at least according to Patrinos, Frazier, and Branscomb, the DOE *had* brought Collins and his staff into the loop. But that didn't seem to matter now.

"We told Marvin what he'd done was outrageous," said one of the participants. "It was as if Belgium had showed up at a NATO meeting saying 'Look, we've negotiated an agreement with Russia!' The DOE hasn't contributed shit to the Human Genome Project. The scientists working hard on this project should have had their interests and feelings taken into account. It offended everyone royally. If this was an attempt to bring about peace, it wasn't a well-thought-out one. It was dead on arrival."

Afterward, Frazier called Patrinos from a phone booth. Patrinos had never heard him sound so traumatized. "The MOU is dead," Frazier told him. "It was pretty rough. I don't mind being the fall guy. What hurts is I failed to convince them to go forward."

"Where do we go from here?" Patrinos asked.

"They want you to try to broker a new agreement with Celera, one that includes NIH and the Wellcome Trust. But they're not going to accept any terms other than that Celera holds to the Bermuda Accord."

To Patrinos, it sounded like Collins was just giving him another chance to save face. Celera would never accept such a condition, since adherence to the Bermuda Accord would deplete the commercial value of its data. The next day, he took the news of the defeat to Martha Krebs. She told him that if he chose to push forward with the MOU in spite of NIH's opposition, she would back him. After all, the DOE was an independent agency. But Patrinos knew it was time to quit. He was philosophical about it. As a Greek, he understood what he called "the fate of small nations." If you want to survive, you learn early the value of accommodation over confrontation. "These were extremely smart people with big egos and a lot at stake," he later said. "I'm just a public servant trying to do my job." He sighed, and for a moment stared out the window of his office at the DOE. "Sometimes, though, I feel like a little sailboat sailing next to an ocean liner, trying not to get sucked into its wake."

Not everyone at Celera was upset that the MOU had been killed. Peter Barrett, for one, couldn't have been more pleased. For months he and his team had been negotiating with Amgen over the terms of their early-access agreement. The contract was complex, intricately structured, and—if some of its terms became public—highly volatile. It wasn't a mutual back-scratching, it was business: Celera agreed to provide Amgen with unrestricted early access to all its data—fruit fly, human, and the mouse genome, too, eventually—and in exchange Amgen agreed to provide Celera with $5 million a year. In addition—a provision not made public—Amgen would get first shot at developing any genes coding for "secreted proteins," which are coveted drug targets. In exchange, Celera would be entitled to receive payments from Amgen at designated milestones in the development of a pharmaceutical from such a gene, including royalties on any drug that went on the market.

Barrett insisted on another clause. Anyone who wanted early access to Celera's pot of genomic gold, including Amgen, had to agree to sign up for five years. Larry Souza, Amgen's vice president for research, was balking at that requirement, and so were Novartis and Pharmacia Upjohn, the two other companies with deals in the works. But Barrett was a tough negotiator and he wasn't going to budge. As an added incentive, he had set a deadline. Anyone wanting early access had to sign on the dotted line by midnight on December 31; otherwise, the deal was off. Barrett knew the deadline was both stick and carrot: all the companies involved had had good years, and it would be an advantage for tax purposes to be able to book the expense in 1998. Still, it was hard enough getting an agreement under this kind of pressure without Venter and his new policy director simultaneously offering to spoon-feed data to a government agency for free.

Even with the DOE deal blessedly dead, the early-access contracts were going to come down to the wire. As Christmas approached with a mess of issues still unresolved, Barrett's team took to staying at Celera through the night. For one four-day stretch, Barrett can't remember sleeping at all. One night, in the middle of a snowstorm, the heat in the building went out. They brought in some blankets and huddled under them while they kept working. On the afternoon of the last day of 1998, a fax arrived from Tom Zindrick, the point person on the deal at Amgen. His team had been undergoing much the same ordeal out in California. The fax was dated "D day." "I will provide the [agreement] shortly," Zindrick wrote, adding, "Sorry for the delay—I forgot a commitment at home last evening. On the bright side, at least I'm still married."

That evening, a line of cars passed through the security gate and down the long drive of Venter's Potomac tract mansion for his annual New Year's Eve party. Marshall Peterson, the hardware expert who would be running Celera's supercomputer, arrived in a new Porsche, looking a little embarrassed by it. Hamilton Smith and his wife came in the rusty Mercury. A uniformed valet whisked the car away. Inside, the house was nicely done up for the holidays and glowed with candlelight, but the underlying décor still seemed oddly un-broken-in—more a display of lifestyle than a consequence of it, as if the owners had filled the place with their tastes and interests but just hadn't gotten around yet to living in it. Venter and his wife greeted people in the chandeliered vestibule.

Barrett and his team were not among the guests. They had scattered for the holiday but were still working frantically by fax and conference call to slam the early-access contracts through before the deadline, now just hours away. Their Celera colleagues stood about in the Venters' sunken living room, scientists and spouses awkward in their tuxedos and gowns, like overage promgoers, vying for space with the great stuffed couches and nibbling the passed hors d'oeuvres. The three poodles mingled. Smith, in his tweed sports jacket, cocked an ear down to conversations, trying to get in on the small talk, before disappearing downstairs to shoot pool. Gene Myers followed his wife around the room with his eyes, smiling proudly at how splendid she seemed in her gown. He looked bronzed and handsome in his new tux, with the diamond stud flashing in one ear. "I got him to go to a tanning salon," his wife, M'Liz, confided to a friend. "Not for the tan, but to try to improve his mood. Very un-Gene, but he did it."

Wine flowed and the party began to loosen up. Venter moved easily among his guests, telling ribald stories and cracking jokes. In the dining room, he put his arm around his wife and described a White House dinner he and Claire had attended a few months earlier, at which President Clinton hadn't been able to keep his eyes off her cleavage. "Craig, would you *stop*?" she said.

Just a few minutes before midnight, Venter mounted the landing of his stairway above the candlelit gathering and clinked his champagne glass. "I just got a call from Peter Barrett," he announced. "We've just signed our first contract with Amgen for twenty-five million dollars, and the other two have signed letters of intent. That's seventy-five million dollars in revenue, before we've sequenced a single base pair!" Then he added, grinning, "Now we're *really* in deep shit!"

Everyone in the room knew what he meant. For all the talk of Celera as a new paradigm, a great experiment, a moment in history, it was now explicitly a business, with customers who had paid millions for a product whose value depended on the speed with which the product could be delivered. Of course, a profitable enterprise was precisely what Perkin Elmer was investing $300 million in Venter to achieve. Still, one can't help but wonder how differently events might have unfolded later if the business contracts hadn't been signed quite so soon, and if the DOE deal had gone through instead.

CHAPTER 13
VENTER UNITS

The terms of the early-access agreement called for Celera to start delivering data to Amgen by the end of March. To deliver data the company needed a data-delivery system, and it could not have a delivery system without a computer to store and process the data in the first place. At that moment, the first of January 1999, Celera's computer infrastructure, which Venter was touting as the most powerful civilian system in the world, consisted of an empty room with some spools of fiber-optic cable lying about and the first installment on $80 million worth of hardware from Compaq stacked in crates in an adjoining hallway. Moreover, Celera's information technology department consisted of one person, Marshall "Mad Dog" Peterson. Under normal circumstances, it would take a team four to five weeks to get the system installed, tested, debugged, and networked. Because of the Amgen deadline, Peterson figured he had four or five days at most. After Venter's midnight announcement, Peterson waited a polite interval of time, said goodbye to his hosts, and went home to Alexandria to sleep for a couple of hours and clear the champagne out of his head. Then he packed a toothbrush and a change of clothes and drove up Interstate 495 to Celera. At four o'clock in the morning on New Year's Day, he had the Beltway mostly to himself.

Peterson understood computers very well, but he did not like to talk about them much. In his view, even the most sophisticated system was still merely a tool, and his job was to match the tool to the task at hand. In this case, however, he was being asked to build a system without knowing what the task really involved. "How much power does it take to assemble the human genome?" he had asked an interested reporter, as if inquiring into the sound of one hand clapping. "Craig doesn't know. Gene doesn't know. *I* certainly don't know. Nobody's ever done it before."

A typical supercomputer—say, a Cray—built its muscle by stringing together enormous numbers of processors. But the all-at-once assembly of the human genome could not be solved by lots of little processors; it required gross amounts of active memory that could handle a single process, very fast. Virtually all computers on the market at the time employed a 32-bit architecture, which was limited by physics to 4 gigabytes of active memory. The assembly algorithms alone were estimated to need 20 gigabytes of RAM, at the same time that the computer would be servicing all the rest of the company's needs. Compaq had won the Celera contract because its new supercomputer, the Alpha 8400, was built upon a 64-bit architecture that could handle 128 gigabytes of RAM—four thousand times the active memory of an average desktop machine. Peterson had tested the Alpha 8400 and a competing machine from IBM by seeing how long it took them to assemble the DNA fragments of the H flu genome. Four years earlier, TIGR's 32-bit Sun mainframe had completed the assembly in seventeen days. IBM's new supercomputer finished the compute in three days, fifteen hours. Compaq's Alpha 8400 took only eleven hours. Celera ordered a dozen Alphas, to start.

Peterson's other concerns were those that any other information technology manager might have, only magnified. The requirements for hard-disk storage capacity, for instance, was estimated to be 10 terabytes. A terabyte is 1,000 gigabytes, or the equivalent of a stack of paper six miles high. If the data stored in Celera's computer were written on paper instead, the stack would reach into the ionosphere. With data costing about a million dollars a day to produce, the reliability and security of the system were obvious imperatives, too. Hackers, freeloaders, competing companies, and crazies had to be shut out. Peterson had hired a computer security expert to design the company's firewall and private virtual networks, and a second expert to try to hack through anything the first one put up to keep him out. Just to limber up, one day the hacker had broken through Perkin Elmer's electronic defenses, signed up with a user name and password, and strolled around for a while in the parent company's top secret databases, like a kid going through his father's locked desk drawer.

By nature and conditioning, Peterson took a keen interest in more conventional threats to the operation, too. "That strip of trees behind the building should be cut down," he recommended at a senior staff meeting in early January. "It's just fifty yards away from your office windows, and a convenient nest for shooters."

"Maybe we could make everybody wear a bald wig instead," Venter said. "That way they won't know who to shoot at."

"Or we could mine the woods," Sam Broder added. "*Boom! Boom!* All day long, whenever a squirrel jumps down too hard!"

"The executive offices should be moved to the second floor, at least—if not the fourth," the ex–Vietnam helicopter pilot said. He wasn't smiling. "You want to be able to slow people down who are coming to get you."

"Now you're scaring me," said Venter.

"I mean to," he replied.

The meeting was interrupted at this point by the arrival of a stack of a brand-new, special edition of *Time* devoted to the genetic revolution in medicine—including a provocative write-up on "The Gene Maverick" at Celera. Exploding squirrels and other security matters were forgotten while everybody flipped through the magazine. It featured an eerily lit photo of Venter standing between two rows of gleaming sequencing machines, looking down on the viewer with his legs spread wide, hands placed commandingly on his hips. "This guy looks like the Übermensch," Broder laughed. The photo caption read "ROBOT ARMY. . . . If Venter wins the genome race, it will be largely thanks to automated sequencers like these."

The caption was inaccurate. The photograph had been taken at TIGR, not Celera, and the machines flanking Venter were not ABI's new capillary machines but the old Prism 377s. Only 11 of the 230 promised Prism 3700 machines had been delivered and none was operational yet. Mark Adams described some of the problems his group was encountering. The first dozen samples they'd tried to run had come out totally blank. Solving that problem had only exposed a cascade of others, with no rhyme or reason to the failures. Some of the machines had mechanical problems: robot arms that gestured uncertainly, as if ambivalent about participating in the greatest science project since the atom bomb; syringes that were supposed to inject a polymer gel into capillary tubes but blew their contents onto the operator's lap instead. On still other machines, the laser beam kept wandering out of its calibrated path. "This isn't all bad," Adams observed sheepishly. "Once the rumors get out to Francis about the problems, he'll stop worrying about us, and slow everything down. That should buy us an extra week or two."

Celera was now officially in a "quiet period." Perkin Elmer had scheduled a stock offering for the new enterprise for April 1999, which meant that Venter and the other company officers were forbidden by SEC regulations from engaging in press interviews, presentations to market analysts, and other public relations activities that might stir expectations for the stock into a froth. A quiet period is usually followed by a "road show," when these same officers are granted a few days to crisscross the country in private jets, broadcasting their message to as many investors as they can in an eighteen-hour day. But for now, no more market analysts came by for the grand tour of the still-empty spaces. For weeks Paul Gilman kept a *Nova* film crew at bay. Venter was forced to ignore the journalists who kept calling. Whatever he said to the contrary, his love-hate relationship with the press was mostly love. Now he watched a pile of pink phone messages from reporters mount up on his desk, eyeing it like a man on a diet gazing at a piece of cake.

It was quieter now throughout the building. The reality of what had to be accomplished had made everyone a little grim. People who were lucky enough to have offices with doors shut them. Marshall Peterson, who had spent the first seventy-two hours of 1999 entirely at Celera, had gotten the supercomputer installed with the help of some Compaq engineers. Within two weeks, it became obvious that some departments had far underestimated the computer power needed to do their job, so he was working nearly around the clock again to install new processors and redesign the system. Mark Adams, trying to outpace the to-do list on the PalmPilot forever clutched in his hand, moved with a stiff-legged efficiency through the halls, his upper body gliding along as if it were a piece of furniture being moved on a dolly. Hamilton Smith stayed up in his wet lab in a corner of the fourth floor, teasing his bits of DNA into their bacterial hosts with his big speckled hands. Most of the rest of the scientific staff touched biology only through their keyboards. In a roomful of cubicles, the combined sound of their tapping was like the rustle of gypsy moth caterpillars munching on leaves.

At the same time, Sam Broder huddled in his office with a couple of staff, sketching out Celera's ultimate product: a web portal where drug companies, biotechs, academic scientists, clinical researchers, family doctors, and laypeople could find everything that was known about the functioning of the human animal, from base pairs to behavior. Click on an organ, and up would pop its gross anatomy of tissues and vessels, its proteins, metabolic pathways, the genes expressed in it, and the diseases that could tear it apart. Or search from the bottom up instead: enter a sequence of base pairs and the site would deliver prospective genes with similar sequences; drag your mouse along the sequence and a window would pop up at the point where a mutation causes a disease or is associated with some behavior. "We want to catalogue every human trait," Venter told Broder's team. "No matter what you hear about—whether it's eye color, hair loss, schizophrenia, greed, or no matter what else, we want it."

The previous summer, Venter had boasted that Celera would be "on the forefront of everything." But being on the forefront means you have no precedents to use as bearings, and each department was struggling to figure out how to get to where it wanted to be. The problem was especially acute for the software and gene discovery teams, who had to start delivering a useful product to Amgen in less than ninety days. The uncertainty played out in the work of the construction contractors; a gutted space would turn into a maze of cubicles, then a week later the cubicles were resorbed and replaced by walls defining offices meant for a completely different purpose. Only Venter seemed able to think of the operation as a whole, and he kept himself updated by sitting in on the small brainstorming sessions that moved things ahead.

But even Venter seemed to lose his grasp sometimes. "Marshall is the enigma to me right now," he said one day in February, as he looked for a free table in the basement cafeteria. Between the employees and the construction teams still working on the top floors, the sound of the room at lunchtime had changed from a low murmur to white noise. "To be honest, I can't tell you what the hell he does."

The communication problem was especially acute between the Celera scientists and the Compaq system designers. They would speak in their separate languages, then sit and stare at one another, hoping for some sign of comprehension in the others' eyes. To try to break through the impasse, Peterson organized a Celera-Compaq summit. He hired a county-owned mansion in Rockville for the retreat, and an enthusiastic management consultant in a black turtleneck to lead the meeting. The consultant understood neither language but presumably knew how to get people talking. He asked everybody to break up into smaller groups so they could get to know one another better. Each group would retire to its own room, where an easel of paper and some Magic Markers were waiting. "I want you all to put your heads together and try to *draw* the future of the Compaq-Celera relationship," he said. There was a glum silence, as if the facilitator had come by his abundance of positive energy by sucking it out of everyone else.

"What if we don't know how to draw?" somebody ventured.

"Just use stick figures."

Amid a lot of shuffling, groups of Celera and Compaq people, in roughly equal parts, formed and the drawing began. One team sketched some little stick men tangled up in giant helices of DNA, screaming for help. An initially more hopeful drawing showed Celera and Compaq as a lighthouse above a stormy sea, guiding to safety a fleet of boats named after pharmaceutical companies. But somebody else in that group added a giant shark labeled "Incyte" and a sinking ship called "Drug Failure," with drowning stick figures all around it. A third group depicted Celera and Compaq as two figures walking beside each other up a mountain toward a great light. From a distance it appeared they were hand in hand, but a closer inspection revealed them to be linked with handcuffs.

One drawing was a little more optimistic. Vivien Bonazzi, co-director of Celera's gene discovery team, had started it by drawing a mountain of sand on one side of the paper and on the other side a half-built sand castle. The mountain, she explained to her group, represented the unspeakably enormous piles of data that the Celera scientists were going to have to contend with. The sand castle was the functional *knowledge* formed from that data and delivered to the world, starting with Amgen. "The question is, how do we get from one to another?" she said. After a moment, a Compaq engineer took the marker from her and hesitantly sketched in a bulldozer carting sand from the mountain to the castle. "The bulldozer is Compaq," he said. "See, we're helping you haul the sand to where you need it."

"Good," said Bonazzi. "That's a start."

• • •

On February 8, 1999, Celera's scientific advisory board convened for the first time. Venter might act like a maverick, but his track record had enabled him to recruit some heavyweight members of the scientific establishment to serve on the board. Richard Roberts, the chairman, won the Nobel Prize in 1993 for his illumination of the internal structure of genes. Arnold Levine was president of Rockefeller University and renowned for his discovery of the p53 tumor suppressor gene. Melvin Simon, who headed the Biology Division at Caltech, had invented the bacterial artificial chromosomes (BACs) with which the Human Genome Project was sequencing the genome one bit at a time. Victor McKusick of Johns Hopkins University, a towering, gentle, elderly man in an elegant dark pinstripe suit, was widely regarded as the father of modern medical genetics. Norton Zinder was a professor emeritus at Rockefeller. In the glory days of molecular biology in the 1950s, Zinder had discovered how a virus could transfer genetic material from one bacterium to another, which had led to a new understanding of the location and behavior of bacterial genes. Finally, Arthur Caplan, of the University of Pennsylvania, was one of the most distinguished bioethicists in the country, often called by the government or the media to voice his opinion on human cloning, stem cells, and other ethically sensitive issues.

The meeting provided an opportunity to take stock of how things were going. Tony White came down in his private jet from Perkin Elmer headquarters in Connecticut, and Michael Hunkapiller flew in from California. Everybody gathered in the Mediterranean Room. (Venter, ever the sailor, had named all the conference rooms after bodies of water.) After some talk about the business model—"Celera's mission is to become the definitive source of genomic and medical information in the world," Peter Barrett proclaimed—the company's senior scientists reported on their progress. In spite of the problems getting the machines to work, the discussion was mostly upbeat. The agreement with Gerry Rubin in Berkeley to collaborate on the *Drosophila* genome had gone through without a hitch; Hamilton Smith was making good progress on the fruit fly DNA libraries, and he reported that he was also beginning to collect samples of DNA for the human genome itself.

Marshall Peterson followed with a dizzyingly detailed overview of the world's "second most powerful" supercomputer system. (Celera had conceded the top spot to a Department of Defense computer in Los Alamos used to model the effects of nuclear explosions.) "In practice, I want to be a utility," he concluded. "When you turn on a light switch, you don't have to think about whether the light will come on. It's the same thing here. I want computer infrastructure to be the dullest part of the Celera story."

"Well, you're doing an excellent job so far," said Venter.

Gene Myers got up next to report on how his assembly team was coming along. Tony White had never met him before, but he had heard that Myers was essential to the success of the enterprise and regarded him now with a sort of grim curiosity. Myers was unshaven and as usual had his wool scarf wrapped around his neck; in fact, he had vowed not to take it off at work until he had proved that the shotgun method worked. He had also taken to wearing his sunny orange Polarfleece practically all the time, like a cheery breastplate to ward off his own dark moods.

On this day, he had encouraging news to report. Instead of leading with it, however, he took the opposite tack. As if he were exorcising demons, he spent the first fifteen minutes discussing all the things that might possibly go wrong with the assembly. His biggest worries concerned the consistency and quality of the sequences coming out of the machines—the raw material that his computer algorithm would have to assemble. "If the data were always perfect, there wouldn't be much to worry about," he said. "But there are always going to be a certain amount of sequencing errors, contamination, and so forth. The trouble is, as you introduce more and more noise into the equation, you get to the point where the problem becomes statistically impossible to solve. So, we have to design something that can work with what we can realistically ask of the biologists. We'll know in about three or four weeks whether we've nailed it." There was a moment's silence.

"You will let us know, won't you?" White asked sardonically. "I'll give you my pager number." Turning to Venter, he added, "Remind me not to take this guy with us on the road show."

White and Hunkapiller left after lunch, and the august members of the advisory board donned hard hats for a quick tour of the Celera operation. Venter led them down to the basement, where in a windowless room the eleven sequencing machines were temporarily installed. The room was too small to accommodate the party, so the scientists took turns peeking through the glass panel in the door; with the boxy machines arranged in rows, what they saw looked a little like a high-tech

Laundromat. Next Peterson guided them through the data center, including an empty room that would house separate mainframe computers for the three pharmaceutical clients, protected by a firewall even from Celera itself. The lock on the door would be fitted with a retinal scan; no Celera eyeball, not even Venter's, would be able to open it. Finally the group bundled into the elevator to view the football-field-size spaces where the sequencers would be permanently housed. The one on the fourth floor was nearly ready, with rows of power outlets along the floors and hanks of cable tucked in amid the duct work.

Somewhere along the way, the tour had picked up another member. Bringing up the rear of the column was a man who did not look like a member of any board, unless he was representing the interests of some other planet. He was very tall and thin, and his gangly body was draped in a loose-fitting luminescent-copper sports jacket over a billowy black shirt and milk-chocolate stovepipe trousers. The hair spilling out in a ponytail from under his hard hat was a coppery color, too, and so was his sketchy beard. He had marked cheekbones and deeply sunken eyes, like a wizard in young middle age. It was Robert Millman, late of Millennium Pharmaceuticals, who three years before had tried to patent TIGR's *Haemophilus* genome on William Haseltine's behalf. Millman was Celera's patent attorney now. His eyes, already unnaturally bright, seemed to catch fire as he looked over the vast expanse where the sequencing machines would soon be humming. Of course, he was well aware of Venter's vision for a business based on open research. But that was not what he was thinking about. "I've got the choicest job there is in biotech," he confided to someone else taking the tour. His voice was raspy, as if he'd been talking all night in his sleep. "This is a patent attorney's wet dream."

In spite of Gene Myers's baleful presentation to the advisory board members, his assembly group was making more progress than anyone else at Celera. He had discovered a latent talent for management that would have been forever dormant in academia. Not only had he put together a tightly knit team, but he had become so adept at getting what he needed for his group and protecting them from interference that they were now regarded as a kind of elite corps—not without resentment from some of the other software engineers with less glamorous assignments. Myers proudly referred to his people as "the geek group."

"I thought everybody at Celera was a geek," said Venter.

"Yeah, but my group, we're the geek samurais," Myers replied. "We're geekvana."

"So what happens in geekvana?"

"You don't want to know."

But Venter very much did want to know, and a week after the advisory board meeting he sat down with the assembly team in a perpetually overheated conference room off the cafeteria. Myers looked haggard, but he was determined to make up for his gloominess at the advisory board meeting. After introducing his geeks, who were a bit in awe of Venter, whom most had never met before, he launched into a review of the team's successes. The assembly algorithm consisted of several discrete stages, or subroutines. The first good news concerned the Overlapper, the big, number-crunching stage of the assembly, in which the supercomputer would compare each fragment of DNA from the Prism 3700 sequencers against every other fragment that had been sequenced to see if and where their base pair order matched up. When the computer found a fit, it would utter a fleeting, nanosecond "Aha!" and, at least for the moment, join the two pieces into one larger piece. In genomics, this joined-together piece was called a contig—quick-speak for "contiguous fragment."

Myers had lured an efficiency expert, Art Delcher, from the University of Maryland to write the initial code for the Overlapper. Delcher's main concern was cutting down on the grotesque amount of memory the program would require once the bulk of the human genome began to pour into it. For practice, he was randomly breaking apart the genome of the worm *C. elegans,* which the public program scientists had finished a month before, then putting it back together using Celera's computer. The simulation was performing even better than Delcher had hoped. "That's one scary barrier we've gotten through," said Myers. "It kept us happy for at least three minutes. But we've got something more."

Moving animatedly back and forth in front of the whiteboard, he sketched out the next stage of the assembly. For a perfect imaginary genome that never repeated itself, the Overlapper would be all that one would need to put together the puzzle, even a puzzle consisting of millions of pieces. But, of course, both the *Drosophila* and human genomes were riddled with identical repeating sections. As a consequence, all too often a fragment of DNA would overlap with two, three, or even a dozen other pieces, because the overlapping sequence was represented in more than one place in the genome. At all costs, the assembly program had to avoid making false connections, which could mislead researchers for decades to come. For the next step of the assembler, then, Myers had written a program that essentially broke apart most of what the Overlapper had put together, keeping only those joins where a piece went together with only one other piece. Myers called the result of these uniquely joined pieces a "unitig." In effect, rather than attacking the problem of a complex genome's thousands of repeats, the Unitigger stage dodged the issue for the moment, telling the computer, "Let's just assemble the pieces we *know* go together correctly and throw the rest into a bin to deal with later." The Unitigger was an act of genius, in one stroke reducing the complexity of the puzzle by a factor of 100.

"Once you hear it, it sounds so obvious," Venter said. "What this means is that two reads that go together on Day One may get broken apart on Day Two, when another fragment arrives that could overlap with them as well."

"Right," said Myers. "The program is constantly readjusting. It has to be able to change its mind with more evidence."

"I wonder if there's not another way you could reduce the overall problem by an order of magnitude," Venter offered. "If we knew which chromosome a particular fragment came from, that would simplify the assembly by twentyfold."

Myers stopped pacing, and his face darkened, as if a cloud had passed over it. "Then it wouldn't be a whole-genome shotgun assembly anymore, would it?" he asked.

"Well, no, of course not," said Venter. "So maybe that kind of information shouldn't be part of your program but just a way of checking the assembly after it's done."

"I wouldn't have a problem with that," said Myers.

"So when will you know if the algorithm works?"

"My hope now is in about a month."

"Great. Just before the stock goes public. Not to put any pressure on you guys, but if this team fails, we all fail, no matter what. We're betting the whole company on you. And I was thinking, just in case we needed a secret backup plan, we might consider putting the public program's raw data into your assembler, too, as it becomes available on GenBank."

Myers's face darkened again. "We're not going to need to do that," he said, "because we're not going to fail."

A few days later, Venter was sitting in his office when Marshall Peterson walked in. Peterson had been working closely with Myers and the two had become good friends. He closed the door behind him.

"I'm really worried," Peterson told his boss. "Gene looks awful. He doesn't know how to turn himself off. I'm afraid he's going to crack."

"It's my fault," Venter said. "I shouldn't have pushed him so hard." He thought for a moment. "Claire and I are taking some time off next weekend to race *Sorcerer* in the Caribbean. Maybe Gene and his wife would like to come along. It might get his mind off his work."

With a little prodding from M'Liz, Myers agreed to go on the trip. Venter had also invited Paul Gilman, an avid sailor, and his wife, along with some friends from TIGR. The event was the annual Heineken Challenge Cup regatta off Saint Martin. Myers left his scarf behind and seemed relatively relaxed. The weather was magnificent, and the atmosphere of a racing yacht regatta sponsored by a beer company did not lend itself to brooding. At the registration desk, each boat received a welcoming kit containing a six-pack of Heineken, a bottle of Saint Martin's own guavaberry liquor, a tube of sunscreen, and six condoms.

The race was a three-day affair, to be won by the yacht having the best combined time. Venter was considering it more a frolic than a real contest. There would be no staggered start, and since several entries could easily outrun the heavy-hulled *Sorcerer* in practically any wind, there was little chance of winning. Still, he hadn't completely put aside the possibility of an upset. On the first day, Tom Motley, *Sorcerer*'s captain, persuaded him to attempt a "ballsy move" at the starting line that, if successful, would put them on the opposite tack from the rest of the fleet and crossing ahead of their right-of-way. The gambit failed, however, and *Sorcerer* rounded the first mark in eighth place out of twelve boats. *Attitude,* a low-slung sloop generally considered the favorite, was already half a mile in the distance, her pretty herringbone sails making the most of the light breeze. To have any chance of catching up, *Sorcerer* needed wind. Instead, what breeze there was began to die away and soon the yacht was all but becalmed. There wasn't much for anybody to do. Venter broke out the Heinekens. Myers passed. Nobody talked much for a while.

"So Gene, what are you thinking about, right now?" Venter asked, breaking the silence.

"Large structural variations," said Myers. "If we can't resolve them in *Drosophila,* then there's no way we can do human."

The wind began to pick up a bit, and *Sorcerer* finally crossed the finish line, last. Venter didn't seem to care. In the evening he and most of the others took the dinghy into shore for a little shopping on the glitzy commercial avenue running parallel to the beach. Myers had brought his laptop along, and he stayed behind in the boat to work.

"To be honest, the thing I worry about most is Gene's worrying," Venter confided to M'Liz Robinson, as they walked together along the crowded sidewalk. "We can't afford to have him blow."

"It's just his way," Robinson said. "He brought his computer with him on our honeymoon, too. At first I was offended. Then I realized that he couldn't help himself."

Back on *Sorcerer* after dinner, they found Myers in the cabin in the same position that they'd left him, his muscular shoulders bunched over the laptop. When they came in, he didn't even look up.

The second day of the race dawned with a haze that refused to burn off, along with a crisper wind that gave some hope that *Sorcerer* could make up some of the ground she'd lost the day before. The yacht crossed the starting line smartly at the gun and kept pace just behind *Attitude* to the first mark. On the long downwind leg, however, she began losing ground again and was soon well back of the leaders. "There's no way we can keep up on this tack," Motley said.

"Let's take a flyer," Venter replied. "We could head out more to windward, then tack back and surprise some of these guys at the mark."

While the other boats kept along the shoreline off to port, Venter pointed *Sorcerer* out toward the open sea. When the island was almost lost in the haze, he brought the boat about. The prow cut deep into the waves as it gathered speed. By the time Gilman spotted the mark in the binoculars it looked as if the tactic had backfired; the other boats were much closer to the target. But *Sorcerer* had a straight shot without a tack and rounded the mark ahead of four or five yachts that had previously been showing their sterns. *Attitude* was still well out in front, but there was at least hope of a high-place finish. Then, halfway through the tack around the third mark, disaster struck. With an amateur crew handling the sails, the tack had not been crisp, and the boat was slow in responding. Just behind, a small yacht flying a Swiss flag was bearing straight at *Sorcerer* on the starboard tack. *"A droit!"* its captain shouted, demanding his right-of-way. Venter pulled the helm hard to port. Nothing happened. *Sorcerer* was in irons, momentarily dead in the water. The Swiss boat kept on its course. Now everybody else on her was yelling, too, except for a young woman in a brown bikini sunning on the bow, who seemed as indifferent as if she were carved from wood. Defiant assertion of one's right-of-way is expected in a sailboat race, but clearly the Swiss captain was either too obstinate or too inexperienced to realize *Sorcerer*'s predicament. Venter and Motley started shouting for him to give way, but it was too late. He had passed the point where he could avoid ramming *Sorcerer,* and unless he altered course, Venter's boat would be T-boned by the headlong progress of his bow. At the last second he pulled hard to starboard, crashing into *Sorcerer* and swiping her hull with a grinding whine, sending sprung stays and pieces of her railing flying in the air. Claire Fraser screamed. Everybody on both decks was cursing everybody on the other.

"Asshole!" Motley shouted.

"Connard!" spat back the woman in the bikini, who had been forced to rouse herself to keep her legs from being crushed.

"Swiss bastard," muttered Venter as the other boat pulled away. "Raise the protest flag." The Swiss boat had already raised theirs. Both captains would have to appear at a hearing before the race committee to decide who was at fault. Motley leaned over the side to inspect the damage. "In all my years of sailing, I've never seen anything like that," he said.

At the hearing that evening, the race committee decided that both boats had been at fault. They were disqualified but would be allowed to race the next day. For *Sorcerer,* however, the regatta was over. Whether from the collision or some unrelated cause, the housing to the bow thrusters beneath the waterline had been damaged, making it impossible to raise the thrusters once they'd been lowered. There was no way *Sorcerer* could compete with this obstruction dragging beneath her hull. Other problems had developed over the weekend. The engine battery wouldn't charge, the automatic mainsheet winch was broken, and the head refused to flush. Venter and Motley spent the morning diving under the boat, trying to free the thrusters, while everybody else swam and lazed about on the deck. *Sorcerer*'s record for the three-day regatta: Last, Disqualified, and Withdrawn. Venter took it in stride. In late afternoon, Motley ferried most of the guests to shore, where they could get a taxi to the airport.

"My only regret," Venter said as they climbed into the dinghy, "was that I promised you all a good race and didn't deliver." As the dinghy pulled away, Gene Myers called back to him. "I'm giving a talk in France on Wednesday," he said. "Is there anything I shouldn't say?"

"I'd really appreciate it if you don't mention that we can't get the sequencers to work," Venter shouted back, grinning.

He spent the next morning tinkering with his crippled boat. The calamitous weekend seemed barely to have dampened his spirits; in fact, he seemed even more buoyant than usual, as if adversity were like the loading of a springboard that would fling him even higher. *Sorcerer* would have to motor back at half speed to a yard in Fort Lauderdale, where the hull could be repaired and repainted. He figured the work would cost between $5,000 and $10,000, but he could afford it. Compared with the problems back at Celera, these were trivial matters. But he acted as if the problems at Celera were also trivial. They would work themselves out one way or another. The sequencers would be made to function, for instance, for no other reason than that they had to.

"If we were buying them from a third party like Amersham, I'd be nervous," he said. "I'd be selling my house and moving down here. But PE is betting a 5.4-billion-dollar company on our success. If their machines don't work, they lose three to four billion in evaluation overnight. Tony can't afford that. He knows that if we fail, he'll be the first one fired."

Tony White could not induce dysfunctional machines to start spitting out DNA, of course, even if it meant his job. Venter was basing his confidence on how he saw the immediate future developing from the present circumstance, like a chess player surveying the position of the pieces on the board and thinking several moves in advance: Tony White was Michael Hunkapiller's boss, and if Hunkapiller's machines were performing at just half their expected capacity, then White would order him to double the number of machines Celera was receiving from ABI, and the operation would get back on schedule. Venter made it sound as if this outcome were a conceptual inevitability, like the development of a fetus from an embryo. He appeared equally sanguine about the competition from the public genome program. Just for starters, the Human Genome Project's devotion to the map-first, clone-by-clone approach meant that its scientists would have to prepare twenty thousand separate DNA libraries, one for each cloned hunk of the genome, and each one prone to possible errors. For the whole-genome shotgun strategy Celera was employing, Hamilton Smith needed to make only two. In Venter's reckoning, Francis Collins was a mild annoyance right now, but he would not be a factor in the future. The public Human Genome Project would disappear within a year or two, or just as soon as he proved that Celera could succeed. "This isn't about a race with them," Venter said, "and it isn't about making money, either. It's about looking for meaning in having existed. To call what I'm doing a success, we have to actually change society." Beating the public program to the genome by a few months was not going to make that kind of difference. His actions would have had no impact, and therefore his life no meaning.

"I want to be judged on having changed things by a substantial period of time," he said. "What was Einstein's impact in terms of knowledge and information? How much longer would it have taken us to get to where we are, had he not existed? You could quantify that as an Einstein Unit. So how much is a Venter Unit worth? Just getting the genome done would be stopping way short of what can be accomplished. It's the utilization of the information that matters. That's why I hate the race analogy. It's the wrong scale. It says, 'Maybe I can get there one second sooner.' I'd like a Venter Unit to be more than a second."

The cabin was growing hot, and with nothing more to be done for the boat, Venter decided to go for a swim.

"Of course, the whole thing depends on whether we can do *Drosophila*," he said, putting on his bathing suit. "The good news is, we have the world's attention. The bad news is, we have the world's attention. If we fail it will be one of the most spectacular burnouts in history."

On deck it was a still, gorgeous day, the air so clear that the yachts, the figures on the distant beach, and even the leaves in the trees overlooking the harbor seemed wrought with a fresh new clarity, as if one had previously been looking at the world through a milky lens. A flock of snow-white terns circled overhead, the underside of their wings stained blue-green by the light reflecting off the water. Venter dove over the side, breaking the glass-calm surface as if it were a mirror scattering into shards. With a few lazy, muscular strokes he was fifty yards away. He turned around and swam back, then lay on his back beside *Sorcerer,* his image bobbing gently in her polished blue hull. The owner of another yacht saw him swimming and called out.

"Dr. Venter!" he said. "How's the water?"

"Great!" he called back. "Everything is just great."

CHAPTER 14
WAR

While Venter and the others were racing his yacht in the Caribbean, decorators were putting the last touches on his permanent office suite. The carpet was a warm bright blue, like a tropical sea in sunlight. Framed copies of articles about Celera and Venter's earlier exploits hung in the approaching hallway and continued in an uninterrupted parade through the spacious anteroom. Leaning against the wall, waiting to be hung, was a three-foot blowup of a recent cover of *USA Today*'s weekend magazine. It showed Venter sitting cross-legged in a blue-checked shirt and khakis, looking at the viewer with the surprised anticipation of a boy who's been told that his essay has won the free trip to Disneyland. In the white space around him floated images of Copernicus, Galileo, Newton, and Einstein, and the question "Will this MAVERICK unlock the greatest scientific discovery of his age?" A bookshelf held an array of other Venterabilia, including his honorary medals and awards. One reporter later wrote that approaching Venter's office made a visitor "feel a little like Dorothy at the door of the great and powerful Oz." Venter resented the implication. "It's not like I put this stuff up myself," he grumbled. "Besides, everybody appreciates it. It reminds us that we're part of something bigger than any of us."

The outer chamber had ample workspace for Lynn Holland, Venter's personal assistant since his years at TIGR, and Chris Wood, Paul Gilman's assistant. In the middle of the room was a small round table with a few chairs, a thermos of coffee, and sometimes a box of Krispy Kremes that Holland put out for whoever happened in. People rarely sat in the chairs, and the plain laminate table itself had the aspect of a hurried afterthought in the decorator's mind. Even so, it quickly began to serve as a welcoming nexus, a place for people to hang about and gossip for a few minutes. Hamilton Smith would often wander in while an experiment was running, or Sam Broder might come down the hall just to see what was up. Through the glass walls of his office on the left, Venter could see who was coming in from the hallway. If it was Ham, Sam, or some other familiar, he might come out to chat and joke around. Even if he was in the middle of a meeting, his face would lift when he saw them and he'd give a wave, like a school chum who had only a little more homework to do before he could join his friends on the playground.

Paul Gilman's office was on the right, smaller than Venter's, but a great deal more comfortable than the windowless hole he had occupied in the temporary suite down the hall. He had a glass interior wall, too, so Venter could beckon when he needed him. Gilman was as good an audience as anyone for his boss's bad-boy antics, but when the situation demanded reserve, Gilman's preternaturally professional demeanor acted on Venter like Ritalin on a hyperactive child. Venter believed Gilman was also one of a small coterie of people in the company who truly "got it"—who understood his vision of an enterprise that would break down the wall between business and academic science and rebuild it into a platform, like scaffolding in an orchard, upon which the whole world could stand and pluck discoveries now beyond its reach. The scientists in the trenches—even his closest associates, like Mark Adams and Ham Smith—didn't quite "get it." In Venter's view Peter Barrett really didn't get it; he was more interested in making money than in breaking new ground. That was his job, of course, and Venter acknowledged that Barrett's efforts had brought Celera millions in revenue before it had sequenced a single base pair. But the chief business officer's corporate-bred office politics and bottom-line thinking made him uncomfortable. In turn, Venter's big-sky visions made Barrett nervous. After sharing the temporary suite with him, Barrett had relocated with his business development team to a corner of the fourth floor in Building II. It was about as far from Venter as one could be and still work at Celera.

While Barrett's team tried to entice more multimillion-dollar drug company clients and the scientists scrambled to provide a product for the clients they already had, Paul Gilman kept his focus on the ultimate target: the academics, doctors, clinicians, teachers, and eventually ordinary folk who would make up Celera's "hundred million customers." Are you among the one-third of the population who can stave off heart disease with daily doses of aspirin? What kind of diet is proved to work with people of your genotype? Come to Celera and for a nominal fee take charge of your own health and quality of life.

Gilman acknowledged that most people were still more afraid of their genetic information getting into the wrong hands than they were curious about it. But he had a plan to help change that. "We'll start with innocuous traits," he explained to Gene Myers over lunch one day. "A person goes to the web site and decides which genetic trait they want to test themselves for. They send in a cheek swab, we analyze their DNA and let them know whether they're a morning person, have perfect pitch, whatever."

Myers stared back at him, his dark brow knitted up over his black eyes. He seemed to be giving the idea hard thought. Or possibly he was still thinking about large structural variations.

"Maybe they get a certificate back with the results, declaring them a bona fide Morning Person," Gilman continued. "OK, maybe there isn't any commercial value in it just yet. But it would draw people to the site and gradually get them used to more serious uses for this information. And who knows, maybe it would catch on, like Pet Rocks."

Barrett's business development team, preoccupied as it was with securing $25 million contracts from drug companies, did not devote a great deal of time to the Pet Rock notion. Gilman had more immediate concerns as well. Francis Collins and other scientists in the public program were continuing to broadcast the message that Celera's assembly technique would fail. The statements were having a damaging effect. "Every time we make a pitch to a prospective client," Barrett complained, "we get the same question: 'OK, you guys tell us your strategy is going to work, but at the same time all these other guys are saying it won't. Why should we believe you?'"

For the duration of the quiet period, there was little Celera could do to respond publicly without violating the SEC rules. But Gilman could still put on his gray suit and figurative white hat and head down to Capitol Hill. Early in February, he arranged an informal meeting for Venter and himself with Congressman John Porter and some of his staff. Porter, a moderate Republican, was the most powerful opinion maker in the House on biotech matters. He chaired the congressional subcommittee overseeing appropriations to NIH and was an enthusiastic supporter of the Human Genome Project.

"We should point out to him that in the same time period that NIH has spent 1.85 billion dollars on the Human Genome Project and gotten seven percent of it done," Venter told Gilman in the car on the way to the appointment, "over five million people have died of cancer." "You can say that," Gilman replied. "But it's exactly the kind of talk that will have everyone in the room thinking you're a dirty fighter."

Gilman had sent Porter's staff some talking points in advance, including NIH's opposition to the attempted DOE collaboration and Celera's concern that NHGRI was channeling funds to Incyte through one of its grantees. In the meeting, however, Venter made no mention of those provocative issues. Instead, after thanking Porter for his support of NIH over the years, he confined himself to a description of Celera's open-research business model and explained the advantage in speed and cost of the whole-genome shotgun method. It was Porter who raised the issue of where the public genome program fit in. "If you really intend to make the information freely available," the congressman reportedly asked, "can you tell me why we should keep the public program going at all?"

"Because there's a lot of things that could go wrong with our strategy," Venter replied. "But after we prove that the whole shotgun technique works, it would make more sense for NIH to spend its money to use the genome to make discoveries, rather than spend hundreds of millions to sequence it all over again."

Gilman left the meeting feeling good. "You did a great job," he told Venter as they rode back to Celera. "I think he got it." At the same time, however, news about what had transpired at the meeting was making its way via a member of Porter's staff to a contact in the Human Genome Project. To the HGP leaders, the meeting betrayed what Venter was really up to—and why he had to be stopped. "That visit wasn't just for a little chat," one of them later said. "It was a planned attack on the competency of the genome project, which was part of their business plan. They wanted to shut us down."

On February 11, a week after Venter's tête-à-tête with Porter, the key strategists in the public genome program met at the Baylor College of Medicine in Houston. Eric Lander, Richard Gibbs, John Sulston, and Elbert Branscomb represented their respective genome centers; Robert Waterston of Washington University, battling colon cancer, was in St. Louis undergoing radiation and chemotherapy treatments but attended the meeting by speakerphone through most of the day. His center was represented in person by Rick Wilson, its deputy director, and John McPherson, an expert in creating genomic maps. The others had brought along lieutenants, too. Francis Collins was presiding.

What a difference a common enemy can make. At the Bethesda meeting of the HGP elite a little over a year before, the genome center leaders had been at one another's throats. In Houston, Collins was still beholden to the will of his generals, but now there were only five left to be beholden to. Balancing the reduction in the number of generals was a remarkable increase in the amount of money they could spend. Before Celera came on the scene, $60 million of NHGRI's total budget in the fiscal year 1998 had been set aside for sequencing. In July, that money had been awarded to seven genome centers. The largest grant, $26 million, had gone to Waterston's operation in St. Louis, and the smallest, $2 million, to Bruce Roe at the University of Oklahoma, a pioneer in the original methodology of DNA sequencing who was highly respected for his ability to train people in the art of sequencing. But since then Collins had consolidated the program even more; with so many hands on the tiller, the great ship of the HGP was simply too ungainly to keep up with Celera's sleek sloop. Roe would not get the money from his grant, nor would any of the other small centers. Instead, the $60 million plus another $20 million was channeled back into a new round of grants in the fall, for which only those who had proved themselves able to sequence at a fast rate were even eligible to apply. By the new yardstick, only Washington University, Baylor, and the Whitehead Institute qualified. Collins did his best to mollify the disenfranchised scientists, and they knew it was for the good of the program. Still, the cuts hurt.

"I'm Bruce Roe, the grandfather of sequencers," the Oklahoma scientist introduced himself to a new acquaintance a few weeks later. "You know—the guy being treated with KY jelly by the NIH."

Judging from past performance and reputation, Waterston's lab would get the largest amount of funding, followed by Baylor, and, last,

Eric Lander's Whitehead Institute. Nobody doubted that Lander was brilliant. ("Eric's mind is simply overwhelming," said another public program leader. "He could squash me like a bug.") But in the past his big ideas had had a way of falling short in execution. Several years before, for instance, as a showcase for the Whitehead's level of automation he had installed a twenty-five-foot system of robots, water baths, and conveyor belts he called the Genomatron, capable of preparing half a million DNA samples a day. It had lots of moving parts and flashing lights, but as a unit the Genomatron had never really functioned; it had wound up in a warehouse in Somerville. Nevertheless, Lander now submitted a proposal asking that virtually all of NHGRI's $80 million go to the Whitehead alone. His intention was not to cut out the other two labs but to send an unmistakable message to Collins: Adjust the size of your thinking. The program needed to scale up by an order of magnitude, and $80 million simply wasn't enough to do it.

A corollary message was that in this contest, the Whitehead had no intention of coming in third. Collins received both messages. The amount was increased to $202 million for the three large centers over two years, with an additional $23 million held in reserve to be distributed later. Waterston's center at Washington University received $78 million, Gibbs's operation at Baylor $31 million, and the Whitehead Institute $93 million. The DOE's genome program budget added another $85 million over two years, and the Wellcome Trust was adjusting its funding to make $77 million available to the Sanger Centre for the project for the coming year alone.

Collins and his colleagues now had more than enough cash to purchase a "robot army" of its own. They also had a pretty good idea how they were going to spend it. There were two capillary sequencers to choose from: ABI's Prism 3700 and Amersham's MegaBACE. In November, the three NIH genome centers had sent their lab managers out to Foster City. After a week of rigorous testing, the managers were blown away by what the Prism 3700 could do. All three decided to go with ABI. So did the Sanger Centre. They had only to wait for the NIH appropriations to move through Congress in March before putting in orders.

But there was still a major strategic issue left to decide—something that Francis Collins knew could set his generals at one another's throats again. The decision to produce a working draft of the genome by 2001 had not been as popular within the program as it had sounded when it was announced the previous September. Robert Waterston and John Sulston intensely disliked the idea. It wasn't that they rejected the draft approach entirely; several years earlier they had proposed their own limited version of a working-draft strategy. (Like all such proposals, it had been drowned out by the mantra of "quality first.") What they didn't like was the notion that all of the Human Genome Project's resources should be devoted to getting the rough draft done, at the expense of abandoning even temporarily the focus on a finished genome. Why pour everything into winning a battle that could ultimately cost you the war? When all the new money became available, they wanted most of it reserved for a "finish as you go" strategy: shotgun each BAC and assemble the pieces in rough order, then shotgun the next BAC while you continued the finishing work on the first.

Washington University and the Sanger Centre were the biggest contributors to the Human Genome Project, and Waterston and Sulston were Collins's most trusted scientific advisors. He had to listen to them. But in his other ear was Eric Lander. Lander had been pressing to sequence as much DNA as quickly as possible ever since Venter had leveled his blow the year before. His voice carried a lot more weight now. At the Houston meeting, he was projecting a capacity for sequencing DNA even greater than his giant NIH grant would allow.

"Jesus, did you do the arithmetic?" Elbert Branscomb of the DOE said to Richard Gibbs in the hallway during a break. "Where is Eric getting the money?"

The answer was that on top of the large grant he was getting from NHGRI, Lander had secured another $38 million by convincing the Whitehead Institute's board of trustees to take out a loan against future NIH sequencing grants. It was a risk, but by taking it Lander had emerged as the most powerful genome center director in the consortium. So Collins had to listen to what Lander wanted, too. The night before the Houston meeting was to begin, Lander confronted Collins in the hallway of the hotel. It was around 11:30 p.m. He was tired of watching Collins play Hamlet. "You're the leader," Lander told him. "You can't go into tomorrow's meeting waffling around. You have to take a position."

Collins didn't say so, and he hadn't even told his own staff yet. But he had already made up his mind what had to be done.

The consortium got to work early the next morning in a brand-new conference room in the Baylor genome center. Soon they were having to shout their words over the intermittent whine of power saws and hammering from the floor above, where the new Prism 3700 sequencers would be installed—a racket that provoked irritated glances in the direction of Richard Gibbs, the meeting's host. "Think of it as the sound of prosperity," he said.

They dug into the logistics, parceling out responsibilities. For the enterprise to work, extraordinary coordination would be required among five separate operations in two different countries. Collins suggested a communication system pinned down by teleconferencing sessions every Friday at 11:00 p.m. until the genome was done. A problem encountered in any one center would be the responsibility of all to solve. It was clear that there would be no more bickering or grandstanding at the expense of one another. To keep the truly international flavor of the project, a dozen smaller centers in Germany, Japan, France, and the United States would be asked to contribute some 15 percent of the genome. The rest was up to those present. At some time during the morning session, Lander started referring to the five genome centers as the "G-5." The nickname had muscle and a whiff of militancy about it. It stuck, because it summed up how they felt about themselves. They considered themselves in a war, and they were meeting to decide how to deploy their forces.

In the afternoon, Collins guided the discussion toward the crucial remaining decision. Sulston and Waterston presented their arguments for staying the course toward a highly accurate rendering of the code. Switching to an all-out draft would spawn even greater logistical problems, they said, since the labs would all be entering new territory even as they tried to coordinate their efforts. Even with the draft approach, someone was going to have to map the BACs on the chromosomes and decide which center was responsible for sequencing each one. Otherwise there would be a chaos of redundant effort, while other hunks of the genome were neglected. Washington University had the most experience in mapping and cloning BACs. In John McPherson they also had the best genome mapper in the business. But even McPherson couldn't perform miracles.

Lander swooped around the mapping argument. A map wasn't needed up front, he said. Better instead for the centers to grab clones randomly and crash them through the pipeline in a "map as you go" strategy, since the chances of two centers picking the same clones were remote, at least in the beginning. The chances of duplication would increase as the sequencing proceeded, but they could worry about that later. The important thing was to get the machines installed and running nonstop as quickly as possible. Waterston then raised the bugaboo that had haunted the Human Genome Project from the beginning. "If we go all out for a draft, aren't we killing the motivation to finish?" he asked. "Will the federal agencies be willing to pay for the complete product? Finishing is tedious work. Will our people leave for more interesting research?"

"We just don't have a choice to think like that anymore," Lander responded. "If Venter wins, we're not going to be able to finish in any case."

"Even if Craig can do what he says, it's going to take him at least two years to get to 10x," said Sulston. "If we put just two-thirds of our resources into the draft—"

"You're missing the point," said Lander. "Celera isn't going to wait to the end to declare victory. They will issue a press release when they get to 1x, another when they finish 2x, and so on. There will be a drumbeat of press releases. And that means we'll be forced to respond."

Collins had been quiet through most of the discussion, but now he took the floor. "Bob, you know I'm as committed to going all the way as you are," he said. "I will never let that goal out of my sight, not for a minute. But Eric is right. We are in mortal danger of going out of existence. Think what's going to happen, early next summer. Celera could have ninety percent of their genome done. If we continue trying to finish as we go along, we'll have a third of ours. We can shout ourselves blue in the face that our one-third is a hundred times better quality than their ninety percent, but it won't make any difference. Folks won't understand the difference. They'll just be looking at who won a race."

He wasn't through. Before leaving Washington, Collins said, he had gone over some numbers. He had added up the capacity each sequencing center would have once the funds were appropriated—the number of machines, the number of reads each machine could produce every day, the predicted success versus failure rate, and so forth. He had checked and rechecked his figures and assumptions, just to be sure. Now he presented his final argument to the group. To complete a draft genome of 5x—90 percent of the code assembled, if not completely ordered—would require approximately 36 million reads. According to his calculations, the G-5 could do just about that much in a year. "Allowing for some time to implement the ramp-up," Collins said, "I think we could complete the draft by next spring." He looked around the table. Even Lander seemed taken aback.

"Next *spring*?" said Branscomb. "You mean 2000, not 2001?"

"I know it's audacious. But if you add up the numbers, that's what you get."

There was more talk. Concerns were voiced, devils were seen lurking in details. Someone raised the point that no matter how much code they sequenced, they were still bound by the Bermuda Accord to daily release, which meant that Celera could grab their data off the web like everyone else. The faster they went, the faster their enemy could go. Collins ceded the point. But gently he kept bringing the discussion around again to the peril they were facing: either they went after a draft or risked being shut down. Gibbs was for it, and so was Lander, of course. Branscomb thought the spring 2000 deadline was unrealistic but he was for the effort anyway; after all, large-scale centralized sequencing had been the DOE's conception of the project from the beginning. But Waterston and Sulston were still reluctant.

Collins was most worried about Sulston. Waterston often followed his lead. More important, the Sanger Centre didn't have to worry about being shut down by Congress or anybody else. Sulston could decide to back out and go his own way at this very moment, and the Wellcome Trust would back him proudly, with a chorus singing "God Save the Queen" in the background.

"Well, John?" Collins said, when it seemed they'd touched every base. Sulston's thick beard clouded most of his face. Collins was trying to read his eyes behind his glasses.

"Everybody in my center is going to hate this," Sulston said. "But it's the right thing to do."

At the end of February, Collins gave testimony defending his budget request to Congressman Porter's subcommittee. His written remarks, distributed beforehand, included the statement that Celera's shotgun strategy was unlikely to produce a complete, highly accurate sequence.

"For him to keep on attacking us in public like this is just plain slimy," Venter told Gilman. "Go down to the hearing. Maybe you can persuade him to lower his rhetoric."

Gilman rushed down to the Rayburn House Office Building and followed the familiar corridors to the subcommittee's hearing room. He caught Collins just as he was going in. "I appreciate your reservations about the shotgun method," Gilman said, "but don't you think you could give us a chance to try it before you tell everybody in public that it won't work?"

"Gosh, Paul," Collins said, "whether it works or not is a scientific matter, isn't it? Are you suggesting I shouldn't mention that?"

"If it's a scientific matter, then we should sit down and have a good technical discussion about what we're doing," Gilman answered. "But if you keep on criticizing our method in the absence of such a discussion, we'll have no choice but to counter what you're saying."

Collins stiffened. "For now, I guess I'll just have to trust my own experts on this," he said.

"Well, I tried," Gilman said. He walked away and took a seat in the gallery. When it came time to give his oral testimony, Collins didn't mention Celera at all. Gilman was relieved. Maybe he'd gotten through to him after all. Then Chairman Porter asked his first question. "Private sector companies . . . claim to be able to sequence the human genome at a faster rate and a lower cost than the federal government," the congressman began. "The president of one of those companies came in to see me and laid out their timetable, and obviously this leads me to the question, Why should the federal government pay for this if it is going to be done in the private sector and be made available to the public at a faster rate and at apparently no cost to the public directly? This is your opportunity to tell me why they are wrong."

Gilman's relief vanished. It was an obvious setup to elicit a well-rehearsed answer. Collins had even brought posters along comparing the two methods, which his policy director, Kathy Hudson, a Hill veteran and former molecular biologist in her late thirties, with a freckled face and a tight helmet of curls, held up so the subcommittee could better see them. If I'm ever caught holding up posters for Craig Venter, Gilman thought, I hope somebody will shoot me. The first showed the human genome as a book with many pages. The public program's strategy, Collins explained, was to take each page of the book, one at a time, shred it up into bits small enough to sequence, then "tape" the pieces back together again with a computer program that matched up their overlaps, before moving on to the next page. The next poster Hudson displayed compared the expected results. On the top was a straight line representing what a piece of the genome would look like sequenced with the public program's page-by-page method. Underneath was what the world would end up with thanks to the whole-genome shotgun technique. It was a broken black line interrupted with plummeting red zigzags, like the paths of dud fireworks.

"The strategy that is proposed by Celera is to skip over the step of pulling out one page at a time and basically take the whole book at one time and put it into the shredder and then try to reassemble what the sequence must have looked like," Collins continued. "You can imagine that is a substantially more difficult process. . . . Taking the whole genome at once, you would undoubtedly, at least in the view of most scientists who have looked at this issue, run the risk of ending up with pieces that are misassembled. That is what the red squiggles are—pieces that came from other parts of the genome, or gaps that have not been closed, or orientation which is incorrect."

In spite of these problems, Collins acknowledged that the whole-genome method did have the advantage of getting a lot of sequence done in a short time. But he pointed out that this posed a problem of a different sort. Celera, after all, was a private company that needed to make a profit. Gilman braced himself. Here comes the part, he thought, where he talks about how we go out at night and snatch little children off the streets.

"Many in the scientific community are concerned," Collins said, "about a circumstance where large amounts of this critical information might, in some way, be constrained from utilization by everybody who wants to use it. It is such *basic* information, and the notion that it would, in some way, be moving out of a public domain enterprise into a single private company has raised some cautions in the minds of many of my advisors."

Gilman left the hearing feeling that the chances of a collaboration with the Human Genome Project were growing increasingly remote.

In early March, after the money was secured from Congress, the orders from the HGP's genome centers started pouring into ABI for the new sequencing machines. Sulston's operation wanted around fifty, at a third of a million dollars apiece. Waterston's group requested thirty to forty, and so did Richard Gibbs down in Houston. (The DOE would need machines, too, but hadn't decided yet whether to use the ABI's Prism 3700 or the MegaBACE.) The largest order by far came from Eric Lander's group. The Whitehead Institute had secured enough outside money to buy a hundred machines. Delivery was to begin that summer.

Many people later came to believe that this surge in demand for the new sequencers was Perkin Elmer's hidden agenda all along—that Celera was just an ingenious marketing tool enabling Michael Hunkapiller to sell more machines. But Hunkapiller claims he was just as amazed by the size of the orders as anyone else. He had assumed that the public program labs would be sticking mainly with ABI's older slab-gel machines, the 377s. Though slower, the 377s had certain technical advantages that came to the fore in the sequencing chemistry the public program employed, and it seemed uncharacteristic for the conservative Human Genome Project to risk putting its faith in technology that had yet to prove itself in production. Hunkapiller wasn't about to complain about the size of the orders, of course. He didn't tell Venter about it, and Venter didn't think to ask.

CHAPTER 15
THE IDES OF MARCH

On Monday, March 15, Venter woke up in a sweat, gasping for air. Ever since returning from the Caribbean the week before, he had been plagued by asthma attacks and a flame of hives up his back and sides. He sat up and reached for the inhaler in the night table. His asthma attacks were rare but fairly serious. Lynn Holland, his assistant, blamed the current series on some chemical in the new carpeting installed in his office. He thought about staying home. But there was too much to get done. He took another draft from the inhaler and got up.

"You look terrible," Holland said when he came in around nine-thirty. "I bet you anything it's these carpets." As usual, she had printed out his schedule on a little slip of paper, which every day she stuffed into Venter's breast pocket in the futile hope that it would keep him on track. Today he had a meeting with Kathy Giacalone, Celera's human resources director, and immediately after that he was expected over at TIGR for an interview with Brazilian television. In the afternoon a film crew was arriving to tape his acceptance of an "Outstanding Alumnus" award from the American Association of Community Colleges. Then at 3:30, he had a phone call scheduled with Francis Collins. Following the collapse of the DOE collaboration attempt in December, Harold Varmus had assured Venter that the NIH would soon be back to Celera with an MOU proposal that included all of the members of the Human Genome Project, not just one agency. But there had been no word since, and today's call with Collins had been scheduled for a comparatively trivial purpose. On Thursday, the two rival genome project leaders were due to speak back-to-back at a breakfast gathering hosted by the Technology Council of Maryland. The council had assured them that it wanted to avoid rather than precipitate any confrontation. But both men wanted to go over the ground rules with each other first, just in case.

Venter also wanted to find time to call Michael Hunkapiller. Two scientists at the Sanger Centre were about to publish a review of ABI's new capillary machine in *Science,* and Venter had obtained a copy. It was hardly a rave. According to the British scientists, the Prism 3700 produced shorter "reads"—sequenced fragments of DNA—than either Amersham's MegaBACE or even the ABI slab-gel machines it was designed to replace. And while it ran twice as fast as the slab-gel machines, it cost twice as much, resulting in no net gain in sequencing speed for the same amount of money. The scientists didn't mention that they had evaluated the machine using a beta version of its software, which hampered its performance. But Wall Street analysts were hardly going to take that into account. The Sanger Centre was part of Venter's competition, and he was convinced the review had more to do with politics than with science. He wanted to know if Hunkapiller was going to make any response. He was not yet aware that the Sanger Centre had recently ordered dozens of the very machines they were giving such poor marks. But he soon would be.

"Mark also says he knows you're really busy but really needs a minute," Holland said. "Sam knows you're busy but really needs a minute. Gene knows you're busy but . . ."

"I know you're busy but I could really use a minute," Paul Gilman said, coming out of his office.

"OK, but I've got hives all over my body and it's highly contagious," said Venter.

"There have also been a lot of calls from reporters," Holland continued, handing him a bunch of pink slips. Venter glanced through them. Nicholas Wade from the *New York Times* had called, along with science or business writers for *Science,* the *Washington Post, USA Today,* and the *Wall Street Journal.* "These people all know we're still in a quiet period and I can't talk," he said. "Francis must be up to something."

Gilman went into his office and clicked on to the NHGRI web site. There he found a press release that had been issued that morning. "Based on experience gained from the pilot projects," read the lead, "an international consortium now predicts they will produce at least 90 percent of the human genome sequence in a 'working draft' form by the spring of 2000, considerably earlier than expected."

Gilman could hardly believe it. He'd been expecting for months to hear from the NIH regarding a plan to collaborate. Instead he was learning from a web site that the Human Genome Project had radically altered its sequencing strategy in order to publish a quick version of the genome a year or more ahead of Celera's. The tone of the announcement had all the signs of a planned media blitz; they had even garnished the release with an enthusiastic endorsement of the program from the vice president. Gilman printed out the release.

"What's the matter with you?" Holland said as he passed through the outer office. "You look like you just ate something gross." Without answering, he went into Venter's office and closed the door.

"This is pure bullshit," Venter said when he read the lead. He took the inhaler out of his pocket and took a hit from it. "Where would they get the resources for that much sequencing?"

It didn't take long to find out. The public program had declared war on Celera, and the dealer supplying the arms was none other than Celera's own sister company, Applied Biosystems. Venter's first urge was to call Hunkapiller at his home in California. But it would be stupid to panic. He would talk to Hunkapiller, but it would be better to have more information first. By late morning he had found out the approximate size of the orders for the machines. He asked Lynn Holland to schedule an emergency meeting of his senior staff as soon as possible. "See if you can get Mike on the phone now," he said. "Tony White, too."

She reached Hunkapiller around one o'clock and put the call through to Venter. Gilman stood off to one side, listening with his head bowed and his hands in his pockets. Venter didn't waste time on pleasantries. "NIH has issued a release saying they'll have ninety percent of the genome done by this time next year," he said. "They're basing that on buying a couple hundred sequencers from you. Maybe you can explain all this to the press, because this is going to fuck us."

"No, it's not," Hunkapiller shot back.

"It certainly will in perception," said Venter. His eyes darted to the pile of pink slips on his desk.

"Let the press have their fun in that regard," said Hunkapiller. "Obviously, the government is trying to prevent you from walking away with the laurels. I don't think it's going to work."

Holland interrupted to say that Tony White was ready to join the conversation. There was a snowstorm in Connecticut, and he was the only person around in PE's headquarters. He hadn't heard anything about the public program's announcement.

"We're fighting fires down here like crazy," Venter said, after filling him in. "Mike is saying he can get them their machines by August."

"Mike, let's address Craig's problem," White said. "How much of a lead are we giving him?"

"I told Francis that we'd fill orders as we received them," said Hunkapiller. "We've all discussed that before."

"We need more sequencers *here,*" Venter interrupted. "We need a numerical superiority."

"We're already planning that you're going to need another seventy by early summer," Hunkapiller said. But Venter had squeezed out a spoken commitment, with their mutual boss listening in. He pulled away from the speakerphone and took a big draft on the inhaler. When he came back, his tone had changed. "There's a *huge* upside to this news, too," he said. "There's no way they can do what they say they're going to—but even if they do, all it does is save us a hundred million dollars."

"I'm always interested in hearing how somebody can save me a hundred million dollars," said White. He already had a pretty good notion. You didn't have to have a Ph.D. to see that the public program was caught in a Catch-22. They were accelerating their project to beat Celera, but since they put all their data in the public domain where Celera could download them for its own use, the faster the Human Genome Project went, the faster Celera could go, too.

"Craig's right," said Hunkapiller. "I think we should take this NIH announcement as a very positive development."

"I don't have a problem with it, as long as we can get up and running before they're up and running," Venter said.

"Let's be clear," said White. "They are not going to be given priority over Celera. I guarantee it, Craig. The government's not going to beat you. The bottom line is, we didn't just get our fair share of the market.

We got it all. We came out with a new platform, and the government bought it."

White and Hunkapiller hung up. Holland called through the door that the *Wall Street Journal* was on the line and Nicholas Wade had phoned again. The film crew for the taping of the community college award had arrived and was waiting with their equipment in the outer office. Venter looked at his watch. "Let's talk to Nick first," he said.

"Be careful," said Gilman. "The SEC reads the *New York Times.*"

"We're allowed to speak to the press in the normal course of business," said Venter. "The NIH is trying to destroy us. Sounds like business as usual to me."

In the next hour he spoke to half a dozen reporters, pausing to catch his breath between each call and fidgeting constantly with a pen or some other object on his desk as he talked. Over the phone, however, his voice came across as calm, even insouciant. Each call was a variation on the central theme: the public program's ramp-up was the best thing that could have happened to Celera. "We're delighted," he told Wade. "You understand this better than the rest of the press—any data they put into their draft approach, we go that much faster. I just hope their quality is good enough for us to use. In any case, getting the genome done is just the start of our business. We have one of the biggest computers in the world, the best bioinformatics team on the planet, and we've designed our business model to take advantage of anything the public program does."

Gilman grabbed a piece of paper, frantically scrawled the words "Quiet Period!" on it, and waved it in front of Venter's eyes. He nodded.

"Nick, I've just been reminded I'm not supposed to say things like that right now, so I'd appreciate it if you would keep that off the record. And off the record, too, it's a little sleazy that the Sanger Centre is trashing the capillary machine in *Science* at the same time they're putting in this big order for them."

"It does seem unusual," said Wade.

"Well, they can't help themselves, they're British," said Venter, ignoring or forgetting that Wade was British, too. During the next call, the film crew bundled in with their cameras and booms. They were on a deadline and couldn't wait any longer. Venter managed to sit still while a young woman made up his face. He was talking on the speakerphone at the same time. "This is nothing but good news," he told Elizabeth

Pennisi, a reporter for *Science*'s news section. "I'm struggling to find what the bad news is."

"I didn't say there was any bad news," said Pennisi.

At 3:45, Holland interrupted again to say that Collins's assistant had her boss on the line.

"Tell her I'll pick up when he picks up," Venter said.

The two men had not communicated since Collins's visit to Celera back in November. That meeting had not gone well. The only subsequent contact between the NIH and Celera had been the encounter between Collins and Paul Gilman just before Congressman Porter's appropriations subcommittee hearing in late February. That hadn't gone well, either.

"It's not going to serve your interests to get into it with Francis over this press release," Gilman now told Venter. "This call is to work out the protocol for Thursday's breakfast event. Try to keep to that."

Collins's jaunty voice issued from the speakerphone. "Hi, Craig," he said. "Sorry to be a little late calling. It's been a pretty exciting day for us!"

"I know, Francis. I've been getting calls all day. I have to say I'm disappointed I had to learn about this from the press."

"I really am sorry about that. I didn't know the press was still that interested in genome stories." Gilman glanced at the press release on Venter's desk, with its second-paragraph endorsement from Albert Gore. *Sure,* he thought. *And you didn't know children were still interested in Santa Claus, either.*

"In any case, the news is terrific," Venter said. "We couldn't be more pleased. Though, frankly, I don't see how you're going to do what you say you're going to do. For your draft-quality genome, you'll need to cover the genome five times. That's thirty-five to forty million reads. Celera can't do that many in a year. I'm just curious how you can do that."

Gilman wrote THURSDAY in block capitals on a page in his notebook and held it against his chest.

"Well, it's certainly going to be a stretch!" answered Collins. "But when you add up all the institutes, thirty-five million reads is about the capacity they can deploy. It just seems like an opportunity that shouldn't be missed."

"But you're starting in August, right?"

"That's right."

"Well, over here, we're trying to keep our statements to the press on a reality basis," Venter said. "I know how easy it is to get caught up in politics, but both of us—"

"Our press statement is *very* reality-based," Collins interupted. "We wouldn't say we can get the working draft done by next spring if we didn't think we could. I'm sure you feel the same way about your own rough draft."

"We're not *doing* a rough draft," Venter said, "not unless you consider *C. elegans* a rough draft."

The completion of the genome of the tiny roundworm *C. elegans* by the Sanger Centre and Washington University had been announced with much fanfare the previous December—the first genome of a complex, multicellular organism.

"Well, as far as *C. elegans* goes, there are certainly gaps that have not been closed," Collins replied. "The *Science* paper was quite clear about that."

"If you took the number of gaps you have in *C. elegans* and extrapolated to human, there would be fifty or sixty thousand."

"Gosh. That does sound like a lot. But we never said it was complete."

"You said in the paper that it was 'essentially complete.' So, which is it—essentially complete, or a rough draft?"

Collins didn't take the bait. "Obviously, we need some new terminology," he answered. "We don't need a war of words."

"If we could stop the war of words, I'd be very happy," Venter said. "Every time you tell Congress our assembly strategy is unlikely to work, we lose market value."

"Well, you have to admit, Craig, most everybody would agree that the whole-shotgun could run into some problems."

"So you're saying that since we don't *know* whether we can do it, we *can't* do it."

"I'm saying that until it's done we should consider this a scientific issue and stop taking potshots at what people are doing or saying."

Gilman held up his THURSDAY sign again. But Venter would not, or could not, let go. "What are you basing your conclusions on about our science?" he said, beginning to raise his voice just a little.

"Expert opinion. That's all either of us has. Maynard Olson laid this all out at the congressional hearing last summer, and there's been no new data since. The only way to demonstrate that whole-genome shotgun will work on a mammal is to do it, and that's what you're proposing, and that's great. But Paul Gilman came up to me right before the appropriations hearing, and, I think rather inappropriately—"

"I don't consider Maynard Olson telling Congress that we'll have catastrophic problems much of a scientific discussion," Venter interrupted.

"And I don't consider your policy person threatening me before a congressional hearing a scientific discussion, either," said Collins. "You can present a model, but until you've demonstrated it, it's always open to question. That's what science is all about."

"Look, if you're uncertain about our chances, fine," Venter said. "But when you use that to build up your own budget—"

"That's an unfair statement," Collins said. He detailed some of the perils that the human genome presented to a shotgun approach. Venter began rebutting them, but Gilman, waving his sign, finally caught his attention.

"Look, if you can do 5x of human DNA as you say you can, great," Venter said, in a more friendly tone. "We can do 5x, too, and between us we'll have 10x by June next year. Maybe we can combine our efforts and the whole world can celebrate."

"That would be wonderful," Collins said. He sounded relieved to pull back from the abyss where the conversation seemed to have been leading. "Some time in the next month, I'd like to have a real sit-down and see if we can thrash out an MOU."

"I sort of think of your announcement today as a de facto MOU," Venter said. There was a smile in his voice, and Collins laughed, but he could not have thought it very funny. Venter was making no secret of his intent to exploit the public program's Catch-22. "But look, we should be talking about Thursday. We certainly want to avoid a circus. I'll be a gentleman and let you speak first. You set the tone, and I'll try to behave."

"Or you can set the tone," said Collins, "and I'll try to behave."

With that, the phone call ended. Venter took a pull on the inhaler. His face still looked puffy, but the bad-boy grin was back. "We certainly have some exciting days around here," he said to Gilman.

"Sure beats working at the National Academy," Gilman replied.

While Venter was on the phone, Mark Adams, Hamilton Smith, Marshall Peterson, and the rest of the senior staff had gathered in the outer office for the emergency staff meeting. Gene Myers was standing over by a window by himself, very still, his jaw clenched. His orange Polarfleece was the brightest object in the room, but his aspect was so grim that the garment and the man canceled each other out. A month earlier, he had cringed when Venter suggested he use *any* of the public data to shore up his shotgun assembly. What he was hearing today was far worse. When Venter came out of his office, he went directly over to Myers. "I know what's on your mind," he told the mathematician. "Let's talk about it. I've got to make a pit stop before the meeting. Let's take a walk down the hall."

Myers followed him, walking with an aggressive rigidity. "Are we doing a whole-genome assembly still, or not?" he said. "I didn't come here to be a write-off so PE could sell machines."

"The science always comes before the business," Venter replied, as they turned into the men's room. "This doesn't change that. We can still have it both ways." He stood next to a urinal. Myers took the neighboring one.

"My people have been working their asses off to write a program that depends on 10x of data," said Myers, into the wall. "What are we supposed to say to everybody if we only go to 5x? 'Hooray! We sequenced fifty percent of the human genome'?"

"Look, Gene, proving that shotgun works on *Drosophila,* then human—that means as much to me as it does to you. I'm not losing sight of that. But if they're going to put all this data out for free, we have to use what will help us."

"It's not the same as doing it all ourselves."

"Your algorithm is still what puts it all together. Whether the data is internal or external, it'll still be your assembly."

"We have no control over the quality of their reads. If they're not up to our standards, and I doubt that they will be, it's garbage in, garbage out. There's nothing our assembly program can do to change that equation."

"I totally agree," said Venter, turning to the sink. "Ideally we still do it all ourselves, without using any of the public data. But we don't want to be stupid. A hundred million dollars is a lot of money. Think about it. With the savings, we could do mouse. We could do human and chimpanzee, maybe find Francis somewhere in between. For now, we'll proceed as if we're going to 10x. But we've got to be ready to rethink the issue month to month."

Myers's embedded frown didn't loosen. They walked back to where the others waited around the conference table in Venter's office, their dour faces making clear that each had been weighing what the day's events meant for their prospects. With his gleaming pate, Venter looked like a switched-on lightbulb among dim shades. His breathing had gotten better, and though he fingered the inhaler throughout the meeting, he rarely used it.

"As you know by now, there's been a total role reversal," he began. "A year ago, NIH was saying they were doing the complete genome, and they were calling ours a rough draft. Now they're doing a rough draft, and we're the ones who'll have a complete product. I wonder if anybody out there sees the irony in this."

"It doesn't matter," said Hamilton Smith. "You have to remember that in the eyes of the larger community, whatever they do is good."

"It worries me that we haven't ramped up our sequencing yet," Myers said.

"Don't forget, they've got to get their machines running, too," Venter replied. "I doubt very much that NIH can get close to meeting the deadline they've put on themselves."

"They remind me of one of those blowfish that puffs up to scare the enemy," Smith chimed in.

"Yeah, well don't forget what happens if you eat that blowfish," said Marshall Peterson.

"Couldn't PE make it, like, harder for them to get the machines than us?" somebody asked.

"If you do that, they could sue PE for giving Celera an unfair advantage," said Robert Millman, the patent attorney. "Next thing you know, we'll be handing over all our patents and intellectual property on the data as compensation."

"That's not the point," Venter said, a little curtly. "Look, I have the same frustrations. Our own company is providing the weapons for NIH to go to war with us. We have to work harder now to win. But I don't think any of us would be comfortable if we won by somehow slowing down the sales of instruments to these people. Especially if their having them could help somebody make major discoveries."

"We need to be able to document how long it will take them to ramp up," Peterson said. "You've got to know your enemy."

"We need to focus on getting ourselves operational, and stop worrying about them. We need three hundred machines running at full capacity by the beginning of June. Mark, how many are up now?"

"Fourteen," Adams replied, scrunching up his mouth as if he had just bitten a lemon.

Venter acted as if he hadn't heard, not letting the number sink in. "Look at all the things we've done brilliantly," he said. "The jobs you've all done, it's little short of amazing. OK, nobody thought we'd have trouble getting these machines running. I sort of thought that would be PE's responsibility. But by next month we'll have ten new engineers here to help."

"That reminds me of another risk this announcement exposes us to," Peterson said. "The NIH is going to have trouble getting their machines to work, too. Where are they going to look for people who can help with that?"

"Are you suggesting they're going to raid us?" said Sam Broder.

"Now you're getting paranoid," said Venter.

"It's not an issue of being paranoid, it's being prepared."

"Marshall's got a point," said Kathy Giacalone, the human resources director. "If these people are coveted, we've got to protect them."

"Listen," Venter said. "Whatever they do, however well they ramp up, they'll be dumping that much more raw data into GenBank. They're doing our work for us! Maybe Francis intends to take our raw data, too, combine it with theirs, and declare victory. But he's made a slight miscalculation. The one small detail I didn't think it worth mentioning to him today was that our human data won't be downloadable. The finished product will be downloadable, just as we promised. But not before then."

"So what's your best-case scenario?" Broder asked. "What happens to our business plan?"

"What's different about it? Amgen, Novartis, and now Pharmacia Upjohn have signed up knowing damn well the data was going to be in the public domain in two years anyway. They didn't want to wait for it."

"If the government pulls this off, the data will be in the public domain now in one year," said Broder. "One year is a lot different than two."

"Will everybody just stop sweating?" Venter said. "All they've done is double our pleasure and double our fun. OK, so maybe they halved the time we've got to work with. But as long as we do our job well, we're guaranteed to win."

Myers got up to go. "I'm going back upstairs to write some code," he said. "I'm going to bury myself in code, and pretend I never heard any of this."

Three days later, more than two hundred biotech executives, researchers, and government people showed up to hear Collins and Venter speak at the Tech Council breakfast. Neither one of them kept his promise to behave. Collins displayed his chart showing Celera's broken genome with its red bits falling off, this time blown up on a big screen. When it was his turn, Venter referred to the Human Genome Project's forthcoming working draft version of the code as a "patchwork quilt." He sarcastically thanked the NIH for their recent investment in Perkin Elmer. In the question-and-answer period, things got worse. They were standing on the stage together, about twenty feet apart. Venter accused his rival of purposely misstating Celera's patent intentions in order to justify his budget. "Francis is making an appeal for your tax dollars," he said. "I'm not making an appeal for anything. We're using private dollars to give the genome away." Collins retorted that it was Venter who was deceiving the audience. Would Celera's human code be made freely available every three months, as promised? Or would drug companies be getting a privileged first look? What about SNPs? Would they be free for all researchers to use, or swallowed up in intellectual property constraints? Through the exchange, Collins and Venter had been edging closer to each other, their voices growing more heated.

"Celera is not going to act like a public charity, if that's what you mean," Venter answered.

"You don't have to phrase it that way," Collins snapped back. "Just answer the question."

"Our data release policy hasn't changed from when we stated it in *Science* last year."

"Does that mean the data will be available on GenBank, or only if you pay a subscription? Which is it?"

"Neither. Not on GenBank, because we think it's an outmoded model. But anybody can come to our web site and do a search. Even you."

By this time, they had closed to within four feet of each other. A collective tension hovered over the people in the audience, born from the exciting realization that they had been forgotten. Before the two men got any closer, the moderator, a blond woman in high heels, stepped up from the back of the stage. "Clearly we need the energies of both the public *and* private sectors," she said with a bright smile as she inserted herself between them.

Still, things could have gone even worse. At least the press wasn't there.

CHAPTER 16
HE DOESN'T GET IT

On Tuesday of the following week, the cafeteria tables in the basement were cleared away to make room for Celera's second all-hands meeting. Things had changed since the first one. Back in September, a few dozen people had gathered on the ground floor and gotten high on the promise in its moldy air and on the great adventure to come. Now, six months later, the rows of chairs stretched almost to the back wall. Most were occupied by people with jobs, not visions. Due to the sheer weight of numbers—Celera now had almost 350 employees—a point had been reached where people's primary loyalties were directed toward their own departments rather than the company as a whole. Body language defined the edges of bailiwicks, and for the most part dress defined role. Secretaries came into the meeting wearing pumps and skirts. Asian software engineers sat in calm attentive rows, their button-down shirts neatly tucked in. Pale, pierced techies in T-shirts and jeans blinked and twitched from being so abruptly yanked into an analog world. Robert Millman, the patent attorney, was the exception, erect and alone at the edge of one row, wearing a fractal-patterned brown silk shirt over corduroy trousers that were as subtle as chartreuse can be. In his long face and bright sunken eyes was the look of a Sufi who had come down to the meeting from some mountaintop. Gene Myers, huddled with the other senior people in the front, had spruced himself up. His hair was combed, his citrus fleece zipped up and squared off against his pants, and his scarf crossed neatly on his left shoulder.

Everyone at the gathering was about to become not just a Celera employee but a shareholder, just as soon as the company's stock went public in a couple of weeks. Tony White had come down from Connecticut to explain what that was going to mean. He had brought along Dennis Winger, his chief financial officer, and some other brass from the parent company, who had dressed down for the occasion by taking off their suit jackets and ties. Peter Barrett stood with them, in a sweatshirt with CELERA and the company's new logo—a tiny dancing figure whose tapered limbs formed a subtle double helix—printed discreetly on the breast in blue. Boxes full of the sweatshirts were piled by the door, one for everybody on the way out.

There had been major changes in the parent company as well in the past few months. Perkin Elmer's analytical instruments division, its heart and soul since its founding in the thirties, was about to be sold for $338 million to a conglomerate called EG&G. Along with it went the Perkin Elmer name. From now on, Celera's parent, renamed "PE Corporation," would be devoted exclusively to life sciences, with two divisions. White took out a marker and on a flip chart nearly as tall as himself drew two great overlapping circles, like a Venn diagram. He labeled one circle "PE Biosystems," which included Applied Biosystems and some other, smaller entities that sold real, hard stuff—instruments and reagents, primarily—to the research community. The other circle he labeled Celera Genomics. People in the back craned their necks to see. "This one," he said, pointing to the PE Biosystems circle with a thick finger, "makes money." The finger moved to the right. "This other one spends it." This distinction, White explained, had great bearing on what was about to happen to the company's stock.

There had been several possible options on how to repackage PE Corporation in the marketplace. The company could continue to trade as a single entity, for instance, or it could split into two distinct enterprises, each with its own board of directors and stock. On the advice of Alex Lipe, the merger-and-acquisitions specialist from Morgan Stanley, White and his board had chosen a third, more creative alternative. Pending approval by its major shareholders, Celera would trade as a "tracking stock." It would have its own ticker symbol on the New York Stock

Exchange, but there would be no new offering of shares. Instead, for each share an investor already owned of the old Perkin Elmer company, he, she, or it—for most of the investors were institutions—would now get one share of PE Biosystems stock and half a share of Celera. The two entities would thenceforth go their independent ways in the market. The beauty of the scheme was that investors who preferred a steady, profitable company wouldn't sell off en masse now that Celera and the burden of its enormous burn rate were half of PE's business. They could simply shift more of their money into PE Biosystems. New investors attracted to a high-risk, potentially high-return venture—and with the dot-com frenzy at its height, there were plenty of these—could place their bets on Celera.

"Celera's movement is probably going to be down, at least initially," White told the new shareholders. "A lot of pension funds and other institutions that own Perkin Elmer now can't hold stock in a company unless it turns a profit. You all won't be doing that for a few years. These people are going to be forced to sell their Celera stock as soon as they get it, which will drive the price down. Some newly listed tracking stocks lost thirty, forty percent of their value right away. That's why we've decided to issue only twenty-five million shares of Celera, instead of the fifty million we were going to. We want to keep the stock from falling into single digits. Stocks trading down there are considered junk. Once a stock is in single digits, it's a hell of a lot of trouble digging it out. Go below five, and you're forced off the exchange."

"Does this mean we'll get only half the number of shares we were expecting?" somebody asked.

"Yeah," said White. "But they'll each be worth twice as much, so it comes out the same. We expect to be trading in the twenty- to forty-dollar range."

Another hand went up. "What happened to the three hundred million from the sale of the old Perkin Elmer division?"

"You're sitting in it."

"Can we buy more stock if we want?"

"Be my guest," said White. "But you'd better not buy stock because you just made some breakthrough in your lab. For more on that, let me introduce Bill Sawch, our chief legal counsel."

Sawch was a trim blond forty-something dressed in khakis and a white oxford shirt. He went over the formal definition of insider trading—carefully noting first that his remarks were in the nature of a favor, since the SEC held the individual, not the company, responsible for illegal trades.

"OK, if we know something, we can't buy our own stock," one of the Asian software engineers said. "But can we short Incyte?"

"No. If the information is something that an average outside investor would like to have to make his own trading decisions, you cannot trade on it, period."

"What if my brother-in-law shorts Incyte?"

"That's insider trading, too. The SEC is very good at tracing your relatives, even your neighbors."

When all the financial questions had been asked and answered, Venter took the handheld mike. The talk of downward dips and insider trades had left a lot of people looking a little off balance, and he wanted to end the meeting on an upbeat note. "I think it's important for everybody to be good at something in life, and we're good at spending money," he said. "If the stock price goes down for a few days, don't sweat. It won't take long for the market to understand all the great things that are happening here. We've got a deal with another major drug company about to be signed, and there are some people interested in us doing the rice genome, and—"

Peter Barrett literally grabbed the mike out of his hand. "Well, now you're *all* insiders," he said. "But seriously, please don't make that information public. And please don't ask Craig any more questions that will make him say things he's not supposed to."

Later, after everybody had collected their Celera sweatshirts and gotten back to work, Barrett could be seen in the hallway, gently banging his big square forehead against the wall. "People have told me I'm the only adult around here," he said. "It's true. And Craig is my biggest child."

Peter Barrett had been banging his head on the wall a lot recently. For weeks he had been struggling to get Venter to focus on the road show, the whirlwind sales pitch that would precede the stock offering. Now there were only a couple of weeks left. "There's all this negative information out there, and the business model hasn't jelled yet," Barrett complained. "Without an understandable evaluation of the business model, we'd be better off not taking this on the road at all. We'll be killed."

Venter didn't see what the fuss was about. Providing the world with the complete recipe for the making of a human being was a pretty strong business model, and he was sure he could present it with style. He was constantly in demand as a public speaker. During the quiet period, he had tried to steer clear of talking up his current project. But there was plenty of interest in his past accomplishments and his visions for the future. The audiences came to hear about genes and medicine, but left with the impression of a man who also knew how to live. Here was a serious scientist who did not take science or himself that seriously, who glowed with self-possession and wealth but seemed invitingly vulnerable, a fifty-something veteran of Vietnam who had fought with people in power throughout his career but still exuded the now bashful, now boastful charm of a high school heartthrob. Now, with the quiet period ending, he had the greatest story in the world to tell. Why worry?

It was just this attitude that Barrett worried about most. "Craig is so proud of the grandeur of all this," he said one day in the parking lot, sweeping his arm to indicate the buildings, gleaming white from a recent sandblasting. "He loves to talk about how we've got the biggest computer in the world, the biggest sequencing facility, and so forth. But unless you tie all that to the 'What's in it for me?' question, it's a loser."

Barrett knew that a road show wasn't about impressing investors the way you wow coeds and science writers. It was about exciting the lust of potential investors. A road show is the finance community's equivalent of foreplay. The purpose is to arouse in the market a desire for your company so hot that it can be quenched only by getting a piece of it. You could tell them that Celera had the inside track on the biggest revolution in the history of medicine, but this was like whispering sweet nothings in their ear. They aren't that interested in your fame or your vision of the future: they want to know how they are going to make some money by lying down with you, right now. You have to show them what nobody else has, and do it again and again. You make your pitch to Fidelity in New York; pack up your videos, PowerPoint presentations, and other toys; fly to Geneva for a one-hour meeting with a private Swiss bank; and fly back again to New York for a quickie before bedtime with Soros Management in your suite at the Trump Tower. While the rest of the world lumbers on in some other dimension, you are propelled through a vortex of flights, faxes, limos, and elevators, and when the bell tings and the doors open on the twenty-third floor, there is another group waiting for you. Welcome to Denver, how was the flight, dim the lights, let's see what you've got. You must always sound confident and fresh, even when the whole sweaty business starts getting a bit tedious. You have to grab them so they grab you and won't let go.

A road show usually precedes a new company's initial public offering. Celera did not need to raise start-up funds—the sale of the old Perkin Elmer had seen to that—but it was under no less pressure. Even before the restructuring could go through, the major investors' boards had to vote their approval of it, and the road show was PE's chance to convince its representatives why they should. More important was how the market at large would respond. PE was 90 percent owned by big institutional investors, like mutual funds and pension funds; Fidelity Investments alone counted for 10 percent. If a lot of these heavy hitters sold the shares of Celera stock they were automatically to receive under the recapitalization, as expected, somebody else was going to have to buy them up, or the price would plummet into dangerous territory in spite of the precautions PE had taken. The market had to be convinced that Celera had something that no one else did.

Whatever its scientific distinction, Celera, as a latecomer in a crowded market, was not going to be an easy sell. Since Francis Collins's March 15 surprise, Celera's business plan was being squeezed in a vise. Before, it had been relatively easy to articulate how the new company differed in a financially promising way from its commercial competitors. Incyte and Human Genome Sciences may have had a head start on the patenting of genes, but Celera wasn't just about genes, and it wasn't about patents. It was the beckoning prospect of having the whole shebang: the genes, their regulatory regions, their interactions and mapped locations relative to one another, and the information tools needed to navigate through this brave new landscape and pluck out the gold. But now investors were aware that, at least in its rough outlines, the whole shebang would soon be available to everybody. The Human Genome Project had Celera caught in a Catch-22 of its own: To differentiate itself from its seasoned competitors, the company had to downplay the value of intellectual property. But if it downplayed the value of intellectual property, what did it have to offer that couldn't be gotten elsewhere for free?

To make matters even more dicey, the public program had just forced Celera's hand on another front. The company had been counting on the pharmaceutical industry's hunger for information on SNPs—the single-base-pair variations seen as the key to understanding why one person got sick and another didn't. A big advantage of the whole-genome shotgun sequencing method was that the location of SNPs would pour like golden coins from Celera's sequencers into its pay-for-view database. This was information that Venter had never promised to make public, and was thus a major incentive for drug companies to sign up. But in April, ten of these same drug companies teamed up with the Wellcome Trust to form a nonprofit consortium that would go after SNPs and place them in the public domain. It wasn't just the Wellcome Trust's involvement that betrayed the direct link between the so-called SNP Consortium and the Human Genome Project. The scientists directing the work would include Eric Lander, Robert Waterston, and John Sulston. The resulting database, which the consortium hoped would contain some 300,000 SNPs, would be run out of Cold Spring Harbor Laboratory. Venter waved away this threat as too little, too late: by the time the consortium was on its feet, Celera would be well on its way to gathering millions of SNPs on its own, with the consortium's offerings tossed into the database as gravy. He sounded very sure. "Our message to the Street stays the same," he said. "We welcome what other groups are doing, because it only adds to our own database."

Like it or not, however, the message *had* changed. All around Celera, even in Venter's own office, the human genome sequence was now being referred to as the company's "loss leader." When it came to wooing customers and investors, Celera was no longer just a portal to the code of human life. More important, it was the transforming substance that would make sense of what you found inside: the human genome, the SNPs, the fruit fly and mouse genome and rice and cow and chimp and all the genomes to follow, all gathered in one place, with software searching tools that would uncover drug targets like divining rods. Venter's new favorite analogy was to the Bloomberg financial news web site. Someone looking for financial information could traipse around the web on his own, visiting dozens of different databases—or he could go to Bloomberg and find all the information gathered and organized for him already. Likewise, Celera was one-stop shopping for the genomic age. Sure, in a year you could get some of the same data from GenBank—maybe. You could also drink tap water instead of the bottled kind. But if you want the kind of safe, reliable, rustfree, abundant, disease-defeating information you need to quench your thirst for profit, come to Celera. If you don't, your competitors will.

Bolstering that message was Pharmacia Upjohn. True to Venter's little slip at the all-hands meeting, the huge company had signed up with Celera just a few days before the road show was to begin. It was a blessed, $32 million shot in the arm, including fees to gain access to the future mouse genome sequence, even though that was at least a year up the road. But Peter Barrett was still nervous. They would have only half an hour to make the road show pitch to investors. He and Charles Poole, PE's investor relations director, labored over the story—what Craig would say, what Tony would say, the order of slides and talking points, what to put in the video, when to play it. He brought in consultants from Morgan Stanley to help shape the presentation and other consultants to critique their work.

"This is garbage," said one, as Barrett went through the PowerPoint slides. "You're ten minutes into the presentation, and I'm still not sure what Celera *does.* I mean, would *you* buy this stock?"

"Sure," said Barrett. "Because what we do is going to be better than anyone else, and what we do better is different than anybody else."

"None of that is coming through."

Tony White was nervous, too, especially about how his star scientist was going to perform. Venter was proving harder to manage than he'd expected. White hated it every time a magazine ran another story on the "gene maverick" who was going to tackle the human genome all on his own. Anybody who talked about this project in the first person, White thought, was missing the point. He also hated it when Venter waxed on about the 100 million expected customers. On Wall Street, exaggeration isn't sexy. If you say you'll have 100 million customers and end up with only 99 million, you are judged a failure. Venter, on the other hand, didn't think he was exaggerating at all. He was simply looking ahead. "Tony doesn't get it yet," he said. "He doesn't understand his own business model."

A week after the all-hands meeting, White sent Dennis Winger back to Celera for an emergency brainstorming session, along with Noubar Afeyan, who had suggested sequencing the entire human genome at his first board meeting at PE and was now the company's chief business officer. There had been at least a little progress in the meantime. With the help of the consultants, Barrett and Venter had found a phrase to describe what Celera alone had to offer. They were calling it "the logic of biology." It summed up the company's wedding of biomedicine and computation. Beneath the squishy surface of life there was an internal order as precise as that on a microchip. Where there was order, there was the possibility of control. And where there was control, there was the possibility of curing cancer and other diseases, and with it, the possibility of making lots of money.

"How are we different from those who've come before us, like Incyte?" asked Winger, anticipating what he expected to be the first question anybody would ask on the road show.

"That's just it," Barrett answered. "They don't have the logic of biology. All they've got is a bunch of genes. Genes are just the component parts. We have what holds the whole thing together."

"You can't cure disease by looking at a parts list," Venter added, "because the problem is in the integrated circuit. Take the cystic fibrosis gene. When the gene was discovered, it was assumed that if you have the mutated form, you get the disease. Now it turns out that mutations in the gene can lead to all sorts of conditions. Sometimes they cause male sterility, liver disease, or inflammation of the pancreas. Sometimes they don't cause any health problem at all. People have been aborting children because they had that gene, and all the time there might not have been anything wrong with them at all. The people advising them didn't understand the logic of biology."

"That's wonderful. That certainly shows the need," said Afeyan. "But if I'm sitting here with my money hat on, I'm thinking, This will generate value in fifteen years, so I make a note to buy the stock in fourteen."

"The bottom line is that people are paying thirty-two million for our product," Venter said.

"The bottom line is it's costing us three hundred million to generate that product," said Winger.

"Amazon.com is burning a lot more money than we are, and their stock just went up another sixteen points yesterday," Venter said. "Maybe we should be losing *more* money. The real danger is breaking even sooner than we want to."

Winger looked across the room at him, trying to figure out whether he was serious or not. "You let me worry about that," he said. "OK?"

The road show was set to begin on Monday, April 19, 1999. Morgan Stanley had arranged the schedule. Barrett had rehearsed Venter as much as he could, when he could get him to sit still at all. By Friday, the video was in the can. From its opening command to PREPARE YOURSELF . . . A CHANGE IS HAPPENING, the images sizzled, the music throbbed, the message sang. To anchor all this exotic new energy to the proven value of the mainstream, Barrett had added an endorsement of the company from Eckhard Pfeiffer, the silver-haired, power-jawed CEO of Compaq. But on Saturday, unfortunately, Compaq's board of directors voted Pfeiffer out. Barrett got the news by page on Sunday morning. He banged his head against the wall again. The road show video might as well include an endorsement from Milli Vanilli.

By 7:00 a.m. Monday, when he met Venter, White, Winger, Lipe, and Poole in Boston's Logan Airport for one last huddle, Barrett had a new, Pfeiffer-less version of the video tucked under his arm. They split up into two teams. White and Venter would present for the Red Team, Winger and Barrett for the Blue. Red Team arrived at their first appointment without incident. But Blue Team discovered their route blocked by the imminent start of the 103rd Boston Marathon. They abandoned the limo and edged their briefcases through the horde of runners waiting on the starting line, wrapped in aluminum blankets. It felt surreal.

Red Team meanwhile was working on Essex Investment Management. It couldn't find their G-spot. Everyone but Venter had worried that he was going to ignore the carefully worked script and fly off on his own, maybe even start in about the 100 million customers. Instead, he was sounding wooden. White could tell he was reading verbatim from the prompts on the PowerPoint slides that the Essex managers couldn't see. At the next stop he took more of the time for himself, and a little more at the next, waiting for Venter to find his rhythm and half afraid of what would happen when he did.

"The good news is, Tony's beginning to get it," Venter told Paul Gilman on the phone during a break. "That's the bad news, too. He keeps butting in on my presentations."

On Thursday, April 22, while the others were in Detroit, Venter was back in New York, momentarily on his own for a breakfast lecture to an audience of venture capitalists gathered at the New York Athletic Club. On the walls hung dim-lit oils of bruised, leather-helmeted football players, hockey stalwarts in baggy woolens, and other athletes from yesteryear, before sports became all about money. A couple of hundred boom-happy male and gym-toned female financiers had shown up to hear him talk about "How Gene Mapping Will Reward Investors' Patience."

Venter rose to the subject and beyond, without bothering to linger too much along the way. He talked about Celera, of course, and the revolution in biomedicine that would come once the genome was sequenced. He explained the company's vision as the combination of the best science and the best of all possible business models, one that would reward its customers handsomely even as it saved the "next Epogen" for Celera itself. He talked about individualized medicine, microbes in space, and the possible origin of life on Earth from the flush of an alien spacecraft's toilet. For a man whose substantial livelihood depended on the accumulation of genetic information, he was passionate in his concerns about its misuse. He talked about the "so-called cystic fibrosis gene," which he said was really just part of a much more complicated genetic pattern. What if some simple-minded law enforcement agency thought a "gene for pedophilia" could be used to ferret out child molesters? "They could genotype all the pedophiles in prison," he said, "find a common pattern, then use that profile to look for pedophiles at large before they strike. It might catch some child molesters. But it might brand you or me as one just as easily. Genes can't predict behaviors. They can only predict general traits that might lead to some behavior, or might not."

"What would you propose to guard against that happening?" somebody in the audience asked.

"My solution would be to change the Fourteenth Amendment," Venter said. "Instead of saying you can't discriminate on the basis of race, creed, gender, or whatever, it should just say you can't be discriminated against on the basis of genetic information, period. Because that covers everything else."

Whether his talk convinced people to buy the stock and be patient about the rewards wasn't clear, but it certainly got them interested. For half an hour afterward they stood three-deep around him in the lobby, while he continued to answer questions on subjects ranging from the ethics of cloning to the concept of a virtual cell. Finally a handler tugged at his sleeve; they had only a few minutes to get to the airport for a flight to San Francisco. Venter let himself be led to the elevator and crowded in with some others. As the doors began to close, a diminutive Indian gentleman in an impeccable blue suit slipped in too. "So, Dr. Venter," he said, as the elevator descended. "One last question. What is the role of God in all this?"

"Well," Venter said, not missing a beat, "he's been a big help so far."

. . .

A week later Venter was back up in New York. Celera's scientific advisory board was convening at Rockefeller University to discuss one of the most sensitive issues in the company's agenda. Board member Arnold Levine, the university's president, had offered to host the meeting in an elegant conference room near his office. The matter at hand was the question of who, precisely, who was going to become "the" human genome. The public program's piece-by-piece method of sequencing the genome ensured that their completed product would be a mosaic of many anonymous individuals, since a different person's DNA, theoretically, could be used for each piece. But Celera's all-at-once, whole-genome shotgun strategy dictated that the DNA of only a handful of people—perhaps five—would be used. One individual in particular would contribute a majority of the sequence, with the other four used to provide redundancy and to contribute more information on how and where one person's genome varies from another's. The individuals would be selected from a larger pool of volunteers asked to donate a blood sample—and, in the case of males, sperm samples as well. Sperm was particularly good raw material. A spermatozoon was essentially nothing more than a little bag of freshly made DNA with a tail. The DNA in reproductive cells, moreover, is as close to the original blueprint of an individual as one can get.

"I asked one extremely well known lady whether she would consider donating," Venter told the group, as the meeting came to order. "When I explained how the samples were taken, she almost fell off her seat laughing. 'This is typical,' she said. 'The men get to whack off and the women get stuck with a needle.'"

Perhaps because he was glad to be away from the road show for a while—or perhaps because he was nervous about knowing something about the donor pool that most of the others in the room did not—throughout the meeting Venter could not seem to resist joking around. But the process of donor selection was a serious matter, studded with legal pitfalls and ethical complications. Just for starters, it was imperative that donors be clearly informed about what they were doing, how their DNA would be used, and what legal rights they had in the transaction. Safeguards to protect their anonymity had to be stringently spelled out and enforced. In no sense must Celera and the donors be seen as having a doctor-patient relationship, where the donor would expect medical advice based on what the company found in his or her DNA. But what if in the process of sequencing, Celera's medical staff discovered that a donor had some serious genetic condition? Did the company have an obligation to tell the person? (The answer here was no.) What if the donor had AIDS and didn't know it? Was it the company's duty to inform the individual? (Yes, and the health authorities as well.) What if a donor later contracted a disease that might have been prevented had he or she known what was in the code? Could the company be sued for not having given a warning? Or what if Celera did warn the donor, who never contracted the disease? Could it be sued for causing unnecessary stress?

The resolution of all these issues and more had to be clearly spelled out for the donors, in order to ensure their informed consent. Sam Broder, who was in charge of the donor selection process, handed out a draft of the consent form that had been prepared after getting input from an internal review board he had convened to advise on the matter. It was over thirty pages long.

"Anybody who could read through all this, masturbate, and give a sperm sample, we don't want his genome," Venter said, leafing through his copy. "He would have to be a lawyer."

"We're doing the human genome, not the shark," Broder said.

"Has anybody given any thought to the problem of the donors' relatives and heirs?" asked Richard Roberts, the advisory board chairman. "They are going to share a lot of the same propensities for genetic diseases as the donors themselves. The donors can't sue us for not disclosing that we had some condition, because they understand the agreement and have signed the consent form. But their relatives haven't."

"Are you going to take a family history?" someone else asked. "Let's say a donor's mother has Huntington's disease, or seven members in her family have breast cancer."

"We can't do that," said Broder. "It looks like you're creating a doctor-patient relationship. Sure smells like one."

Another conundrum in the donor selection process was the issue of race. How could any one person's genome stand in for the diversity of the species itself? Everyone involved, especially Venter, was keen to use the sequencing project to underscore the point that human beings are genetically 99.9 percent the same, with no way of discriminating one person's genome from another's on the basis of race. But if donors were selected to reflect racial diversity, wouldn't that give just the opposite message: that in fact there was such a thing as a representative black male genome or an Asian female one? On the other hand, what were the consequences of *not* taking steps to ensure diversity in the donor pool?

"We want to demonstrate that race is not a scientific concept," said Venter. "But if we do five genomes, it would be fundamentally wrong to end up with five white men." Earlier, he had recommended to the internal review board that the donor pool indeed be racially diverse but that the designation of which "race" someone belonged to was entirely up to the donors themselves. The advisory board approved the decision.

"Have we considered whether there are any physical risks to donors?" Roberts asked.

"For male donors, the risks are very real," said Venter. "Blindness. Hairy palms." The meeting was winding down, but he was winding up.

"Seriously," someone said. "Will the donors receive compensation for research-related injury?"

"Like what?" Venter said. "You get caught in your zipper?"

"What about risks to *us*?" someone else asked. "Let's say somebody makes a donation and then they change their minds, after we've already invested millions in sequencing their DNA."

"We tell them they can't withdraw," Broder said. "They can't hold Celera hostage like that."

"Right," said Venter. "If a guy is going to withdraw, it has to be pre-ejaculation."

Celera's stock began trading on Wednesday, April 28. The opening price was set at $18—a dollar more than the option price for the company's employees. Instead of the predicted dip, it began with a little bounce, not the dizzy leap of a dot-com initial public offering (IPO) in those days but respectable nonetheless. By the end of the day it had risen to $25 and two days later it climbed close to $30. But then it began to slip. By May 11, the day Venter was to speak at a major health care investment forum in Baltimore, it had slid back to $19. Venter was checking the stock every half hour. It wasn't the price he was curious about but the volume. He wanted to see if people were paying attention.

The Deutsche Banc Alex. Brown Health Care Conference was an annual event that drew the most astute of the biotech sector's investors to feel out the industry's subterranean trends. Venter put on his darkest, sleekest suit and took a limo up to Baltimore. White and Barrett joined him. The audience for Celera's session was not impressive, but that turned out to be fortunate. Venter had planned to start things off with the video, but when neither he nor Barrett could get the projector to work, he launched into a number-heavy presentation meant to demonstrate Celera's clear advantage over the competition in sheer sequencing and computer muscle. But the numbers didn't seem to wow anyone, and in a breakout session afterward—a chance for interested investors to pose questions directly to a company's chief officers—things went even worse. The room was stuffy and dark, with heavy drapes on the windows pierced by a single blade of sun that bisected Venter's face where he sat at the head of the table. White and Barrett took seats next to him and brokers and fund managers filled the rest. More stood around the perimeter, pressing up against the walls.

"How do you know this shotgun thing is going to work?" one investor asked.

"We don't know now, but we will soon," Venter said. "The *Drosophila* genome is the make-or-break for us. When it's done, we will have proved our case."

"When will that be?"

"By the end of this year," Venter said. "That's a promise."

"I recall hearing initially that *Drosophila* was supposed to be completed by the *first* quarter of ninety-nine," somebody else broke in. "Did I misunderstand?"

"Yes, you're mistaken," Venter quickly replied. "We were supposed to *start* by the first quarter." (In fact, the original plan had called for an April completion of the fruit fly genome.)

"What about this new competition from the government?"

"We don't regard it as a competition," Venter said, "which is why our partner company is selling sequencers to whoever wants them. We set up our business strategy so that no matter what the public program does, it just helps us go faster and makes more money for PE Biosystems at the same time."

The questions kept coming. How much market share are you anticipating? What are your short-term revenue expectations? When will you be profitable? How are you going to catch up with Incyte?

"Incyte is killing themselves, so we don't have to worry about them," Venter responded to this last question. "They've dragged themselves down with royalty burdens in the contracts."

"I don't think Incyte should be seen as a failure," Tony White broke in. "We have the advantage of learning from their successes and also from some of the problems they've encountered along the way. But they give a fair return to their shareholders, and they're reevaluating their business model. Those are some very smart people out there, and they aren't going away."

"For all you've been telling us," a sharp-faced guy in a buzz haircut asked, "you've got only three customers. So what's your business plan?"

"Find more?" Venter said, with a little grin. "But really," he continued, "this is money, not life. . . . We don't *want* more customers right now, but, yes, we need some clarity. . . ."

Seeing Venter start to flounder, White broke in again and explained how the business model was continuing to evolve. Peter Barrett jumped in, too, and Venter back on top of him, all three men clarifying the business model in somewhat different ways. Around the room there was a faint rustling of keys in pockets, a shifting of feet—the sound of people not getting it.

"Look," said Barrett, as the investors began filing out, "ten months ago we were ten people in a gutted building. Now we have three hundred fifty employees, two buildings, the largest supercomputer in the private sector, and three of the half dozen people in the world capable of writing the algorithms needed to make this work. All we need is a little time."

Afterward, Venter waited for his limo outside the hotel. "I thought that went *terrific,*" he said. "The stock price doesn't worry me at all. In fact, it's a very good thing." The slipping price was good news, he explained, because when it fell to a certain fixed point, a lot of early holders would be forced to dump the stock. This would create a giant buying opportunity for the really savvy investors who got it—of whom there were obviously plenty, or the volume wouldn't be so high. Then the stock would rise again to new heights. "I'm hoping it gets down below my option price," he said. "That way I can buy some more!"

A few days later, Lynn Holland was seen staring dismally at Celera's page on the Yahoo.com financial site. Her boss had gotten his wish. "Nobody *wants* us," she mourned. The stock had fallen to $14 and was pressing close to the single-digit range, the kiss of mediocrity. Nobody in the building was exactly panicking; they were all too busy with problems over which they had at least an illusion of control. Once the sequencers were working . . . once *Drosophila* was assembled . . . once they signed another pharma, the market would respond as effortlessly as a kite catching a fresh wind.

One person did not agree. Four months into his job, Robert Millman still saw Celera as a patent attorney's paradise. But in his mind, it would all amount to nothing unless something could be done about the farce of a business model. You don't lay claim to the richest gold mine in the world and then sell tickets and shovels at the gate. If you want to make a fortune—and what else are you in business for?—you build a thick wall around the mine, gather as much gold as you can carry, and sell it for what the market will bear. He had tried to make Venter understand this, but Venter didn't get it. No matter. If his boss wouldn't listen, Millman would talk to someone who would.

CHAPTER 17
THE HAND OF MAN

On May 11, Millman woke up to the sound of claws clacking against the bathroom tub. His hermit crab was thirsty. The mollusk spent most of its time in the living room, snuggled among the wires behind the TV. But every week or so it set off across the rug, down the hall, and into the bathroom for a drink. Millman did not think of the crab as a pet, to be fed and given water. That would demean it. But he did make some minor adjustments to accommodate the needs of his housemate. Before getting into the shower this morning, he lifted the crab into the tub. It sucked into its shell at his touch. When it felt the water splashing down, it levered itself up on its claws, looked up at him, and drank.

Millman lived with the crab at the end of a narrow, wooded dead-end street in Glen Echo, just outside the D.C. city limits. His long, low house hung over the edge of a deep ravine, like the nest of a predatory bird. A picture window in the living room overlooked the gorge. There were fox, beavers, and raccoons down below, and as he made his way from the bathroom to the kitchen, he peeked into his telescope, just to see what might be moving. He had furnished the place with his collections of nineteenth-century patent models, steam-driven toys, and other curiosities. There were virtually no functional objects in the room that weren't art, and no art that wasn't functional. A two-story clock constructed of bent branches and twigs dominated the entryway; even the gears were made of wood. Over the dining table hung a porcelain trumpet painted the vibrant colors of tropical fruit. Some of the lamps in the room resembled glowing brains, others giant spermatozoa; still others, hanging on the walls, backlit scalps. In the kitchen he opened a cupboard and took out a coffee mug. On the front the artist had sculpted the tortured face of a man whose body tapered around the sides to a reptilian tail that formed the mug's handle, then curled back and disappeared into his own gagging maw. Millman had found the piece at a lesbian art show in Northampton, Massachusetts. It was called "Man Eating the Snake-like Self That He Is."

Millman had not taken the usual path to his position as a corporate lawyer. He had grown up a "cross-functional, dysfunctional, science-nerd jock type" in the San Fernando Valley. After high school, he attended the University of California, Riverside, with the initial intent of becoming a neurosurgeon. When recombinant DNA engineering came along in 1976, he switched to biochemistry for a couple of years before deciding instead to go into emerald smuggling. The career change was inspired by a Venezuelan friend, a medical entomologist, who was studying a particular kind of gnat that had to be imported from South America. The boxes containing the gnats passed through customs unexamined, so it seemed a no-brainer to load some jewels in along with the bugs. For one reason or another, the emerald caper didn't work out, but it had gotten Millman interested in insect biology. He eventually graduated in biochemistry, with an enigmatic flush of entomology courses in his junior year.

After college, he studied plant transformation systems at Washington State University. When his Ph.D. research fell apart, he took his master's and moved to Hawaii for a couple of years, where he taught scuba diving and worked as a street performer doing magic tricks and escape routines. He had been an adept magician and juggler since the age of seven, and now, in the finale of his act, he extricated himself from a straitjacket while blindfolded, riding a unicycle on a tightrope. After a couple of years of passing the hat, he returned to the mainland and took a job with a plant genetic engineering company. While it is tempting to draw parallels between doing magic and conjuring up new life-forms with recombinant sleights of hand, the job was not a perfect fit, since by this point he was competing on the national hang-gliding circuit and needed more free time than his employment would allow. He got a job teaching biochemistry at the University of Maryland, which left the summers open. Two years later, his position disappeared. He started an MBA, but soon afterward saw an advertisement from a law firm seeking help in biotechnology. The interviewer told him, "If you quit your MBA, we'll cut your hours here and double your pay so you can go to law school at night."

And that is how Robert Millman ended up a patent attorney, eventually working in the company that he called a "patent attorney's wet dream." The only obstacle, in his view, was the ludicrous business model being espoused by the company's president. Hamilton Smith, Sam Broder, Paul Gilman, Marshall Peterson, and most of the other senior staff had all been drawn to Celera by Craig Venter's vision. Millman had come in spite of it. When he flew down to Celera for his interview, Peter Barrett's staff first explained to him how people were going to flock to Celera's web site for genomic information the same way they went to Lexis for legal information and Bloomberg for financial data. "[They] told me how everybody was going to spend a dollar a day to see how their genome is changing," he remembered later. "I thought that was really cute. Because genomes don't change." When he met with Venter, Millman told him exactly what he thought: the way to ensure the company's financial success was not to sell access to its database but to jealously protect it as intellectual property. This was hardly the strategy Venter had in mind, but gaining patent protection on a few hundred of the most promising genes passing through the pipeline had been an explicit part of the plan from the start. Venter needed a patent attorney to help with that, and he knew from his own experience that Millman was one of the best.

He offered Millman the job, and Millman said he'd think about it. By the time he got back to Boston, he had decided to take the offer. He couldn't refuse: Celera was bringing dynamite to a cave that Millennium and everybody else were scratching at with pickaxes, and he wanted to be there to collect the gold. Business models can be transformed, or ignored. But he would have to work quickly. Intellectual property is like toothpaste. Once you let it out, as Venter was threatening to do, you can't get it back in the tube.

Six months later, after his shower and breakfast, Millman went into his walk-in closet to decide what to wear. He scanned the racks of bowling shirts, Hawaiian patterns, designer shirts with their fractal patterns and other optical designs. He had several that changed color depending on the viewer's angle—now green, now blue, for instance. He liked to wear one of these when he had some tedious video conference scheduled during the day; the shirt confused the camera, transforming his body into an explosion of white light on the screen whenever he moved. Today he had to dress more conservatively. In the afternoon he was to give an overview of Celera's intellectual property opportunities to a contingent of PE lawyers coming down from Connecticut. Bill Sawch, PE's chief counsel, would be there, and perhaps even Tony White. He owned no white shirts but found a low-wattage green one, which he put on under a dark suit, conventional but for the buttons, which he'd replaced with some obsolete computer chips. To complement the outfit, he chose a plain gray tie. One had to look closely to see that it was made of papier-mâché.

The last decision was the choice of socks. He liked his socks to express his mood or make some comment, even if he would be the only one to hear it. He had Matisse socks, Miró socks, and pairs from Picasso's blue, rose, and cubist periods. He had Betty Boop socks, and a pair showing a drowning man surrounded by sharks. He decided on his "Last Supper" socks. Eleven PE lawyers had signed up to attend his presentation. If White came, that would make twelve.

The first patent in the United States was awarded to Samuel Hopkins in 1790, for a new method he had devised for processing potash. The same four basic guidelines that Hopkins had to fulfill still govern inventors today. The invention must first be original. It cannot have been published before, or be too much like some previous invention. Second, it must be "nonobvious." You cannot get a patent by wrapping a rock in cloth and calling it a no-scuff doorstop. Third, the invention must have a demonstrable function. If you mix silicone with boric oxide and come up with an exceptionally bouncy rubber, you won't necessarily get a patent. Demonstrate its value as a toy, however, and you can call it Silly Putty and make a fortune. The final guideline requires "enablement": The invention must be described clearly enough in writing so that any skilled practitioner in the same trade can read it and fashion the invention himself. A patent does *not* confer ownership. It is simply a contract between the inventor and the government, whereby the former agrees to make public his invention in exchange for legal protection against others making or using it for commercial purposes for the next twenty years. The other way to guarantee commercial exclusivity over an invention is to keep it a trade secret, not allowing anyone access to the design of the invention. The patent system was invented specifically to push originality out into the open, where it can be used to produce more originality.

Considering that these guidelines were thought up two hundred years ago, when no one could imagine inventions derived from life, they have proved remarkably accommodating to bioengineered novelty. Under the first guideline, an organism or any part of one in its natural living state is clearly unpatentable, since it does not originate with the inventor. But in 1972, Herbert Boyer and Stanley Cohen won a patent on the *process* they had invented to manufacture human insulin by cloning its gene. That same year, microbiologist Ananda Chakrabarty applied for a patent on a microbe he had constructed that could degrade crude oil. The utility of his invention was obvious, as was its nonobviousness, and Chakrabarty had no problem writing down the recipe so that any other skilled biologist could produce the microbe. But the patent examiner disallowed the application, arguing that microorganisms are products of nature and therefore not original. Chakrabarty appealed the decision, and by 1980 the case had found its way to the U.S. Supreme Court. In a landmark decision, the court ruled in his favor, on the ground that "anything under the sun that is made by the hand of man" is patentable subject matter, including the specialized life-form he had engineered into being. Mother Nature may have supplied the ingredients, but Chakrabarty baked the cake.

The Supreme Court's ruling provided a conduit for the patenting of other living inventions, but not everything fit through. An oyster with an extra set of chromosomes that rendered it edible even in months ending in "r" was denied patent protection, on the ground that the invention was too obvious. The first patent on a complex living organism was awarded instead to the so-called Harvard mouse, in 1988, which the hand of man had transformed from just another rodent into a transgenic implement for the study of cancer. By this time, there was ample precedent for patenting proteins and genes yanked from their natural environment. In 1987, for instance, Amgen had been granted a patent on the isolated gene for erythropoietin, a hormone found in certain kidney cells that is essential for the formation of red blood cells. The protein had previously been extracted in minute quantities from human urine, but in nowhere near the quantity needed to mass-produce it as a drug. With the isolated gene in hand, Amgen could clone and manufacture erythropoietin from scratch, relieving thousands of patients in renal failure from the constant blood transfusions associated with dialysis. Epogen, the commercial form of erythropoietin, earned the company billions, saved a lot of suffering, and became the paradigm of the successful engineered drug. Biotech fashions came and went, but stirring the dreams of every new scientist-turned-entrepreneur was the hope that his or her discovery would lead to "the next Epo."

Amgen's right to patent protection for Epogen was practically indisputable: its utility was obvious, and the hand of man had moved the discovery from the initial characterization of the molecule, to the sequencing of its gene, to the cloning of the therapeutic protein. But a gene discovery did not have to result in a therapeutic protein to be considered functional. If a useful diagnostic test could be made from it, the gene was patentable. Not even Francis Collins, who co-owned the patent on the cystic fibrosis gene, could cry foul if Celera were to patent a gene with an equally robust demonstration of its biological function and medical use. Likewise, if a gene's protein product could be used as a target for the molecular action of a potential drug—something against which a drug company could test its candidate molecules to see which ones had the desirable effect—well, that clearly had function, too.

On the other end of the functional scale was raw genomic sequence—mere strings of DNA letters, with no knowledge of what they meant—which nobody at all thought worthy of patent protection. Between these two shores, however, stretched a deep, murky lake of uncertainty—or possibility, depending on your point of view. How much do you have to know about a gene before you can say you've discovered it? Is it enough that you know, say, that its protein transports ions across cell membranes for you to claim you know its utility? Or do you have to know *which* ion, in *which* cells, and why? What about *parts* of a gene? In October 1998, the first patent on an EST (expressed sequence tag) was issued—to Incyte. Most experts, however, including those in the Patent Office, were still uncertain over how well the company's claims for the function of the gene fragment would stand up in court.

Until the government could find its own bearings in the murky lake, Robert Millman was going to be aggressive in staking out Celera's territory. He was hardly the only gene patent attorney out there, and he knew that outmaneuvering his competition required breathless footwork, high invention, even a little magic. But he saw nothing sinister about it at all. Protecting the intellectual property his company discovered was simply his function, and he intended to fulfill it by doing whatever it took. If Venter had wanted a pet, he shouldn't have hired a Doberman.

"What we do first," Millman was telling the PE lawyers, "is go for the low-hanging fruit." He whisked off his dark suit jacket and let it fall on the chair beside him. He loved giving presentations. It made him feel like he was back onstage. The tables in the Atlantic Room had been arranged into a big square, the lights dimmed. Hook-nosed, tall and thin, Millman stood on one side, the light from his laptop casting crepuscular shadows in his red-gold beard. Even in his business attire he looked like a wizard in young middle age. With Peter Barrett, Sam Broder, and some other Celera people joining the PE attorneys, there were perhaps two dozen people in the audience. The lawyers were a mix of types: a silver-haired doyen, a freckled twenty-something with a red sponge of hair on his head, a tall woman whom Millman greeted with a courteous bow when she entered, as if he knew of her wisdom from one of his earlier incarnations. Bill Sawch sat directly opposite him, looking relaxed and alert in a way that suggested he was invariably relaxed and alert.

"Celera's business strategy," Millman continued, "is based on the fact that we have the largest gene pipeline on earth." His voice had hoarse edges, as if it had fought its way up from somewhere deep. "The first product of the company is the database itself, which we sell to subscribers. But our second core asset is internal gene discovery, beginning with the stuff within easy reach. The low-hanging fruit."

"What exactly do you mean by 'discovery'?" asked the silver-haired lawyer.

"Discovery," Millman replied from the semidarkness, as if the word were an ancient riddle that answered itself. But the man pressed.

"Like, discovery of a gene, or something farther down?"

"Discovery."

There were no more interruptions. Human Genome Sciences, Millman pointed out, already claimed to have patents filed on 90 percent of the genes in the human genome, but Incyte was making the same claim, and the truth was that nobody really knew where the competition stood. Keeping in mind, too, that the government's view of gene patents was still evolving, the key was for Celera to be proactive, to grab as much potential intellectual property as possible and sort out later who really owns what. Celera was getting a late start, but it was in a good position. Every Wednesday at 4:00 a.m., the company's supercomputer belched out the previous week's accumulation of data. Once the pipeline was fully operational, more sequenced DNA would be flooding into its gene discovery group every week than other companies would see in months. In a perfect world, this would be breakfast for one. Unfortunately, there was a hitch: under the terms of its early-access agreements with Amgen, Novartis, and Pharmacia Upjohn, Celera was obligated by contract to release the data simultaneously to those companies as well. "What this means," Millman told the group, "is that if we were to miss a receptor gene and one of our subscribers gets a patent on it, PE Biosystems will have to get a license from them to develop a diagnostic test from that gene." The point hit home. Our fruit, in somebody else's basket. Toothpaste out of the tube.

Fortunately, there was a way to get some of it back in. If Celera simply spat out raw, untouched DNA to its three customers, Millman explained, then anything of value that those companies found in it was theirs to keep. But what if Celera did a little tidying up of the data first—just a few quick pats with the hand of man? The gene discovery group was not allowed a single day to see the data before its partners. But there were some routine procedures one might apply to the data even before it was sent to gene discovery—quick automated searches for "genelike" patterns in the fresh batch of code. It was common practice, for instance, to throw new information in bulk up against public databases of known genes, looking for similar sequences, or "homologs." Function tended to follow form: if a new DNA sequence closely resembled that of a known gene, then the two likely performed similar functions in the cell. DNA sequences are the blueprint for amino acid sequences, which in turn determine the way a given protein folds into shape. If there were codes in the new gene sequences for compelling shapes in a protein—for instance, an indentation that matched, like a lock to a key, some protrusion in another protein known to be involved in disease—that information might also strengthen a patent application.

"Is this enough to get a patent?" Millman asked. The lawyers awaited the answer. "No. But I am showing the Patent Office that through a ten-million-dollar-a-month sequencing effort, I'm getting something of value."

Especially valuable were the genes that might code for so-called G-protein coupled receptors, or GPCRs for short. These are proteins embedded in a cell membrane, like cloves in a ham, that serve as docking stations for some other protein carrying a message to the cell from outside. When the messenger docks on the part of the GPCR protruding on the surface of the cell, that event causes a change in the GPCR's shape at its other end. That transformation in turn allows a G-protein inside the cell to grab hold of the message and pass it on, via other proteins, to the nucleus, where it becomes a directive to the genes waiting inside. The message might be "Make more of this hormone!" Or it could just as well be "Hold up, we've got enough now!" Since the difference between health and disease depends on maintaining a normal balance in the body's biochemistry, all of the proteins involved in such a cascade of interactions are vital. But GPCRs, stuck in the cell membrane, are a fixed target, which is why the majority of today's therapeutic drugs hone in on them. Prozac is one. So is Claritin.

The new GPCR genes spilling out of Celera's pipeline weren't all going to lead to discoveries with the value of those drugs. But a tiny handful might, and since no one knew which ones they were, why not tuck a little Celera placeholder on every sequence that resembles a GPCR, just in case? For a mere $150 apiece, Millman said, he could file a "provisional" patent application covering hundreds of potential genes. The provisionals would give Celera breathing space, a year to decide which ones merited further development and protection with a regular patent application. Some people might justify a quick grab like this as a "defensive patent"—merely a safeguard to keep somebody else from doing it first. Millman disdained the term; in his view, every patent should be exploited to its full potential. Even if Celera never did the further work needed to discover the gene's specific function, claiming it as a homolog might prove useful later on. If some other company were to discover the gene's specific biological utility, hit them with your previous claim and see if you can extract a royalty from them.

Of course, this strategy depended on getting a provisional application into the Patent Office before Celera's customers could do the same thing. To that end, every Wednesday morning Millman had been coming in early to work. So far, all the data were *Drosophila* sequences, and, with the sequencing operation still gagging on technical glitches, precious few of them at that. But even a fruit fly homolog to a human GPCR gene contained a morsel of potential value, not to mention anything passing by that might look interesting to the insecticide industry. First he made sure that his instructions to tidy up the new release of data had been carried out—that the sequences suggesting possible GPCRs, ion channels, secreted proteins, and other gene families had been noted and catalogued before being sent to the customers' servers. Then he downloaded the release into his own department's computer. There he worked a little automated alchemy to add some boilerplate, change a few keywords, and otherwise transform the information into the exact format of a provisional patent application, right down to the stringent requirements for type font, tab settings, and margin spacing. Then he clicked Print. When it was finished, Millman loaded the box of paper into his Subaru and rushed it down to the Patent Office in Crystal City, Virginia. If he left by 11:20 p.m., he could still get it stamped in before the office closed at midnight, locking in Celera's claim while its customer-competitors were still scrolling through the data.

So much for the low-hanging fruit. The next step, Millman told the assembled lawyers, should be to climb up the tree. In the dim light, he seemed to grow a little taller as he spoke. Before starting Celera, PE Corporation had acquired GenScope, a small California genomics company. "Celera West," as it was now called, had the expertise and facilities needed to create full-length cDNAs—laboratory reconstitutions of actual genes—from the mere strings of letters in Celera's database. This was the first step in validating the real biological function of a gene, and when it was completed Millman would file a second round of provisional patent applications on any promising leads. After that came other steps, each one getting closer to defining a gene as a true drug target, winnowing out more chaff and leaving only kernels of increasing value. Let's say you start with a thousand GPCR-like gene sequences in your database. Most will prove worthless. You develop half of them into cDNAs and find that out of those, say, a hundred are expressed in human heart cells. You still don't know their function, but now the drug companies are paying closer attention. Among those hundred, maybe you find ten that are expressed only in the tissue of patients with congestive heart failure. Any big pharmaceutical company in the world would pay millions for the right to use those genes as drug targets, and you've got intellectual property on them. You've reached the high-hanging fruit.

"This is what God created giraffes for," Millman said, and for a moment his head seemed to arc up high on his long hairy neck. Getting to this point would of course take time, money, and lots of work. In the meantime there were plenty of other browsing opportunities. In the process of gearing up to assemble the genome, Celera and PE Biosystems were leaving behind a trail of collateral intellectual property. Gene Myers, for instance, had written a computer program that compressed thirtyfold the deluge of raw data coming off the sequencing machines. This would save the company millions of dollars in computer storage space; either keep the program a secret or protect it with a patent. In the DNA prep room upstairs was a carousel-like device that someone at ABI had designed to stir samples by using the action of magnets on tiny beads placed in each little vial. Visitors shouldn't even *see* the thing before its IP was safe. And what about Myers's assembly program itself? If a computer program can be invented that puts together the 3 billion letters of the human genome, it can do the same for mouse, rat, cow, rice, corn, and dozens of other species with commercial value. Don't let it out the door.

Then, of course, there were SNPs. The SNP Consortium had announced its intention to defensively patent the human genetic variations it found before placing them in the public domain. "So, we have to patent them first," said Millman. "This presents an incredible reality. How do you patent three million variations?" Finally, there was the human genome itself. Obviously, it could not be patented in its natural state. It was Mother Nature's "prior art," in the language of the Patent Office, and, unlike a single gene, could not be embalmed in some artificial molecule. Even if Myers's work was an unqualified success, the genome would still exist only as a bolus of organized information on a computer disk. Databases of information could not be patented, and while they might be copyrighted in Europe, they could not be in the United States. But there was time to think of some other approach. Perhaps the system of *using* the assembled genome could be "physically embodied" in some way. "There's all kinds of tricks and games I can play," Millman concluded, flicking on the lights. "Any questions?"

At first, silence. The lawyers were taking the measure of suspiciously good news, as if they'd just been told that chocolate ice cream is, after all, very good for the heart. Even Bill Sawch needed a moment to gather himself. "I wonder," he said to the others around the table, "if Robert had made this presentation to outsiders—like say, NIH—what would you expect their reaction to be?" Some people laughed, breaking the spell.

"Whatever we do, Collins will say we're the bane of the world," Sam Broder said.

"So, we get a patent on annoying the NIH," said Millman.

"We could patent Craig," said Peter Barrett.

"Seriously," said the young attorney with the spongy red hair, "we need to get across that we're not building an evil empire. This is for the benefit of mankind. Sure, we'll make a little money along the way, but this is not to gouge the public. That's for the drug companies to do."

"But does everybody agree with the general approach?" asked Millman. "A broad patenting strategy, with steps taken to knock out patenting by our subscribers. OK?"

"As long as you couple it with a policy that doesn't outrage everybody," said the red-haired lawyer.

"Nobody expects Incyte and Human Genome Sciences to have a conscience," said the tall woman. "They were commercial operations from the get-go. What's different here?"

"What *about* Incyte and HGS, Robert?" asked Sawch. "They've got tons of patents already in the works. Do you foresee a World War III on gene patents?"

"Definitely," Millman answered. "They'll be waiting for us with everything they've got."

"So the more artillery *we* have, the better off we'll be, right?" the silver-haired one inquired.

"That seems to make better sense than giving the artillery away to the competition so they can blow our heads off," said Millman.

The meeting adjourned. Millman headed back up to his isolated office on the second floor. He felt encouraged by the impression he had made. Sawch called him back a few days later. He complimented Millman on his insights into where Celera should be going, and offered a suggestion. If Venter was against his "broad patenting strategy," Millman might consider talking to some people higher up, if the opportunity arose. Tony White would be interested, and so would the PE board of directors.

The next day Millman wore his brightest, happiest socks, the ones with incandescent lightbulbs bulging out over his anklebones. They were hidden beneath his olive-green corduroys, but he knew they were there. "Nobody here understands the land grab like me," he said. "I'm going to be Francis Collins's worst nightmare."

CHAPTER 18
EVIL BOY

Around the same time, Collins and the other Human Genome Project scientists were gathering for their annual meeting up at Cold Spring Harbor Laboratory. The mood was very different from the desolate affair of the previous year. The decision to outrun Celera had galvanized the academics into a cheerful, righteous militancy. The weather was brilliant. Francis Collins and Eric Lander, the architects of the plan, strode about the sunny flagstone terrace like officers on the deck of a British man-of-war. Venter's name was on everyone's lips again, but now he was perceived as more of a cartoonish malevolence than a real threat, as a man more to be ridiculed than feared. In the shade, the elderly, near-translucent James Watson held court in a bright yellow sweater, white trousers, and floppy-brimmed tennis hat, lobbing sardonic gibes about his nemesis into the air over his smirking listeners' heads. During one lecture the speaker used his PowerPoint zoom button to turn press photos of Venter into comically exaggerated portraits of menace. The audience howled.

The *New York Times* had again published a provocative feature by Nicholas Wade on the "maverick implementer" out to beat the government, just a day before the meeting commenced. This year no one at Cold Spring Harbor was cowed. "Craig should not be portrayed as a maverick," said Bruce Roe, the sequencing specialist at the University of Oklahoma. "He should be portrayed as an opportunistic maniac and a leech."

Having consolidated most of the public program's power and money into the G-5, Collins used the occasion to rally the spirits of the smaller American labs and international partners. They did not require a lot of rallying. In a closed-door session, the 15 percent of the sequencing not already allotted to the G-5 was divvied up among eleven smaller sequencing centers in the United States, France, Germany, and Japan. The "G-16," as Collins dubbed this broader coalition, emerged with a press statement reaffirming its commitment to the map-first, sequence-later approach and to a "vigorous spirit of international cooperation," deploring "the trend towards treating human genome sequence as a commodity." Two days later, Collins stood before the full assembly of scientists, postdocs, grad students, and technicians packed into the main auditorium. Latecomers sat in the aisles and stood three deep along the walls. Somebody had found a chair for Watson and placed it on the far end of the stage, where he sat still, profile to the audience, as if he were a tribal icon on display. He was still wearing his tennis hat.

"I hope this doesn't sound corny or grandiose," said Collins, "but I feel this is an historic moment. This is the most important scientific effort that humankind has ever mounted." He repeated the sentence, drawing out each word. "This is far bigger than going to the moon. It will change biology for all time." The hall was dead quiet. Nobody questioned why that particular moment was historic, and if Collins's doughty diction made his words sound even more grandiose, they were that much more effective in conveying their sentiments. Beneath his thick glasses and home-cut hair, he seemed not just the leader of the program but an unapologetic embodiment of its ideals. As for those who would treat the human genome as a commodity, Collins summed up their values in a single image flashed on the screen behind him: a giant spiral of DNA, made entirely of money.

No one needed to raise a hand to ask who these deplorable commodifiers were. Apparently they did not include Randy Scott of Incyte, whose company had been eagerly and openly privatizing human genomic data for years. Scott was right there in the audience, waiting to give his own invited talk in a later session on gene patenting. The jug-eared, affable, always deferential Incyte leader had never drawn much indignation from the public program scientists. He was a businessman doing business, who knew to stay on his side of the wall and away from theirs.

Back in Rockville, the news that he was being hanged in effigy up on Long Island hardly surprised Venter. "It's not that I'm going to break my promise to release the genome that frightens Francis," he said, his voice full of contempt. "What he's *really* afraid of is that I'll keep it."

Venter nevertheless seemed hurt by the reports of the ridicule he was receiving, as if he still harbored some yearning to be accepted by the same community he was publicly disparaging. He had to admit, too, that in spite of his own openness to the media, Collins was winning the public relations battle. Some reporters loved Venter's "bad boy of science" image a little too much, and others clearly did not love it, or him, at all. He could dismiss the opinions of his enemies as jealous hypocrisy, but articles and editorials in the United States, Europe, and Japan demonstrated a much wider distrust of his intentions. In the Netherlands, somebody had written an avant-garde play about the history of civilization, bringing in Venter near the end in the role of the devil. A part of him delighted in the reach of his notoriety, but another part seemed mystified and stung by the intensity of the enmity directed at him, as if it were all some awful global misunderstanding. "I don't mind being *Bad* Boy," he said to his wife over lunch one day late in May. They were sitting on the patio outside the Celera cafeteria, Venter with his eyes cast down to avoid the sun. "I just don't want to be *Evil* Boy."

There were more pressing worries. The sequencing of fruit fly DNA had begun on April 8, around the time it was supposed to have finished. To begin work on the human genome in time to compete with the Human Genome Project's spring 2000 deadline, the sequencing part of the *Drosophila* project, if not the assembly, had to be completed by the end of July at the latest. Celera was now halfway through that allotted time frame but only 3 percent of the needed fly DNA had passed through the machines. New machines arrived from ABI every week, but there was not enough time to install them, because the ones already online were constantly breaking down. Some, in fact, had never worked—"dead since arrival," Marshall Peterson called them. In business development, meanwhile, Peter Barrett hadn't managed to sign up a single new customer. With the public program's "working draft" on the horizon, the drug companies were more reluctant than ever to spend millions committing themselves to a Celera database subscription when they might be able to get serviceable data for free just by waiting a few months.

Celera still had the hypothetical advantage over the public program of a grander database that would eventually include the all-important mouse genome—and the promise of more sophisticated, user-friendly software tools to mine the data for biomedical value. But until the machines began to churn out sequence, there was no database to speak of. The software teams were floundering. They didn't have time to design the programs needed to attract new customers and simultaneously satisfy the demands of the current clients. Anne Deslattes Mays, whom Venter had brought with him from TIGR to head the software teams, desperately needed more skilled people, but there simply weren't enough computational biologists on the planet to hire.

Then one day, Deslattes Mays's whole world collapsed. She got a phone call from a hospital in California. Her husband Randy, hiking alone in Yosemite Park, had suffered a stroke. Another hiker had found him lying by the side of the trail the next morning. He was in critical condition. Tony White sent the PE jet to pick him up, and Sam Broder used his medical contacts to arrange for Randy Mays's care. But he had suffered serious damage to the language centers of his brain, and even with constant care he might never recover. Venter gave Deslattes Mays leave for as long as she needed. The people reporting to her directly, who had never been quite sure of their roles, were now left to fend for themselves.

"This isn't a company," said Gene Myers, who was relying on Deslattes Mays's teams to provide the clean, preprocessed data he needed to make the assembly work. "It's a train wreck."

Myers had bonded tightly with his own group but had become irritable and impatient with people outside it. He was still struggling with the idea of incorporating the public program's data into his human genome assembly. He reluctantly admitted that it would save time and money—but only if its quality met the standard of his algorithm. Since he often referred to the sequences downloaded from GenBank as "crap," "piss-poor," and "utter garbage," it did not seem as if he thought it would. He was far from sure that the quality of Celera's *own* data would be up to the task. Early test runs using Celera's fruit fly DNA augured a total disaster. Myers brooded, barked at his colleagues, and muttered about quitting. People understood the pressure he was under and tried not to resent his complaining.

There was one encouraging sign. A handful of universities were showing some interest in taking out a database subscription at Celera's academic rate, a few thousand dollars a year, rather than the few million charged to drug companies and biotechs. The income would be trivial initially, but if one institution could be induced to sign up it might escalate to a flood of academic customers. A team from Vanderbilt University in Tennessee seemed the most likely to break the ice. Their representatives came up for a visit in early June. Venter, Barrett, and Gilman had just sat down with them when the lights went out and an alarm started to sound in the hallway. Venter went out to see what was going on. In the corridor, water poured from the ceiling. A pipe had burst above the lobby, which was itself directly above the main electrical switch room. To prevent a fire, facilities manager Bob Thompson had ordered the power to the building cut off. Red alarm lights were flashing and men in hard hats were running around. Venter grabbed Thompson as he hurried by with a bullhorn. "What's the damage?" Venter asked him.

"It looks like we lost the lobby," said Thompson. He was wet, and had the glazed look of someone besieged by bad luck.

"What about the data?"

"No problem. The data center wasn't affected."

Venter turned to his prospective clients, who were trying to find a dry place to put their feet. "This is great!" he said. "You're getting a real-time demonstration of how well our backup systems work."

Vanderbilt decided to wait awhile before signing up. But there wasn't much point in having customers anyway, unless you had a product to sell them. Michael Hunkapiller had trained and sent a dozen technicians to Rockville. Along with Mark Adams's staff, they were doing everything they could to get the sequencers up and running. There was no consistent pattern to the failures. On some machines, the laser beam wandered out of its path, causing the instrument to automatically shut down. But what was to blame—the laser, the software running it, or vibrations from the drilling and hammering still going on on the floor above? For a week, a milky wash appeared across the neat columns of red, blue, yellow, and green base pairs tripping down the monitors attached to sequencers, rendering the DNA unreadable by the computers. Then it disappeared as mysteriously as it had come. Tubes clogged, filters blew. Adams kept his tension lidded, parceling out just the amount of attention to each problem needed to keep it from exploding into crises. But just behind his right ear, a dime-sized patch of gray had appeared in his dark hair, as if he were corralling all his stress into one out-of-the-way spot. He was driving the others to work eighty-hour weeks until the problems were solved. Since the public program's announcement, his own people were even more willing to put in the hours, but the ABI technicians felt like conscripts in somebody else's war. After all, ABI was supplying machines and service personnel to the Human Genome Project centers as well. They worked side by side with the Celera biochemists, but in the lunchroom they kept to themselves.

Late one Tuesday afternoon, Venter called his senior staff down to the Atlantic Room to take stock of the situation. He usually gathered energy as the day went on and it drained from those around him, but today even he seemed weary. "Another shitty day in paradise," he said as he walked in. Shadow, his favorite poodle, followed him and settled with a sigh under his chair. They had a guest at the meeting. Marshall Peterson, whose military and corporate background made him uncomfortable with the casual way Venter ran his meetings, had invited a time-management expert to join them: a chipper blond woman in high heels, pearls, and a blue-and-white polka-dot skirt. She was there to tell them how to "get more out of a meeting." Contrasted with her audience—Myers, hunched over his laptop with half his face buried in his scarf; Ham Smith slouching almost to horizontal in his chair—she looked like an emissary from some better organized, perkier planet. From a box, she drew and handed out laminated cards listing "the twelve tips to a more effective meeting." The scientists passed the cards around like dutiful schoolchildren. Tip number one was "Have an Agenda."

"No problem there," said Venter. "We all have agendas here. We just don't share them with each other." He scanned down the list, holding the card at arm's length so he could read it without his glasses. Some of the other tips included "Allow Small Talk at the Beginning," "Play by the Rules," "Watch Body Language," and "Utilize Technology and Room Design."

"Sounds more like a dating guide," he said.

The time expert gave him a big smile. "Good for you!" she said. "It's *good* to have fun in meetings. See? That's tip number twelve." After reviewing all the other tips, she demonstrated tip twelve by reaching into her box and distributing some yo-yos and other party favors. "Enjoy!" she said. The toys got passed around the room. Robert Millman dourly handed each one one in turn, showing no interest. Sam Broder played with a yo-yo, too tentatively to get it to rise. Mark Adams was examining a soft plastic baseball decorated with a happy face. Somebody put a rubber dinosaur on Myers's laptop. He frowned, passed it on, and kept on typing. Venter was fingering a squishy sphere about the size of an orange, filled with some glowing gelatinous substance. "This feels like a breast implant," he said, and tossed it across the room to Peterson, who gave it a squeeze, too hard. It burst and the Day-Glo goop inside trickled out between his fingers.

"No wonder you can't get a date," Broder said. Peterson blushed and wiped his hands with a handkerchief. Then Adams bounced his happy baseball on the table, tripping a digital voice mechanism. *"Holy Cow!"* it said. *"What a clout!"* Under Venter's chair, Shadow perked up his ears and got up to investigate. When Adams put the ball down, the dog nosed it off the table. *"Strike three!"* it cried as it hit the floor. Shadow took it tentatively in his mouth and carried it back to his place. The consultant packed up her things and left hurriedly, as if she were afraid of catching an infection. The meeting got down to business, and without an agenda.

"Mark, how many sequencers are there here as of today?"

"About ninety," Adams said. "We'll be running fifty tonight."

Shadow gave the smiley baseball a nip. *"Going . . . going . . . GONE!!"* it bellowed.

"Why only run fifty, if we've got ninety?" asked Venter.

"There's some weird stuff going on," Adams replied. The Prism 3700s were designed to complete four sets of DNA samples before needing to be reloaded by a technician. During the final run, Adams explained, the yellow signal for the DNA letter "G" was fading away. He suspected that as time went on, formamide, the chemical used to keep the DNA suspended in solution, was interfering with the bonding of the G nucleotides to their yellow fluorescent dye. Theoretically, the DNA could be just as efficiently suspended in water as in formamide. But Adams had tried water, and it didn't work either. "Go figure," he said. "Maybe these machines need Evian."

"What is Hunkapiller doing to help with these problems?" Hamilton Smith asked.

"Nothing," Venter replied. "He blames us. First Mike says we don't have the machines level. When we convince him we do, he says we need to bolt the feet down. He thinks we should fix the environment here before asking for any more instruments. He's looking for ghosts in the alley."

"Technology this sophisticated is bound to have bugs," said Smith. "We just need a little more time."

"There *is* no more time," Venter replied. "Not with the federal government trying to put us out of business. If thirty-five percent of the instruments are down on any one day, then PE is just going to make up the difference." He turned to Bob Thompson. "Where are we storing the nonfunctioning ones to return to ABI?"

"I found some space off the cafeteria serving area," Thompson answered.

"We shouldn't have machines sitting in the cafeteria," said Venter. "When is the third-floor sequencing room going to be ready?"

"June first, at the latest," said Thompson. "The last major project is the installation of the air handlers on the roof. We took delivery yesterday. For some reason, nothing went wrong."

"That's great. So it'll be ready only six months after it was supposed to be."

Thompson didn't reply. Under Venter's chair, Shadow bit down hard on the baseball. *"Yrrrrrr OUTTA there!"* it said.

"I wish I was outta here," said Venter.

CHAPTER 19
CHESS GAMES

A couple of days later, Mark Adams's team worked out a fix for the fading yellow Gs. The mystery was traced to the high evaporation rate in the solutions containing the DNA, caused by the intense heat generated by the power-hungry sequencers. Adams reduced the evaporation by putting little squares of tinfoil over the trays containing the samples. This required reprogramming the movement of the machine's robot arm to prick the foil before it transferred the DNA from the trays into the capillary tubes. When that operation turned out to jostle the trays out of position, the sequencing team took to fastening them down with a bit of Scotch tape. Fixing a $300,000 machine with tinfoil and Scotch tape might not be the most elegant solution, but at least it solved the problem.

In Craig Venter's view, the biggest threat to Celera that spring was neither the truculent machines nor the heightened threat from the government program. It was Tony White. Since the Human Genome Project's announcement that it would issue a rough draft of the genome by spring 2000, the PE chief was taking a more intimate interest in what was going on down in Rockville. This should hardly have come as a surprise, given that White had bet his own company on a gambit that was starting to unravel. But Venter considered the intrusions dangerously meddlesome. He thought of himself as engaged in a high-stakes game of chess with the public program, demanding an artful manipulation of scientific strategy, internal politics, and public opinion. The last thing he wanted was to have White telling him which chess pieces he was allowed to move or, worse, trying to move them around on his own.

Venter also suspected that White was wavering on the open-research business model. He was telling Venter that he should "break bread" with Randy Scott over a possible merger with Incyte, a company that owned more gene patents than any other company in the world and released its data only to the big drug companies with the budgets to pay for it. It wouldn't take much pushing at all, Venter thought, for White to reverse the business plan and deliver the genome only to those who could pay for it. "We're talking here about a guy who sold catheters and surgical gloves for a living," he said, referring to White's career at Baxter International. "He wears his capitalism on his sleeve. I'm not going to let this historic opportunity get away because some short fat redneck needs to prove he's the boss."

White had come a long way from the North Carolina woods, however, and not on his talents as a catheter salesman. He was shrewd, stubborn, and passionate about winning. He also believed resolutely in teamwork and couldn't tolerate people who put their own interests above the company's. Years before, when White was a midlevel manager at Baxter, the company had merged with a firm called American Hospital. A few days after the merger, one of its managers bluntly told him that he'd be damned if he was going to share his operation with White or anybody else from Baxter. "Six weeks later, he was reporting to me," White remembered. "Six weeks and a day later, he was out of a job."

The PE head understood that Craig Venter was a special case, someone who had to be given a lot of room in order to do his job best. If Venter wanted a Nobel Prize, that was fine with White. But he was turned off by Venter's unquenchable need to personalize the venture. He was spending far too much of his time, White thought, feeding his image to the media and lecture halls full of titillated admirers. He was appalled, too, by the hyperbolic provocations that the Celera president lobbed at the rival government program. This wasn't how you did business, and it distracted everyone from the job they needed to get done. He didn't care much for the articles about Venter hanging all over the place, either. After he complained about them, more appeared.

"Tony's just jealous because I'm getting all the attention," Venter said. "The problem is, he's marginalized himself. I knew from the beginning this would happen. He's gone from being the head of an active Fortune 500 corporation to running a holding company, while Hunkapiller and I do the work."

He was entitled to his opinion, but it was not a good idea to share it with the *New York Times.* In April, Nicholas Wade had come down for an interview with Venter, who gave him the grand tour. The newspaper published Wade's story on May 18, under the headline "The Genome's Combative Entrepreneur." White was furious. Venter, according to Wade, had asserted that the public genome effort had chosen "a flawed strategy that will produce a seriously incomplete DNA sequence." "Both [Michael Morgan of the Wellcome Trust] and Collins are putting good money after bad," Venter was quoted as saying.

It got worse for White as he read on.

"Together with a colleague, Michael W. Hunkapiller," Wade wrote, "[Venter] has in effect hijacked a major company, the former Perkin Elmer Corporation."

"There goes the stock price," White thought. As if his voice mail were reading his mind, it began to fill with investors asking what was going on. He got Venter on the line. "I've got major shareholders wanting to know whether I'm in charge of this company, or you are," White shouted. "I've got a positive reputation on Wall Street. I shouldn't have to be reassuring people I'm the guy looking after their money." To drive the point home, he forwarded the investors' voice messages to Venter's phone. He told Venter that the goading of the government program had to stop, once and for all. Dissing the other guy might be standard behavior in academia, but in business it was an embarrassment, and embarrassments cost money.

After the call Venter appeared in the doorway of Paul Gilman's office, wearing an aggrieved look that broke out into his kid-in-trouble grin. "Tony says everything is *my* fault," he said. "He's ordered me to say something nice about my competitors."

In spite of the conflicts with White, Venter was still fairly confident that the CEO would come down on his side in the argument with Hunkapiller over releasing more machines to Celera. Under the circumstances, however, Venter figured it wouldn't hurt to have some additional support. The next meeting of the PE board of directors was scheduled to take place on June 17, and by coincidence at Celera's headquarters. "We have to use this chance to impress the board," Venter said to his senior staff two days before the meeting.

"We're ten percent done with the *Drosophila* genome, which is three percent the size of human," grumbled Myers. "How impressive is that going to be?"

"I didn't say we had to impress them with how far along we are," Venter replied. "I only said we had to impress them."

"So we blow them away with how *bad* it is here," Ham Smith said, grinning. "That's great! They're not going to let Celera sink."

"We can't just show them a worst-case scenario," Venter replied. "This is the only opportunity these people will have to see the state of things—best case, worst case, functioning, near functioning, chaotic, whatever is really here to see."

"What's the chance that Mike will just turn it around and say that nobody else is having these problems, so it must be something at Celera that's wrong?" Marshall Peterson asked.

"One hundred percent," said Venter. "He's already told Tony it's a lot worse here than anywhere else."

"So what *are* we doing differently from other folks?" asked Smith.

"We're looking into that," said Adams. "It could be the matrix file. If it's bad, the cross-talk between dyes causes bad data. Or it could be that after we installed the new version of the software, we should have done a spatial recalibration."

"Maybe we should hire an exorcist," Paul Gilman said. Nobody laughed, but Adams tried to smile. He looked like a Buddha undergoing torture.

"The good news is that just the eighty machines that we have working makes us the largest sequencing facility in the world," said Venter. "All of these problems are going to be over. Then it will just be up to Gene to do the assembly."

"Just give me the data, and I'll do it," said Myers.

Venter turned to Bob Thompson. "Are the returns to ABI still sitting in the cafeteria?" he asked.

"Yeah," said Thompson. "What, you want me to move them into the lobby?"

"Just leave them where they are," Venter said. "But put up a sign so the board understands that these are nonworking machines going back to ABI."

White flew down the next day, along with Hunkapiller, Noubar Afeyan, and Dennis Winger, PE's chief financial officer. White had no intention of letting Venter and Hunkapiller, his two field generals, get into a blame fest in front of his board. He wanted the matter resolved before the members arrived. On the way in, he nearly tripped over some TV cameras and equipment in the hallway outside Venter's office. "Just once," he muttered, "I'd like to come down here and not find a film crew."

The best way to understand the problems with the sequencers was to hear from the people in the trenches. About twenty people gathered in the Atlantic Room. The tables were arranged in a big square, Venter on one side and Hunkapiller on the other, rarely speaking. White was sitting between them, facing the manager of the ABI technicians in the front of the room, who had been asked to review the status of the machines. His name was John Pollard, a dapper man in his thirties with slicked-back hair. He tried to sound nonchalant, but with his boss and the parent company's CEO both in the audience his body betrayed his nervousness, twitching and twisting as if he were ducking invisible dodgeballs. Pollard displayed a colored bar chart showing a daily count of the machines running, those needing repair, and the ones abandoned as inoperable—"out of the pack," he called it. He pointed to the stubby column for May 31, when a full 60 percent of the machines had been down. "That was the day I started drinking heavily," he said.

"I thought I saw you at the bar I was at," said White.

"It looks to me like we're losing instruments faster than you can install them," said Venter.

"Yeah, but the trend is in the right direction," Pollard answered. Hiring more technicians would help, he added, but there was workspace for only five more, if he could find them.

"How are we going to get even a hundred machines installed at this rate?" Venter pressed. "Our customers are screaming at us that they're not getting any product. We have to get more support from ABI."

"Dumping more instruments into the system isn't the answer," said Hunkapiller, breaking his silence. "This isn't a brute-force issue. You've got to figure out why the problems here are an order of magnitude bigger than anywhere else."

"One thing might be the temperature range on the fourth floor," Pollard replied. "It can get up as high as eighty up there."

"*There's* your answer," Hunkapiller said. "These instruments weren't built to perform under those conditions."

Venter threw up his hands. "We keep hearing from Mike that every single problem we have is Celera-specific," said Venter, "but—"

"No, I never said that," Hunkapiller interrupted. "All I said was you have to take into account what's going on around the machines. We're trying to run them in a construction site."

The argument went on until White had heard enough. "This isn't a matter of who screwed up," he said. "We got a problem, and we have to fix it. *All* of us are in very deep doo-doo if we can't get two hundred and thirty functioning machines in place. If we have a chronic twenty percent failure rate, then we need that many more machines on-site. And the idea that we have to hire only five more people isn't acceptable. I don't care what it takes, or who we have to kill. If needed, we buy a company that has the manpower already in place. This isn't business as usual. This is the full-blown number one priority of this company."

After the meeting, Venter had trouble suppressing his delight. "I'm in a phenomenal mood," he confided. "We've won a major battle."

The board of directors arrived the next day. There was no longer any need to shock them with the negative show-and-tell of dead machines, but Bob Thompson had already carried through Venter's instructions, perhaps a little too eagerly. The dysfunctional sequencers, some with their covers off and electronic guts dangling, had been haphazardly shoved into a dark recess in the cafeteria behind some soda vending machines. Thompson's three-foot painted sign designated the area as a holding pen for dead equipment to be returned to ABI. Nobody noticed it until Hunkapiller passed by at the end of the queue. He was in a dark mood to begin with, and his brooding, thick-shouldered, six-foot-two-inch frame was a little intimidating even when he was in a good one. The sight of the sign put him in a full rage. He tore it off the wall and headed to the elevators, apparently in pursuit of some head to wrap it around. Poor Thompson happened to get in the same elevator. "Who's responsible for this!" Hunkapiller barked at him as the door closed. By the time it pinged open again, Thompson was feeling lucky to have escaped alive, much less still employed. A few minutes later he endured a similar onslaught from Tony White, who had flown into a similar rage when informed about the sign. Thompson admitted that he had put it up but refused to say who had ordered him to do so. His job was saved only after Venter fessed up to his role in the matter.

"How dare you diss my people like that!" Hunkapiller shouted at him in his office. "These guys are working eighty-hour weeks. I'm pulling people off their honeymoons to train them. Do you think it helps to be publicly blaming them for something totally unrelated to their abilities? You might as well put them up in stocks."

Venter, who judged when he'd crossed a line by the commotion caused by the passage, understood immediately that the technicians were not the only ones with injured pride. He apologized to his old colleague, and later to the ABI people upstairs. He made a point of inviting the technicians to the next all-hands meeting held a couple of days later. Still, they lingered outside in the hallway until Venter beckoned them in—big, shy men wearing overalls and size twelve shoes, some with earrings and warrior goatees. They shuffled to the back of the room, where the refreshments were, embarrassed by the applause coming from the gathered Celera employees. "You don't look like the kind of guys who have friends," Venter said, "but if you do have any, please ask them to come work for us, too."

Back in Connecticut, Tony White was not in a joking mood. Even if the extra machines he'd ordered could cure Celera's constipated pipeline, the stock price was still hanging sullenly near its low, no new customers were signing up, and he was fed up with Venter's rebelliousness. White respected his prize scientist's talent and had no intention of firing him, at least not yet. But if Venter was to remain a PE employee, he was going to have to start behaving like one. For weeks, White, through Bill Sawch, had been asking for documentation of Venter's various outside speaking fees and travel expenses. Accepting honoraria for such engagements was against PE guidelines. Charging the travel for these engagements to the company added insult to the offense—especially on chartered planes. Venter ignored Sawch's requests for the records. He viewed his lecturing at colleges, business schools, and health conferences as a superb marketing tool. The honoraria were usually modest, and in any case he was not about to quit speaking to the public just because doing so conflicted with some rigid corporate policy up in Connecticut. White responded by putting the screws on harder, demanding all details relating to fees, expense reimbursements, time commitments, and the scope of Venter's various outside involvements. "I consider this to be very important," he wrote, "and I grow increasingly frustrated at your delays in providing a full response."

Venter complied by sending an inch-thick file of speaking invitations he had received since Celera had been formed. But Lynn Holland, who sent the file, forgot to include an honorarium from Yale University for $150, which Sawch apparently knew about from some other source. He and White viewed the lapse suspiciously: perhaps there were others that Venter was hiding. Holland, mortified, offered to resign. The situation grew uglier. Venter was receiving compensation as a Novartis consultant, even though Novartis was a Celera client. Closer to home, literally, was his continuing appointment as TIGR's chief scientific officer, for which he received a $100,000 annual fee. Considering that TIGR and Celera had various collaborations under way and that its director was Venter's wife, it is hardly surprising that the arrangement raised some eyebrows up at PE headquarters.

Venter protested bitterly that his intention to keep his hand in TIGR had been discussed and approved back when he was hired, as had his other outside interests. But he had nothing in writing to prove his case. He felt he was being subjected to an inquisition. After being called before the corporation's conflict-of-interest committee, he was ordered to curtail his speeches, return the reimbursed travel expenses to PE, and dissolve most of his outside business relationships, including his paid appointment at TIGR. White was not asking Venter to abide by rules that did not apply to any other PE employee. If he was pursuing the matter a little too stridently, so what? Maybe Venter would get the message that a PE employee was exactly what he was.

The CEO had other moves to make. A merger with Incyte was still on the table. Roy Whitfield, the firm's CEO, and its president, Randy Scott, were receptive to the deal if satisfactory terms could be reached. By late spring, the negotiations had acquired a solid if secret existence, with the code name Indigo. Venter had originally been in favor of exploring the merger. By mid-July, however, he had come to think of it as a "dumb idea." While he liked and respected Scott, he believed Incyte to be a dying concern whose acquisition would only dilute the value of Celera's stock. But as soon as Venter changed his mind, White, who had initially been cool to the merger himself, had turned in the opposite direction and gave Indigo his blessing. There was, of course, an excellent business rationale for acquiring Celera's only major competitor in the market for genomic information, especially since Incyte had successfully signed up virtually every major drug company to its database, had solid revenues, and owned more intellectual property on genes than any company on earth. Celera still had zero revenue, zero gene patents, and three unsatisfied customers. So White's reversal may have had nothing to do with pressuring Venter. His offhand suggestion that perhaps the two of them and Scott should jointly lead Celera until they sorted out who should remain in charge—well, perhaps that was just another turn of the screw. But he was acting the way any nervous CEO might with a company burdened by a $175 million burn rate and a feeble stock price. He was keeping an open mind, exploring his options.

The same could be said of his actions on July 8. Venter had taken a couple of days' vacation to go sailing off Cape Cod, and in his absence White slipped back down to Rockville in his plane to interview every person who reported directly to the Celera president. White viewed this as sound business practice. A CEO had to reassure himself that there was a consensus behind the leadership of his various divisions, and the best way to do that was to solicit the opinions of the leader's senior associates when he wasn't there. One by one, Mark Adams, Gene Myers, Sam Broder, and the others—even the venerable Hamilton Smith—were called down to White's rarely used office in Venter's suite. White asked how they felt about the way things were going, and what might be improved. No one had anything disturbing to report, until it was Robert Millman's turn. He walked in and closed the door behind him. "You've got to stop the bleed," he told White. "You're giving away all your jewels." All the resources of the company, he said, were being wasted on developing an information business that was doomed to fail. For support, one had to look no further than the online Encyclopaedia Britannica, which survived on advertising alone because nobody wanted to pay for its information. "If an encyclopedia can't get sixty-nine dollars for its information online, do you think grandmothers are going to buy their grandchildren access to the human genome over the Internet?" he asked.

"This isn't about grandmothers," White said. "It's about supplying the health care industry."

"We shouldn't be releasing sequence information to *anyone* until we've got the human genome finished and protected. Not even to subscribers."

"And in the meantime?"

"Maximize your IP. Leverage gene discovery. Patent the hell out of everything you find. If you hold all the jewels, you can dole them out any way you want."

"Have you put this forward to Craig?" White asked. "I assume he's not exactly thrilled with this idea."

Millman shrugged. "I can't even get in to see Craig," he said.

"What, he's too busy?"

"He's too whatever."

When Venter returned, he got a call from White. Without mentioning his conversation with Millman, White said he was having second thoughts about the come-one, come-all access to the human genome. Venter tried to persuade him that the business model was sound and just needed a little time to mature. The sequencing operation was getting back on track, he said, and once the whole-genome shotgun method had been proved on *Drosophila,* drug companies and academics would be clamoring for Celera subscriptions.

White did not sound reassured. He suggested that the matter be brought up for review by the board of directors at their next meeting in August. "We'll know by then whether the whole-genome shotgun method works for *Drosophila,*" he said. "If Myers can't pull that off, we've got to think of another way of making money."

Now Venter would have to justify the very essence of his vision of the company to the PE board, most of whom were nonscientists. He was confident he could handle them. He would handle Millman, too, when he got the chance. It was clear from the kind of patent legalese White had been using on the phone just where he was getting his notions. But the whole grimy wrestling match with White was wearing him down. The combined pressures that White was exerting were accomplishing what neither the Human Genome Project's unexpected resurgence nor months of catastrophes at Celera had achieved: they were crushing Venter's optimism. He looked exhausted, pale, sometimes even frightened. "These bastards want me out, and they're trying every devious trick they can to do it," he said.

To make matters worse, he could not share his troubles with his chief confidant. Rather than talk to his wife when he got home, he sulked in front of the television set until it was time to sleep. Claire had never seen him act that way before, but she knew he did not want to talk about what was going on. It was not a conversation either one of them wished to begin. She did not want to have to say "I told you so" about his leaving TIGR to get in bed with a corporation again. She agreed that PE's tactics in extracting information about the speaking-engagement and travel expenses were excessively mean. But she would not have been able to support an argument that amounted to her husband having privileges that no one else in the company was allowed.

One evening in late July, Venter opened the door of his Mercedes convertible parked outside Celera, preparing to drive home. It had been a particularly wearying day. He sat awkwardly, half sunk in the seat, one leg still on the ground outside the open door. He couldn't seem to bring himself to finish getting in and turning on the ignition. He couldn't remember feeling this depressed since his days in Vietnam. If White was going to force him to renege on his promise to release the genome to the public, he could not stay at the company, but he did not want to quit, either. An image came into his head of Celera—Mark, Ham, Gene, their hopes and his—as a child floating away in a flood while he stood on the bank, unable to save it. He started to cry. The light faded to dusk.

CHAPTER 20

HOW TO ASSEMBLE A FLY

Though Venter was hitting bottom, Celera was showing some signs of recovery. The stock price had been inching upward through July. Tony White claimed that it was because the market had finally forgotten about Nick Wade's comment in the *New York Times* that PE had been hijacked by subordinates. Venter said it meant that the smart money was beginning to appreciate his vision. Neither man really had a clue. But whatever its cause, the positive trend eased a little of the tension between them. Meanwhile, the additional capillary sequencing machines were arriving from ABI by the truckload. With three times the number of technicians now on-site to install them, the bugs were being teased out one by one, and fruit fly DNA was finally churning through the pipeline at a respectable pace. On the afternoon of July 28, 1999, as Celera's employees dug into tubs of ice cream to celebrate the company's first birthday, Mark Adams raised his cone and announced that the one-millionth fragment of fruit fly DNA had just been read by a sequencer. Barring any more disasters, it might still be possible to produce enough reads to attempt an assembly by mid-September. That would be just in time for the annual Genome Sequencing and Analysis Conference in Miami. Then the world could see whether the $300 million gamble on whole-genome shotgun had paid off.

The pressure was now on Gene Myers's assembly team. Myers himself had never thought it to be anywhere else. He hadn't yet reconciled himself to the fact that he would probably have to include public-program data in the human genome assembly. But he pushed that issue out of his mind, focusing all his mental resources and those of his team on sequencing the fruit fly genome instead. "I'm banking everything on *Drosophila,*" he said. "Once it gets done, I've proved it. Human will be an anticlimax."

The assembly team had been moved to the fourth floor of Building II, as far as one could get from the corporate Sturm und Drang on the first floor of the main building. There were no high-heeled secretaries walking the hallways—in fact there were no hallways at all: just a bank of cubicles sprouting at the end of a dusty empty space like an experimental garden on the edge of a vacant lot. Marshall Peterson's information technology team, which was responsible for running the supercomputer, had taken root around a corner, and the shared isolation of the interdependent teams had fostered a convivial synergy. The arrangement also made it easier for Peterson to keep an eye on his friend's volatile emotional state. Peterson believed that the best way to cope with work pressure was preemptively, like a doctor pricking a boil. To channel off some of the stress, he had set up a Ping-Pong table in an empty room between the two groups, and the hard *pock* of the ball was a counterpoint to the staccato of keystrokes issuing from the cubicles. A gash in one wall at racquet level paid witness to the intensity of the competition. Peterson also instituted a Monday morning Nerf-gun war between the two groups, sourcing the gaudy plastic weapons at a nearby Toys "R" Us. The point was to get the week off on a relaxed note by pummeling one another with foam projectiles at as close a range as possible.

Myers enjoyed the release of the weekly Nerf battle, perhaps more than anyone. But he never stayed relaxed for very long. There was a grim wariness in his posture and a tight line to the set of his mouth, as if he were expecting at any moment to be pounced on from an unexpected direction. He was relieved that the sequencing operation was finally getting up to speed. But that only brought on another worry. To make up for the delay, the DNA was being rushed on to the sequencing machines as quickly as possible. In Myers's mind, there were too many fragments being moved through the pipeline too fast, by too many people, with too little control. The closer he got to testing the shotgun theory, the more obsessed he was with the horrifying array of things that could go wrong. Fortunately, his angst did not infect the other ten computer scientists on the assembly team. In fact, it made their own lives a little easier.

"We don't worry very much, because Gene does it all for us," said Granger Sutton. "He covers all the meetings and management, too. The rest of us can just work."

"I'm a perfectionist," Myers admitted. "I need to be this way in order to do the stuff that I do. On the other hand, I'm not always great to be around. As my wife will tell you."

Myers, his wife, and their three cats were living in a rented house in a rural town fifteen miles farther out from the city. It was a big, gloomy place, and they were not spending enough time there to fill it up. M'Liz Robinson was used to her husband's obsessive work habits. But back in Arizona there at least had been time to share some meals and go dancing once or twice a week. Now Myers's work worry followed him home like a mean stray dog and curled up possessively at his feet in his basement office. He seemed to have forgotten his dream of spinning round a ballroom in Washington with his beautiful, then-imaginary wife. Since the move, they had not been dancing once. Lately they were seeing more of each other across a conference table at Celera than across the dining table at home.

"We were at a meeting together and I looked over and thought, *That man doesn't look so good,*" Robinson said one day. "He looks sick. His whole face has changed. He's got this lizard look."

A lot of the pressure on Myers came from within. He was determined to show that the assembly of the fly genome was a mathematical rather than a biological challenge. The biologists had made the sequencing stage possible, of course, and once the assembly had been run they would be needed again in the "finishing stage," using laboratory techniques to plug up the myriad little gaps still remaining. But by then Myers's algorithms should already have completed the lion's share of the work. At least 97 percent of the fly's estimated 120 million base pairs should have been revealed in the correct order, with the size and location of the remaining gaps nailed down. No one else, not even Venter, believed that such a pure and austere approach was workable, let alone necessary. Myers conceded that his boss knew more about the practical realities of the whole-genome shotgun technique than he did; in fact, Myers had had no experience with it at all. But the whole point, as Myers saw it, was to circumvent as much practical reality as possible. Celera was about speed, right? The less biology there was in the equation, the faster and more elegantly it would be solved. In short, he was going to use logic to conquer life. "Theory," he told his boss, "is the art of making the real interesting."

What made the problem interesting is that the DNA giving life to a fly, a human, or any other complex organism is not the product of anything remotely like a logical system, but a conglomeration of the blind whims of billions of years of evolution. There is no imperative in natural selection to construct genomes parsimoniously. All that matters is that the life-form they encode be able to survive and reproduce. The result looks a bit like a closet that has not been cleaned out in hundreds of millions of years. The genome contains not only the vital instructions on how to live and replicate but all the dead messages, evolutionary memories, borrowed implements, extra appliances, and wayward detritus that has settled into every nook and crevice over time—some of it useful, most of it not, with no way of telling one from the other. Things that might have once had function might now just be taking up space, like a manual typewriter gathering dust long after its owner has switched to writing on a computer. Most of the clutter takes the form of repeats, which themselves come in different sizes and patterns, including whole families of little bits of DNA introduced by viruses and larger hunks of copied code that at some point in the past hopped from one chromosome to another, perhaps picking up some small but crucial difference later on. The majority of these repeats are functionless, but some are vital to life, and there is no way to tell one from the other without a great deal of scrutiny. Mother Nature is not a perfectionist. In fact she is pretty much a slob. But she gets the job done.

Myers, who balanced his checkbook every week and kept his own closet perfectly ordered, regarded the fact that such a slovenly arrangement results in something as beautiful as life itself with a mixture of awe and alarm. "In nature, anything that is even slightly viable *exists,*" he said one August morning, while he was taking a break in the assembly team's conference room. "It's an amazing system. But it's not the kind a mathematician would design. Give us two or three axioms, and we're grooving. 'Parallel lines do not intersect at any point.'" He gave a fist pump. "*Yes!*"

Since Venter had named the meeting rooms in the main building after the great seas of the earth, the assembly team had named their distant little space the Tranquillity Room, after one of the seas on the moon. Some linoleum tables had been pushed together to make one big one, and a dozen orange plastic chairs were scattered about at random angles. On the table were some empty Coke cans, and a Krispy Kreme box with one deflated glazed doughnut left. Myers bunched his hands around his coffee mug, as if he were trying to get them warm. He hadn't shaved in days or done much with a comb, and his lucky scarf was getting a little shiny with wear. He might have been mistaken for a homeless person who had wandered into the building and found a warm place to sit for a while.

Opposite him was a whiteboard running the length of the room, displaying a summary of the assembly algorithm as it now stood. The program involved several stages, and each stage had been allotted a column with its name on top. In each column was a list of subcomponents, with a check mark next to those whose programming code had been completed. There were a lot of check marks. The interactions between these subroutines were graphically represented by multicolored circles, triangles, and canister-shaped objects connected by arrows. Most of the remaining space on the board was filled in with messier bursts of symbols and equations, where one or another of the scientists had risen from the table and tried to articulate a concept to the others. Scrawled across the top were the words DO NOT ERASE.

The developing algorithm was the creation of the entire team, but it was Myers who had conceived its basic character, and he had endowed it with some of his own. The Celera Assembler, as the program was called, treated nature's disorder as if it were a wildwood full of snakes, swamps, demons, quicksand, and beckoning maidens who, if you responded to their allure, would lead you away and gobble you up. To find its way through this morass Myers had constructed the algorithm to obey a central guiding principle: *Do not make any mistakes.* The repetitive sequences in a genome were like an infinite chain of forks in a path, and the program had to be able to choose the right turn every time and not overlap one piece of code with an extremely similar one that might actually be miles away in the genome.

Previous assembly programs, such as the one written by Phil Green at the University of Washington, coped with the danger of being fooled by repeats by zooming in on every questionable match. Automated sequencing machines were far from infallible, sometimes producing ambiguous readings of the base pairs along a sequence. Green's program calculated the degree of confidence one could have in the identity of each base pair in two overlapping segments, rejecting matchups that seemed too risky. The strategy worked fine on the 150,000-base-pair segments that the Human Genome Project was putting together one at a time. But it was futile against a problem the size of a whole genome; it would be doomed by the sheer number of repeats in the equation. But Myers was looking through the other end of the lens. Instead of zooming in on every possible point of error, his strategy was to zoom *out,* locating and removing all sources of doubt at the start, and then to proceed from the tiny shred of certainty left behind. Thus among the first stages of the Celera Assembler program were two—on the whiteboard in the Tranquillity Room, they were called "The Screener" and "The Overlapper"—containing subroutines designed to winnow out the repeats. First the easy ones, then the more elusive variety were dumped into a sort of quarantine, leaving behind only those fragments that could absolutely go together in only one way. Better that the program join no sequenced fragments at all than put two together incorrectly.

"Repeats leave *signs,*" Myers explained. "We've built an assembler that uses that information intelligently. It knows repeats. It wants repeats. Then it goes out and hunts them down."

The little bits of certainty left behind from this winnowing process were called unitigs, short for "unique contiguous fragments." A unitig might contain two DNA fragments coming off the sequencers, or maybe a third could be fitted on to one of its ends, or a fourth, or more. At some point, however, the Unitigger stage of the program, which followed the Overlapper stage, would dip into the pile of available fragments and come up with more than one that matched up with the letters on one of a growing unitig's ends. This was a sure indication that the sequence represented by the unitig was entering a part of the genome that was repeated in more than one place. At this point, the program would reject the matchup and put the unitig aside—the way someone doing a jigsaw puzzle might cease work on a particular face in the picture when it extended into hair that might belong to any number of other people in the picture. The critical rule was not to force a unitig to extend past the point of certainty. Even then, however, the unitig would be subject to being broken apart again, as additional reads from the sequencers added to the number of fragments available to test against it for a possible match. Just so, a jigsaw puzzler might assume that two blue eyes go together on the same face—until he reached into his pile of pieces and found a piece with another blue eye on it. At that point, he would have to assume that more than one face in the puzzle had blue eyes and would have to await more information to figure out whose eyes were whose.

The Unitigger was a major step, greatly reducing the complexity of the rest of the puzzle to be solved. Still, it could produce only islands of definite code floating in a sea of uncertainty. Gerry Rubin's lab at Berkeley had already finished a rough map of the *Drosophila* genome that could be consulted, like the picture on a puzzle box, to see where each isolated unitig belonged. But to use Rubin's map as a guide would invalidate the work as a test of the viability of the whole-genome shotgun technique. Instead, Myers planned to use the Berkeley map for confirmation only after the shotgun assembly had been accomplished, like looking at the lid of the jigsaw puzzle box after the last piece is in place, just to be sure you got the whole thing right.

Fortunately, Myers and his team had another weapon that, if all went well, might serve to bridge the gaps between unitigs and arrange them in their proper order. Each fragment of DNA off a sequencing machine is 500 base pairs long. In the conventional sequencing approach used by the government genome program, a 150,000-letter BAC was cloned, then shotgunned into many "subclones," each one providing 500 letters for sequencing off of *one* of its ends. But a key part of Celera's sequencing strategy was to read 500 letters off *both* ends of every fragment. Thus each sequence read was one mate of a pair, like the two plastic ends on a shoelace. While the identity and order of the letters between these "mate pairs" might be unknown, you knew that the two end fragments belonged near each other in the genome and at a precisely known distance apart. In his DNA prep lab, Hamilton Smith had constructed two separate sizes of shoelaces: one library of clones 2,000 base pairs long, and another 10,000 base pairs long. Gerry Rubin's lab would also supply a set of "bootlaces," with 150,000 base pairs between the two ends that were to be sequenced. In the next stage of the algorithm, called the Scaffolder, these three sets of mate pairs acted like tiny bridges linking the unitig islands. If the sequence of one mate pair in a clone uniquely matched a series of letters in one unitig, and its mate on the other end uniquely matched up to another unitig, then it was almost certain that the two unitigs were next to each other on the genome, even if all the letters between them were not yet known. Naturally, Gene Myers was not going to be satisfied with "almost certain." The program demanded that each link be confirmed by at least one additional set of mate pairs yoking the two unitigs, reducing the chance of an error to, as he put it, "a quintillion to one." The unitigs could then be joined, and by using additional sets of mate pairs, other unitigs could be added on to the free ends of the joined sequence, and so on to form a chain called a scaffold. There would still be holes between the unitigs, but at least you would have established the linear order of the information in hand. By analogy, a scaffold of the alphabet might look like this:

ABCD FGHIJ LMN PQRST VW YZ.

In this scaffold, you don't have the missing letters E, K, O, U and X, but you do have the rest arranged in the correct order, and you know the size of the missing pieces: here, one alphabet letter each. (Keep in mind that if you blew this alphabet-size scaffold up to the proportions of the human genome, each letter of the alphabet would represent a collection of millions of DNA letters.)

Further steps would be needed in the computation to further resolve ambiguities caused by repeats. But the mate pairs were the heart of Myers's strategy, creating order in the genome without the need for any kind of a priori map. The only problem was that one had to be absolutely sure that the two elements of a mate pair truly belonged together. Myers could not yet be certain of that—and it was this awful doubt, more than anything else, that was drawing in his cheeks, hunching his shoulders, and narrowing his eyes to puffy slits. In order to produce the mate pairs, each sample of cloned DNA from Smith's lab was first divided into two identical solutions. The twin samples then traveled through a complicated prepping procedure involving several transfers from one plate of test tubes to another. If at any point a laboratory technician accidentally swapped one plate for another, or even rotated a plate back-to-front, you would end up with false mate pairs: two sequenced reads that the assembly program would assume to have come off the two ends of the same clone but in fact could be millions of base pairs apart from each other. It was hard enough battling Mother Nature without having to worry about the fallible nature of human beings. Given the speed at which DNA was flying through the prep process, Myers was tortured by the probability of mistakes. "False mates are like time bombs waiting to blow you up," he explained. "They're *liars.* Worse, there is no way to tell that they are liars."

It was not that Myers doubted that the technicians could get it right 99 percent of the time. It was that an error rate of 1 percent could destroy any chance that his math would work. Instead of relying on human beings, Myers wanted an automated bar-coding system to keep track of the sample trays as they moved from one prep station to another. He had been pressing for the installation of such a system for months, but nothing had been done. That August morning in the Tranquillity Room, he had just received a new analysis, and now his worst fears were realized. The analysis showed that the error rate of false mates was approaching 3 percent. "What does it take to get a bar-coding system in here?" he shouted, throwing up his hands and sending the Krispy Kreme box skipping across the table. "Every grocery store has one! This is a major fucking crisis going down, and Craig wants to pack *Band-Aids* around it. He tells me to stop sweating until I can prove there's a problem. Well, guess what—by then it will be too late to fix." Myers shook his head. "This is complete lunacy. It's denial, squared."

Venter was hardly oblivious to the need for a bar-coding system; no one who spent ten minutes in Myers's proximity could be. But it was too late to get one installed in time for the *Drosophila* assembly, so there was no point in agonizing about it. Somebody would come up with a workaround. In the meantime, Gene Myers would just have to get used to how things worked in the real world.

Venter had other matters to attend to, such as keeping his job. He was making a sincere effort to show Tony White that he could accept authority, or at least a sincere effort to pretend that he could. He was still feeling misused by the corporate brass. "Compared to this, Francis Collins is a bee buzzing around my head," he confided. "These people try to push you to the edge to induce you to fail." But there was no sense in risking the whole dream over a battle he could not win. He agreed to cut down on his speaking engagements and reimburse PE for past travel expenses, and he withdrew from most of his extracurricular appointments, including his post as chief scientific officer at his beloved TIGR. He made a point of consulting with White on important decisions. His apparent new attitude, combined with the easing pressure on the stock, put White in a better frame of mind. It still irked him to see the halls and foyers at Celera plastered with Venterabilia, and he worried every time he opened his newspaper, afraid he was going to see another article on the genome race. But if Venter kept his promises, White was willing to recommend to the PE board that they keep him on.

In Venter's mind, there was still the matter of a possible merger with Incyte to reckon with. In late July, Randy Scott flew out to meet with Venter at Celera on a Saturday, when there would not be many people around. The two men had always respected each other and the visit was cordial. They talked for an hour or two, and Scott left hopeful that a deal could be worked out. Scott's patent-driven company did not fit well with Venter's vision of "open research," but he was too vulnerable in his chess match with Tony White to openly oppose the merger. A few days later, Scott flew east again to Connecticut to discuss terms with White, Venter, Peter Barrett, and other PE executives. The PE side wanted to structure the deal as a marriage of equals between Incyte and Celera. But Scott and his colleagues wanted to frame it as an acquisition by the PE "mothership," an arrangement that would command a higher asking price for Incyte. They set the figure at almost 50 percent above Incyte's trading value, which was then approximately $700 million. The high price was justified, they said, because Incyte's high revenues, secured intellectual property, and other assets were undervalued by the market, while Celera, though better capitalized, had no products or much revenue at all. White was appalled at the asking price and let them know with his usual bluntness. Any hope of a deal vanished. "Tony pulled a Khrushchev on them," Paul Gilman, who was at the meeting, later said. "He more or less told them, 'We will bury you.'"

Venter was not totally relieved. In public he had always underplayed Incyte's head start in commercial genomics by portraying them as having gone off in the wrong direction. Sure, they had more gene patents than anyone else—patents based, he would add, on their exploitation of his own EST technique. But gene patents were made of sand; most would never stand up in court, especially those based on mere gene fragments. Celera's open research model was distinct by virtue of its *freedom* from a reliance on patent licenses to make money on downstream drug development. Nevertheless, Venter feared, if Tony White was going to be drawn into a head-to-head war with Incyte, White would want to have every weapon he could at his disposal, including more protection for Celera's intellectual property. Robert Millman's recommendation that Celera hoard as much of the human code as possible for its own downstream drug development was due to come up before the PE board of directors on August 19. With Incyte pulling from the outside and Millman pushing from within, Venter could see that he was going to have a tough time convincing the board to stick with the plan to give the human sequence away to the world. To those who "didn't get it," offering the genome for free would be like baiting a hook with a tuna in hopes that it would attract enough small fry to make up for the cost of the bait. The nonscientists on the board were especially likely to ask the most dangerous question: Why not just sell the tuna instead?

Venter's pitch to the board would be easier, of course, if his open research model were not being squeezed on the other end by the public Human Genome Project. No new Celera customers had signed up, and it did not take a business genius to see that drug companies and academic institutions alike were waiting to see what the government's genome tasted like first. Ironically, Collins's push to get a working draft of the genome finished before Celera could "privatize the code of human life" was pushing the company further in that direction, by undercutting the value of Celera's version as an information resource and forcing the business minds at PE to try to capture value with patents instead. To keep that from happening, Venter needed to get the PE board thinking way beyond the basic human genome sequence. He had to convince them that the human code was just the top layer of Celera's information lode, which would soon dwarf anything either Incyte or the public program had to offer. What counted was the raw speed with which Celera could put together genomic data of all kinds and lay them out like a smorgasbord for Big Pharma subscribers to plunge into, free of the fear of patent restraints.

Most valuable of all—even more important, perhaps, than the human code—would be the mouse genome. In spite of their obvious morphological and behavioral differences, on the genomic level people and mice are amazingly similar. A gene in one mammalian species is very similar to the same gene in another, performs the same function, and often breaks down with the same tragic results. This is what makes mice such superb lab models for cancer and other human diseases: genetically, they are essentially little hairy human beings that can be manipulated in the lab in ways that people obviously cannot.

Equally important, the mouse genome would be a crucial clue to sorting the wheat from the chaff in human junk DNA. Like the human genome, it is mostly composed of repetitive nonsense. But obscured in that barren landscape are the bits that are neither genes nor junk—the regulatory regions that "instruct the instructions," telling genes when to turn on and off or how much of a particular protein to churn out. The regulatory regions are as vital to life as the genes themselves. For that very reason, they have not changed much through evolution, since change would likely have been fatal to the individual organism, whose genetic legacy would thus not pass down to the next generation. Truly useless DNA, on the other hand, has been free to mutate randomly through millennia without jeopardizing the species' chance of survival. Thus the important regulatory regions of two mammalian genomes will look very much the same, while their junk sections will differ, because of all the mutations that have accumulated since the two species diverged. When one genome is laid on top of the other, the important bits will match up, while the surrounding waste will not. Metaphorically, the mouse genome is like a device that can highlight which spots in a barren desert harbor oil. Every oil company in the world would pay whatever it could for that kind of information. It was the same for any pharma looking for genomic information. And if things went as planned, the mouse genome would be exclusively Celera's for years to come.

"The public program has played right into our hands," Venter told his senior staff. "We can use their draft data to finish our human genome sooner, then move on to sequencing mouse by early next year, way ahead of everybody else. We'll be surfing into the beach while they're still getting their boards out. That will put us in an extremely good financial position for a very long time."

That optimistic picture depended on three factors: the scale of Celera's sequencing operation, the sheer power of its supercomputer, and, finally, the viability of the whole-genome shotgun method. The first two prerequisites were finally coming into focus. For the third, Venter needed a proven success; he needed the fly. The assembly could not possibly be finished by August 19, when the PE board was to meet, even if everything worked perfectly. But the prospect of its completion in time for the Genome Sequencing and Analysis meeting in Miami in mid-September could buy him a little time.

Venter returned from the board meeting in Connecticut with a partial victory. He could keep the written promise he had made in *Science* to release and publish the *Drosophila* genome. But the board members were not willing to commit to the same policy for the human genome until they saw how things fared in the market with *Drosophila.* In the meantime, they agreed that Robert Millman should cease his practice of filing patents on mere computer guesses at genes as they issued from the Celera pipeline. The strategy, designed to get a foot in the door on possible intellectual property, made Celera look as though it was trying to compete with its own customers.

"This is a delicate dance," Venter remarked after the meeting. "I'm not going to push too hard now. They seem to be backing my philosophy, but there are events that could take things out of my control. If *Drosophila* is released and the stock goes down, I'm going to have a hard time making an argument." He shrugged. "Probably it will go the other way."

Millman took the news hard. As usual, he allowed his socks to reflect his mood. He walked into Paul Gilman's office, sat down in a chair, and threw a leg up on Gilman's desk. Underneath his red high-top sneakers he was wearing plain white athletic socks. "If Dante had created a hell for patent attorneys, this would be it," he said in his raspy voice. "Abandon all hope, ye who enter Celera."

"I thought you said Celera was a patent attorney's wet dream," Gilman said.

"It was. But in this dream you roll over and realize that the beautiful young woman beneath you is your mother."

Other than Gene Myers, the eleven members of the Celera assembly team were the sort of temperate, self-contained people who might be expected to choose logic as a livelihood—except on Monday mornings. At a few minutes before nine on August 23, the team gathered up their Nerf weapons, as they did at the beginning of every week, and grouped in the common area outside Myers's cubicle to plan their attack on Marshall Peterson's information technology department around the corner. Myers's people were wearing plastic Viking helmets, which looked somewhat fiercer than the multicolored propeller beanies that Peterson's team had chosen as battle attire but which had the disadvantage of flying off with any sudden movement of the head. Exactly on the hour, Myers put

"The Ride of the Valkyries" on his bookshelf stereo and turned the volume on peak. For a little while at least, he had forgotten his worries. "Let's get 'em," he said.

The assembly team had decided on an elaborate attack strategy, but unfortunately Peterson's IT group charged while Myers and his people were still trying to get their child-size helmets on their adult-size heads. Nerf balls rained down, and the assemblers' battle plan instantly switched, as it did every week, to "fire randomly at anybody wearing a beanie." The assembly team quickly got the upper hand, the arrows from their double-barreled crossbows being more aerodynamic than the foam balls expelled rapid-fire from the handheld Gatling guns preferred by the other team. Peterson and his men were soon driven back and pinned down. But the slow reload rate of a Nerf crossbow, combined with the utter lack of lethality of balls of foam, gave the IT team an opportunity to break out. Nerf balls littered the floor, fallen plastic helmets were kicked by running feet. Clark Mobarry, a former Caltech physicist in charge of writing the Unitigger program, mounted a table and emptied his weapon into a clot of hardware scientists. A fair-bearded, broad-shouldered algorithmist named Ian Dew, who actually looked pretty fearsome in horned headgear, shot off a couple of arrows, missed, then charged directly into a hail of missiles to retrieve them and reload. Karin Remington, the demure brown-eyed blonde responsible for the "Repeat Resolver" stage of the algorithm, had donned a plastic breastplate and was swinging madly in all directions with an oversized inflatable mace.

M'Liz Robinson, who had come over from business development to fight on Peterson's side, stalked her husband with a pump-action pistol and caught him in the Tranquillity Room, where she let loose a barrage, perhaps a little too energetically. He tried to block the bullets with his crossbow, making a guttural noise that was close to a laugh. Robinson would later recall it as the most passionate moment they'd shared in weeks. Moments later, Peterson found Myers's scarf in a doorway, seized it, and declared victory for the IT team. It was 9:12. Everybody got back to work.

By that second-to-last Monday in August, things were looking good. Art Delcher, the efficiency expert Myers had hired from the University of Maryland, had nearly finished the Overlapper stage of the program. Its main function was to compare each DNA fragment to every other one, looking for overlaps and discarding the ones that linked a new fragment with a known repeat. He had fine-tuned the routine to a speed of 30 million comparisons per second. As a test, one day the team shredded up the *H. influenzae* genome, which had taken Granger Sutton's original TIGR Assembler a full day to put together, and fed it into the Celera Assembler. The assembler spat out a perfect sequence in five minutes. Clark Mobarry had likewise completed the complex Unitigger program and was daring anyone to find a bug in it. He and his wife were expecting a baby in a couple of weeks, so both his work life and his home life were charged with expectancy. Even Myers was feeling optimistic. He had stayed late one night and by one o'clock had written a little program that sniffed out and corrected potential false mate pairs in the data, more or less circumventing the need for a bar-code system for the time being. One hundred eighty sequencing machines were now installed and data was pouring into the assembly team's coffers. There might still be time to run the assembler on a full tenfold coverage of the fly genome before the GSAC meeting, which was scheduled to begin on September 17. But that might not even be necessary. If Myers had calculated correctly, when the amount of data reached just six times the size of the whole genome, or 6x, the assembly program should have enough redundant, overlapping fragments to be able to put the sequence together into a linear set of scaffolds covering the genome. To be sure, there would still be gaps, but filling them would require nothing more exciting than collecting more data. Just 6x, then, would give Myers the proof he needed to show that the whole-genome shotgun method could assemble a complex animal genome.

Today there was even more good news. Over the weekend, the team had challenged the algorithm with its hardest test yet. They had broken up into random bits the 20 percent of the *Drosophila* sequence that Gerry Rubin's lab had already completed, to see if the computation could put them back together in the correct order. Myers had written a little program that visualized the scaffolds from an assembly run as a broken red line scrolling across the top of the screen, the breaks representing gaps between the scaffolds. When the test run was complete, and with Sutton and some others looking over his shoulder, Myers launched the visualizer on his monitor. At the top, it displayed a unbroken red line running the width of the screen. He had to scroll through several pages before he found a gap. In all, there were only five hundred gaps. Extrapolated to the size of the whole genome, that meant they could expect around three thousand gaps in a genome that now looked to be about 140 million base pairs long. Most of the holes should be relatively easy to plug with some finishing work by the biologists. It was only another simulation. But it was the closest one to real fruit fly DNA yet, and the results were amazing.

There was one long, dark shadow still blocking Myers's view of a triumph in Miami: Robert Millman. The PE board may have consented to release the *Drosophila* genome into the public domain, but the patent attorney was still intent on securing as much IP protection as he could before that happened. The value of the assembly algorithm was obviously something the company should protect. From Millman's point of view, allowing Myers to get up in front of an audience of Celera's competitors in Miami—or *any* audience, for that matter—and blithely describe how he put together a complex genome was a kind of commercial suicide. Ideally the algorithm should be kept a trade secret, which would muzzle the mathematician and everybody else. But Myers was enraged at the mere suggestion. As an academic, he felt that his achievement would not even exist until it was presented before his peers. "All I want is fifteen minutes to say 'I did it,'" he protested to Venter. "That's what I came here for, and I'm not staying if I don't get it."

Millman was willing to compromise. Myers could describe the algorithm in Miami, as long as a patent application on it was filed beforehand, preventing competitors from building one along the same lines without violating Celera's claim. To write a credible application, however, Millman needed to understand the algorithm well enough to describe it. For that, he needed Myers's cooperation, and he wasn't getting it. The issue was still hanging when, on the afternoon of August 25, the assembly team received from the sequencers an upload of new fly genome data that put them over the 6x threshold. The run of the assembly began five days later. It would take about twenty hours. The first person to arrive the next morning was Granger Sutton. Right away he could see there was something extremely unsettling about the result.

Myers came in a few minutes later. "How's it look?" he asked.

"We're getting a kind of weird number," Sutton said, in his imperturbably calm voice. The "kind of weird number" was 802,000—as in 802,000 gaps in the sequence. In other words, the assembler had not produced a sequence at all but merely 802,000 little fragments, like the shards of a vase dropped from a very great height.

Surprisingly, Myers did not panic, either. "That's totally bogus," he said. "There must have been some glitch in the run."

Sutton agreed. He spent a good part of the day inspecting the code, then fed the data back in for a second run. When it was done late that afternoon (it was possible to skip some of the steps, so this run did not take a full twenty hours), the team gathered around the terminal in his little cubicle, some kneeling, others craning their necks to see the results. Nothing had changed. It wasn't just that the assembler had failed to meet expectations. It had failed to make any sense of the genome at all. Everybody was quiet. You could almost hear Mother Nature cackling.

"Something is interfering with the formation of unitigs," Myers said, more bewildered than upset.

"That's not possible," said Clark Mobarry.

"Yeah, I know," said Myers. "But it's still happening, isn't it?"

CHAPTER 21
LINE 678

"I'm not going to make much sense," Myers said, "because I don't know what's going on."

He and Sutton had come down to deliver the news to Venter and the others. A cleaning woman came in to empty the trash, took a look at the grim faces, and left.

"All I know is something's fucking up the unitigs," Myers continued. He had already tossed out some possible explanations for the failure, but none of them seemed very promising. "We know it's not the assembly strategy. There's got to be something in the data."

"The data are good," Mark Adams said, not defensively, but as a statement of fact. "It's awesome data."

"I'm not saying it isn't," Myers said. He pushed his glasses up on his forehead and rubbed his eyes. "Right now I'm not sure what I'm saying."

It was quiet for a moment while everybody thought. Hamilton Smith pushed his chair away from the table so he could stretch out. He looked bemused, as if he were just an observer, curious to see how an interesting story was going to unfold. Adams's lips were bunched up in their lemon-sucking position. Venter toyed with a soda can. His face had a slack, absent look, as if he had left his features untended while he thought. "It's hard to see this through my eyes," he said. "I'm still seeing it through your eyes, piecemeal. You're getting no more order at 6x than at 4x?"

"Less," Myers said. "Something's cutting the unitigs off at the knees."

The basic contour of the problem was simple. With 6x worth of data, the mate pair strategy was supposed to kick in and form large scaffolds that linked the unitigs in an ordered sequence. But the mate pairs weren't even being given a chance to do their work, because the new load of data was breaking the unitigs themselves apart. Instead of filling in the canvas, like a new snowfall settling on ground that was patchy with snow, the additional fragments had come down like rain, melting holes into the patches already formed. It was too early to panic. Any number of things might have gone wrong, and some of them—such as a bug in Clark Mobarry's Unitigger program—might be easy to catch and fix. Mobarry's wife was having a difficult pregnancy, and with the stress at home, a bug could easily have slipped past his attention in a distracted moment. But what if it wasn't a bug? Then the problem had to lie either in the quality of the data the algorithm had to work with or the structure of the computation. It was either "garbage in, garbage out"—or the algorithm itself was garbage.

At least some data problems would be correctable with just a little tweaking, if they could only track the failure to its source. One possibility was that the reads of fly DNA had been insufficiently trimmed. To make his libraries of fly DNA clones, Smith had inserted each piece of fruit fly DNA into a loop of viral DNA, so that the DNA fragment would reproduce along with the virus itself. The bits of virus on each end of a fragment had to be cleanly lopped off before they reached the assembly stage, like bits of excess cardboard clinging to the edge of jigsaw puzzle pieces, interfering with their match. If such "noisy ends" were the problem, then the assembly team had only to do a simple search through the accumulated data for the virus's specific sequence, delete it, and run the assembly again.

"I'll bet you a dinner there's vector in that sequence," Venter said.

"We can do the search," said Sutton, "but you have to wonder why it hasn't shown up before."

"It's got to be *some* kind of noisy data screwing up the compute," Myers said.

"Are you surprised by that?" Venter said. "There's no such thing as data without noise. In biology, we call that 'life.' I thought the idea was to build a program that could work with it."

"That's what we *did,*" Myers protested. He looked more hurt than angry, more bewildered than beaten, like a wrestler who couldn't believe he'd been pinned and right away wanted another shot at his opponent. "To create an algorithm like this is harder than you think."

"I'm sure it is," Venter said, in a more sympathetic tone. "Frankly, I'm almost relieved that we're getting hung up here a bit. Otherwise it wouldn't make sense. Ham will tell you, there hasn't been a single whole-genome project where we didn't hit some kind of wall."

"The walls are where it gets fun," said Smith.

"Yeah, I'm having a blast," said Myers.

"Look, the worst possible outcome would have been complete gibberish," Venter said. "You've proved that you aren't generating gibberish."

"Great. Maybe I can use that as the title of my talk in Miami. 'The Genome Sequence of *Drosophila melanogaster:* At Least It's Not Gibberish.'"

"You need to go home and get some sleep," Venter said. "I guarantee we'll figure out what's wrong tomorrow. I'm absolutely positive about that."

After Myers left, Venter shook his head. "This is what happens," he said, "when the theorists start thinking their shit smells good."

Nobody figured out what was wrong the next day, or the next, or the day after that. And Smith had been wrong: no one was having any fun. For Myers and his people, creating the algorithm had been an exhilarating scientific challenge; poking through its architecture, looking for an as yet indefinable problem was a tedious and anxious chore. Myers was convinced that the flaw lay in the data coming into the compute, but he knew that Venter was right: a certain amount of imprecision was inherent in the process of sequencing. The whole point was to invent a computation that could cope with imprecision. It looked as if he'd failed.

A simple procedure of screening the reads for traces of viral DNA sequence quickly ruled out bad vector trimming as a factor. But another possibility arose to take its place, and when that proved a false lead, something else came up. Every dead end only widened the field of suspicion. Perhaps the problem traced back all the way to the moment when

Smith cloned the fruit fly DNA the previous fall. If too many of his clones were chimeras, formed by falsely yoking together two bits of DNA from different chromosomes, the assembly would choke. Likewise, the size of each clone had to be precise. But Smith was widely regarded as a genius in the physical manipulation of DNA, and if faulty cloning was the problem it would have shown its face at 4x and 5x.

Perhaps the source lay even further back in the process. What if the original vial of raw fruit fly DNA that Smith had received from Gerry Rubin contained an amalgam of more than one strain of *Drosophila,* so that the clones were genetically variable? Asking the assembly program to differentiate between sequencing errors and nearly identical repeats within a single genome was hard enough. Adding genetic variability into the equation would make it too much to handle. Built to be wary of making a mistake, the program would refuse to put pieces together at all. But Rubin was as meticulous a scientist as Smith. It seemed impossible that he would send a sample containing heterogeneous flies.

Whatever the source of the problem, it was clearly wreaking its destruction within Clark Mobarry's Unitigger program. Mobarry was having less fun than anybody. The Unitigger was the most complex, most creative component of the assembly algorithm, and he was as proud of it as Myers was of the complete algorithm. Finding a bug in it would be an embarrassment, but at least it would solve the problem. The possibility that the problem was caused by some more elusive, structural weakness in the Unitigger's architecture was scarier and much more mortifying. Mobarry had no way of knowing where to look for the problem. All he could do was tunnel down into smaller and smaller regions of the shattered assembly, searching for some nonsensical pattern that might betray its cause. The work put Mobarry in a horrible dilemma. He was the only person who knew the details of the Unitigger well enough to recognize a mistake, if one was there. But his wife was due at any moment. Her discomfort with the large and restless unborn infant was hard enough to bear without her husband there to help, especially with another young child to take care of.

Myers wanted to be understanding. He'd driven his team hard, but he had always looked out for their welfare. He could see that Mobarry was being torn apart. But there was too much at stake to let him go. "If Clark's wife needs tending to, somebody else is going to have to do it," he told Sutton. "He's welcome to attend the delivery. But the rest of his time is ours." Myers laughed nervously, as if he had caught the manic tone in his words and wanted to disown it. He asked Mobarry to have his in-laws flown in from California; Celera would handle the arrangements. When Mobarry's father-in-law complained that Mobarry's home did not have cable TV, Myers told his secretary to get it installed. It wasn't what anybody wanted, but it allowed Mobarry to continue staying late.

Everybody else was working just as hard. Borrowing a page from Venter's book, Myers was trying to keep steady for them. "This is my first trial by fire," he confided to an acquaintance. "I've always been one step removed from the real action. Craig says to take it in stride, because genomes never come together the first time. I guess I knew that intellectually. But I wasn't prepared emotionally. To tell you the truth, I'm scared shitless."

Granger Sutton was constructing a way to isolate the Unitigger and test it against a shotgunned simulation of the finished sequence from Rubin's lab. If it failed to put that back together, they would at least know that the problem was in Mobarry's domain. After a couple of tense hours, the results of Sutton's test came back: the Unitigger had easily passed. Mobarry was relieved, of course, but it left them with even fewer leads, and time was running out. When Myers got home that evening, he found nobody there but the cats. He couldn't hold the tension inside anymore, and slammed through the dark empty house, screaming. When his wife came in she found him standing in the kitchen, still shaking. He told her that he'd lost control.

"He didn't have to tell me, because I already knew," Robinson later said. "I could see it in the cats' faces: 'Father's gone insane.'"

The next morning—Friday, September 3—Myers arrived at work looking a little more relaxed. Maybe the screaming had helped. He was wearing a new scarf. It had a more colorful, wilder zigzag pattern. "I thought I'd try a new mojo," he said. The team gathered in the Tranquillity Room. "How's your wife doing?" Myers asked Mobarry.

"Very uncomfortable. No baby yet."

They went around the table, filling the group in on what each had been doing. Art Delcher had begun a line-by-line check on the Overlapper code, in the unlikely event that there was a bug in it throwing off the Unitigger downstream. Ian Dew had stayed at work until 4:00 a.m. and was back in by 8:00.

"So what did you find out?" Myers asked him. Dew looked at him for a moment, dazed. "I can't remember," he said. Once his neurons started firing again, he explained that he'd done a "hard screen run," which theoretically should have mirrored the Unitigger's compute on a small part of the genome and produced the same result. But it did not. "The unitigs came right together," he said. It was the same anomaly that Sutton had witnessed the day before: run a test on any subregion of the genome and it came back with a solid assembly; run the program against the whole genome and it collapsed into crumbs.

"This is getting spooky," said Sutton.

"Maybe we're asking the wrong questions," offered Mobarry.

They scattered to their cubicles and got back to work. Every once in a while, someone would come up with a new idea and bring it in to Myers. Names were bestowed on the best of these flashes of insight: "the pigeon-poop theory," the "double-headed dragon." But none of them held up for more than an hour or two. Sometimes Mobarry could be heard on the telephone, murmuring reassurances to his wife. For the most part, the only sound was the subdued chatter of the keyboards—hopeful interrogatories followed by silence, like the intermittent scuttle of rats looking for a way out of a maze. They went down to lunch, ate, returned. In the afternoon, a horsefly flew in and buzzed around the ceiling. Dew chased it out. Sutton got up and bought a Snickers bar from the vending machine down the hall. Myers paced around. At 4:30 in the afternoon, his wife arrived. There were going to be some big parties at the GSAC meeting, and she had volunteered the two of them to give a quick dance lesson to any Celera employee who wanted to get ready to cut some rug. Myers had forgotten about it. "I have to go fucking dance for a while," he told Sutton under his breath. "I'll be back in twenty minutes."

Three or four people showed up. M'Liz put on some Latin swing music. "The important thing is to keep back on your heels," Myers said, demonstrating. "That gives this little sway to your hips, automatically." He looked graceful but a trifle automatic. Just then the cell phone on his belt rang. It was Sutton, with another possible lead. Myers left his wife to continue the lesson on her own.

In the two minutes it took Myers to walk back up and rejoin the team, Sutton's lead had proved another dead end. The day closed with one last meeting in the Tranquillity Room. There was a brief attempt at brainstorming, but nobody had much left to think with; most of the ideas sounded like somatic responses to the need for more ideas. Everybody looked fed up. Even Myers could see it was a waste of time. "It's been a pretty exciting week," he said, smiling wearily. "We all need to pace ourselves through the Labor Day weekend. I'd advise everybody to work like hell when you're working, and try not to think about work when you're not."

"But right now, we work like hell, right?" Mobarry said.

"Tomorrow is Saturday," said Myers. "Anybody who wants to take the day off should do it. I know I pretty much have to, if I want to stay married. Do the best you can. And keep breathing."

When most of the team left, Art Delcher stayed behind. He wanted to spend a little more time on something curious he'd noticed in the Overlapper: every once in a while, two reads of DNA that appeared to match up were not sticking together. He spent some time comparing two of these truculent overlaps against each other to see what might be going on. He left at 1:00 in the morning, feeling he'd potentially identified the problem, if not the reason for it. He came back later that morning and again on Sunday, poring over the Overlapper code one more time, line by line. Delcher was there alone, at 11:00 on Sunday morning, when he scrolled down to the 678th line of code in the program. He looked up again at the preceding line, then down again to line 678, then one farther down, and back once more to 678. For a moment it felt as if his heart had stopped. He wasn't sure whether to laugh or cry.

CHAPTER 22
DANCING IN MIAMI

Back in 1989, at the first Genome Sequencing and Analysis Conference, at Wolf Trap in Northern Virginia, fifty people had shown up. As Craig Venter's notoriety grew through the nineties, so did GSAC, both in size and in extravagance. In 1998, the year Celera appeared on the scene, the TIGR-sponsored event had outgrown its longtime locus in Hilton Head, South Carolina, and moved into the slightly dowdy grandeur of the Fontainebleau Hotel in Miami Beach, a venue big enough to accommodate the trade show exhibits and swelling list of registrants. Among the featured speakers were pharmaceutical superstars like Allen Roses of Glaxo Wellcome, and SmithKline Beecham's George Poste. Venter invited President Clinton to give the 1998 keynote address; the two men had become acquainted at one of Clinton's Renaissance Weekends, and the president had asked him to serve on a special advisory committee on biological warfare. While initially receptive to Venter's invitation, the White House ultimately declined. In the fall of 1998, the Monica Lewinsky scandal was at full boil—perhaps a meeting on DNA might not be the wisest choice for a presidential appearance.

"Craig's show," as GSAC was commonly known, had returned to the Fontainebleau for another year. Among the scheduled talks was a lecture on forensic DNA technology by the head of the FBI's forensic DNA team, coincidentally the person who had analyzed the stain on Lewinsky's infamous blue dress, and in his introduction Venter could not restrain himself. "The president would have been here last year," he said, "except that he had an appointment with our next speaker." The joke got a big laugh. But it would prove costly.

Almost two thousand people had signed up for the meeting, set to begin on September 17. An even larger percentage than the year before were biotech scientists and Wall Street market analysts. Only a few people associated with the government genome project had shown up. Whether the rumor of a planned boycott was true or not, four of the five G-5 leaders had organized their own meeting—pointedly dubbed "Back to Science"—to be held on Florida's Gulf Coast a few months later, and had bought a full-page ad in the GSAC program inviting scientists to participate "in an atmosphere that is not highly commercial." In contrast, a small army of Incyte employees had descended on the Fontainebleau, all of them dressed in sprightly sea-blue polo shirts with the company logo on the chest. Now that the hope of a merger with Celera was gone, Incyte was making a full-throttle effort to beat its flashy rival at its own game. Last year, Venter had hired his friend Bruce Hornsby and his band to play at the party on the final night. This year's send-off party was scheduled to be a spare-no-expense "Beach Blast," courtesy of Incyte, the company that called itself the "Best view of the human genome." Throughout the meeting "Incyte TV" aired on every television in the hotel, fed material by a video crew working the lobbies and exhibit floors for on-the-spot interviews of the genomic industry's leading lights. Partway through the conference, Randy Scott made a splash by announcing that Incyte scientists had hard evidence that there were 140,000 genes in the human genome—twice Venter's estimate and far in excess of the 100,000 figure cited by most researchers. The announcement ignited a rash of national news stories.

"[This means] that the human organism is considerably more complex than hitherto expected," Nicholas Wade wrote in the *New York Times.* It also meant that Incyte potentially had forty thousand more human genes to sell. Celera had none, but it still had the market's attention. The stock price had taken another spurt forward, clearly in anticipation of some breaking news at GSAC. "If [the fruit fly assembly] pans out," one investor wrote in a popular online stock forum tracking Celera's prospects, "Celera wins the whole enchilada."

"I doubt there will be anything new," somebody else opined. "In fact, I'd say that if there is news to be had at GSAC, it would be bad." Another trader posted a simple question: "Who's Eugene Myers?" he asked.

On the night before his talk, Myers was just another GSAC attendee with a name tag dangling on his chest, standing alone in the glowing blue darkness of the welcome reception. Pangea Systems, the party's sponsor ("an innovator in the application of information technology to the life sciences!"), had anchored giant double helices made of blue and white balloons to the floor. The tips of the columns shivered lightly in unison, eerily in time to the beat of the rock band on the far side of the room. Myers had a glass of ginger ale in one hand and a spicy hors d'oeuvre in the other, which he consumed despite the roaring nervousness bubbling up from his stomach. He could see his wife among the crowd on the dance floor, having a good time. Hamilton Smith came over and stood with him, surveying the scene. On their side of the room it was a little quieter, and the Incyte TV crew was weaving about, looking for somebody to give good sound bite. The team's "reporter," a perky little blonde in tight jeans and an unbuttoned Incyte polo shirt, ignored the stooping, white-haired guy with his pants hitched up too high and the dark frowning dude wearing a Polarfleece to a party in Miami Beach. Instead she reached her mike up to interview a marketing exec in a buzz cut. "So, you're like saying that your new robot picker can cut clone prep down forty percent *without* adding to reagent costs?" she said. "Awesome!"

"These are not the people I grew up with," said Smith, shaking his head.

The party broke up around midnight. By then a lot of helium had leaked out of the base pair balloons and the giant helices were drooping. Myers and Smith had long since left. When Myers's wife got back to their room, she found her husband in the bathroom, struggling with a case of indigestion.

A bug. In the end, the assembly failure had been caused not by noisy data, heterogeneous DNA, or the wrath of Francis Collins's God upon those who would steal the Book of Life, but by a one-sentence mistake in 150,000 lines of code. It was not just a bug, moreover, but a big, fat, dumb-ass bug of the kind that Art Delcher would have been disappointed to see one of his undergraduate students make, but which nevertheless gets made all the time by the best software engineers in the world. When looking for overlaps between a new DNA fragment and each of the others stored in the supercomputer's memory, Delcher had written a line of code asking a question: "Has the stored fragment already been identified as part of a repeat?" If the answer was "yes," then the overlap was rejected as unreliable. But if the answer was "no," the program was supposed to join the two fragments into one larger one, thus creating a unitig. The problem was that Delcher had forgotten to include the "no" answer to the question. Instead of joining the fragments, the program did what every computer program does when it is left hanging for an answer: it defaulted to the answer given to the previous calculation. If that one had yielded a true overlap, there was no problem. But if it had perceived a repeat, then both that match and the next one were rejected. Because of the bug, 30 percent of the perfectly good matches were being thrown out as false ones induced by repeats—more than enough to shatter the assembly into pieces.

"This is Programming 101," Delcher said later, a little dolefully. "Whatever is in memory *stays* in memory until you tell it not to."

Normally when such an egregious error is discovered, there is an undercurrent of irritation at the person who made it. But when you recover your child lost in a department store, your first instinct is not to yell at him for straying. Delcher had made the mistake, but it mattered far more to the rest of the team that he had also found it. When it finished its new run the next morning, the assembler had delivered an ordered, nearly complete genome of *Drosophila melanogaster.* There were still additional fly data in the pipeline to be incorporated into the assembly and a few wrinkles in the computation to be ironed out before the Miami presentation. On the same day, September 7, Clark Mobarry's wife gave birth to a nine-pound, fifteen-ounce boy. The Mobarry baby was perfect in every respect.

Myers still had the problem of Robert Millman. Perhaps it is more accurate to say that Millman still had the problem of Gene Myers. The attorney had put together a gargantuan patent application that treated the *Drosophila* genome sequence, the software to explore it, a microchip containing all of its genes, and various other uses of the information as a single "discovery system." It was an ingenious attempt to get around the fact that the fly DNA molecule itself was Mother Nature's "prior art." The application would not prohibit academics from using the information in their research. But if Millman could get it filed in time, it might block Incyte or anyone else from scooping up the code into their own database and charging customers to use it. The only thing missing from his application was a description of Myers's algorithm. Millman had been trying for weeks to get Myers to write up something explaining how it worked. Without such an enabling description, there could be no patent application on the algorithm, and without an application duly filed, there could be no disclosure. But Myers could not, or would not, understand this simple equation. It was a case of "denial, squared."

It was not until the afternoon of Friday the seventeenth, the very day GSAC was to begin, that Myers agreed to sit down with Millman in the Tranquillity Room. "You realize I've had everybody stop work for this," Myers told him irritably. "Just so you know I'm cooperating."

"Just so *you* know," the attorney replied, "if I don't get something in writing from you guys today that I can mold into a form that doesn't make me sound totally ridiculous in an application, your talk ain't gonna happen. If you'd cooperated twelve weeks ago, we wouldn't be having to do it this way now."

Myers got the point. In a couple of hours Millman had on his desk two-paragraph descriptions of how the Overlapper, Unitigger, Scaffolder, and other assembly processes worked. While the rest of Celera's contingent left for Miami, he set to work. At 10:30 p.m., an hour and a half before the U.S. Patent Office in Crystal City closed its doors for the weekend, Millman walked up to the receiving window. There were some other people in line, but he was the only one wearing a turquoise bowling shirt, yellow slacks, and hand-painted socks, and the only one making a little patent history. The guidelines for patent applications are very stringent, and since the office's inception in the eighteenth century one of the rules has been that all applications be submitted in writing. Printed out, however, Millman's offering—"Primary nucleic acid sequence of the *Drosophila* genome, discovery systems containing the *Drosophila* sequence and uses thereof"—would have taken up tens of thousands of printed pages. The Patent Office had broken new ground by allowing him to file the application on two CD-ROMs instead, which, when it came his turn, he presented to the clerk in an envelope, along with a big, gleamy-eyed grin. She stamped the envelope with the date and time of receipt. Millman drove to Reagan National Airport. He was still grinning when he got on the last plane to Miami.

. . .

The main ballroom at the old Fontainebleau Hotel, where the GSAC lectures were taking place, resembles a boudoir the size of an airplane hangar. Banks of golden chandeliers hang over a couple of acres of peach carpeting, and velvet wallpaper in a darker shade of peach covers the walls. In preparation for the meeting, a dozen gigantic monitors had been hung from the ceiling beneath the chandeliers, so that people on the sides and in the back could see the distant speaker's face. Another giant monitor directly faced the podium, so the presenter could watch himself present. When Myers had first mounted the platform, he found the reflection unnerving. But now he was beginning to like it.

The Sunday morning session began with a talk by Gerry Rubin summarizing the role the fruit fly had played in genetics for a hundred years and detailing the Berkeley team's contribution to the joint sequencing effort with Celera. Mark Adams followed with a video overview of the scientific operation, featuring a scene with Hamilton Smith and his lone assistant trying to act natural for the camera, which drew a few giggles, and some wide-angle shots panning across the rows and rows of ABI sequencers in the production rooms. ("Jesus, these guys are running a damn *factory,*" one academic in the audience muttered to himself.) Adams discussed some interesting features discovered in the fly genome but did not reveal the results of the assembly itself.

Then it was Myers's turn. With Adams's help, he pinned the microphone to his shirt. He opened his laptop and keyed up his first PowerPoint slide. Using data from one of the simulations, he took the audience on a tour through the assembler algorithm, step by step. He paused for a moment, and checked his image on the monitor.

"In case you're already doubting me because this is just a simulation," he said, "let's take a look now at how this works on some real data." If anyone hadn't been paying attention before, they were now. As planned, he said, his team had used Rubin's partially finished sequence not to bolster the Celera assembly but to test its validity in the end. In a series of slides Myers showed that in every place but one, Celera's shotgun version matched perfectly with Berkeley's finished sequence. "On the one discrepancy," he noted with a slight smile to Rubin in the front row, "we have reason to believe that we may be right. There appear to be virtually no errors. We pretty much have it in the bag."

Myers wasn't quite finished. On September 8, the day after the 6x assembly had finished, the first batch of human DNA fragments had been loaded onto the sequencers. Myers had been doing a little preliminary analysis of the data coming through in the meantime, and what he saw was most encouraging. Though thirty times the size of *Drosophila,* the human genome appeared at first glance to be *less* complicated in its structure. "I expect that on human, the repeats are going to be less of a problem," he said, almost casually. "This bodes extremely well. We're ready. We think it's going to go just great."

There was silence when he finished. Then, as Adams came up to join him, the audience burst into applause. Myers looked down at his shoes and tried to suppress a smile. The rear doors opened and the crowd filed out to the lobby for the coffee break. In the murmur of a hundred conversations, every word was about what had just taken place.

"When Venter makes his predictions, he always goes out on a limb," one scientist was saying to his colleagues. He shrugged. "Looks like he came through again."

"The smartest thing he ever did was hire this Myers guy," said another.

"The real question about all this is why they're doing this in the first place," an Incyte computer scientist was telling another group. "Sure, they have a fast program. But where are they going to make their money?"

"Craig is doing shotgun because it's sexy," somebody replied. "Gene is doing it because it's there."

A clutch of the few government Human Genome Project scientists in attendance huddled in another spot, not saying much. One shook his head. "Those fuckers are actually going to do it," he said.

The conference ended on a raucous note two nights later. Bad weather had moved Incyte's Beach Blast into the grand ballroom. The stage show started out loud and crescendoed through the evening. An undulating crowd of genomicists on the dance floor batted scores of beach balls about while their colleagues soaked them with squirt guns. Bare-bellied Latina beauties in hacienda hats swirled in their skirts of red, green, yellow, and blue—was it a coincidence that these were the colors of the four DNA letters on an automated sequencer?—only to be swept off the stage by gyrating dancers on ten-foot stilts. A band of men playing brass instruments and dressed in hoop skirts, leis, and glittery tunics descended into the crowd. A conga line polymerized behind them. Toward midnight, several members of the Celera assembly team tried to attach a couple of dozen helium balloons to Gene Myers's scarf and launch it skyward. They couldn't quite get it airborne.

Myers himself was on the dance floor with his wife, moving to the Latin beat with cool, restrained steps, his arm around her waist, his thigh assured against hers. Amid the flailing limbs around them, they seemed a vision of grace. It wasn't quite like Myers's dream—they weren't in Washington, and he wasn't wearing a tux—but at least they were dancing. She smiled up at him, then noticed, with a little disappointment, that though he was smiling too, his eyes were not looking at her. He didn't seem to be looking anywhere at all. "What are you thinking about?" she asked.

"Drosophila," he said.

"What about it?"

"Just that I wish we'd gotten it to work on the first try," he said. "We could have been gods."

PART THREE

CHAPTER 23

GETTING TO NO

In retrospect, the beauty of the *Drosophila* project—whose finest hour was yet to come—was that its relatively modest objective allowed it to unfold beneath the radar of history. People could interact unselfishly because no one was competing for a Nobel Prize, much less a niche in heaven next to Mendel, Darwin, Watson, and Crick. The human genome was another matter. Craig Venter wanted to be remembered for doing great things, and saw no reason why he shouldn't admit to the aspiration. In contrast, the leaders of the Human Genome Project insisted that their motives were, to use one of Francis Collins's favorite words, "selfless." Collins and his colleagues defined themselves as defenders of the public interest against a corporation, and a man, bent on selling for profit what was rightfully the shared heritage of all humankind. Constantly drawing attention to one's selflessness is a bit of an oxymoron. But if the human code really *was* in peril of being privatized, then it is understandable that the HGP scientists would see themselves as "wearing the white hat." On the other hand, if Venter were to receive the recognition he so desperately wanted, he would *have* to deliver the genome to the world, just as he had promised. So was it the genome that the government scientists wanted to keep out of his grasp, or the credit for capturing it?

At Celera, the reading of the human script began on the morning of

September 8, immediately after the last letters of fruit fly DNA had been flushed from the sequencers' capillary tubes. Mark Adams was on hand to load the first human sample himself, an act he performed with a tremendous sense of relief. The last of Celera's 300 ABI machines had been installed the previous weekend, and DNA was finally flooding through the pipeline at the pace the company's name implied. Nevertheless, the devotion of an extra four months to the *Drosophila* project had cost the company its lead. The day before, NHGRI and the Wellcome Trust had issued a press release announcing that the public program had already sequenced 739 million base pairs, or nearly a quarter of the human genome. The consortium was on track to deliver "the first draft of the genetic blueprint of humankind" by the following spring, as promised.

"It does seem that they're going a lot faster than we thought," Paul Gilman said.

"This convinces me more than ever that Francis has no intention of continuing past the draft and finishing the whole genome," Venter said. "The announcement in the spring will be the big one for the press. We have to combat that. It's time to play hardball."

According to the press release, the Human Genome Project had not abandoned its commitment to a finished genome at all. But Venter was right about the HGP's public relations strategy. By positioning the delivery of the working draft as an historic event, the program had redefined the finish to the race to be exactly what it was capable of producing by the spring, providing everything went smoothly and all the G-5 members could deliver on their promises. Their grants in hand, all had invested heavily in capillary sequencing machines. Eric Lander had emerged as the public program's genome sequencer with the deepest pockets. His genome center at the Whitehead Institute enjoyed first-year funding of $34 million from Collins's NHGRI, the promise of another $55 million to follow over the subsequent ten months, and the kicker of another $38 million secured in a loan to the Whitehead from the Massachusetts Higher Education Authority.

From the beginning of the Human Genome Project, John Sulston and Robert Waterston had been regarded as the quarterbacks of the sequencing operation, with Lander shouting "Gimme the ball! Gimme the ball!" from somewhere on the flank. Now he had the ball and the means to run with it. Sulston and Waterston were ordering dozens of Prism 3700s from PE Biosystems to beef up their existing stock of the old slab-gel Prism 377s. Lander chose instead to scrap all of the Whitehead Institute's slab-gel machines and replace them with 125 new capillary machines.

Michael Hunkapiller was more than happy to fill the order. To get the machines installed and running as quickly as possible, Lander lured away two of Hunkapiller's top engineers on the 3700 project; Hunkapiller was not so happy about that, but he had to admire Lander's chutzpah. Whitehead then set about building an automated operation driven by conveyor belts. In the short term, starting completely fresh meant a lot of headaches. Within a couple of months, however, the Whitehead's pipeline was pouring out DNA much faster than the labs hobbled by the need to manually prep hundreds of slab gels a day. The others were also devoting much of their resources to the traditional mapping and finishing work. For Lander, the game was to concentrate solely on churning out as much raw sequence as possible. It was a brazen blitz designed to out-Celera Celera.

"Eric did a very Eric-ky thing," Elbert Branscomb, head of the DOE's sequencing operation, remembered later. "He understood that in the end it was going to be the total number of base pairs sequenced that determined who was the guy, who was the leader."

There was only one problem: the pipeline that Lander had created had a monstrous appetite, and there was not enough DNA in his allotted portion of the genome to keep it fully fed. The G-5's plan called for Washington University to coordinate the distribution of premapped BAC clones to the various other labs for sequencing, which would ensure that two groups were not duplicating each other's work. But the supply of mapped clones could not keep up with Lander's capacity. He had gone along with the joint decision to "map first, sequence later" only on condition that, should the mapping effort prove too slow, he would be free to pluck clones to sequence randomly from the genome. When he threatened to go forward with random sequencing, the map-centric scientists at the Sanger Centre and Washington University bellowed in protest. But Lander was relentless. His constant demands for more territory led to a crescendo of "heated conversations," as Collins put it, in the weekly G-5 teleconference calls. Collins's diplomatic skills, combined with the ever-present awareness of a common enemy, kept the alliance together. Whether it could still beat Celera remained to be seen.

In the meantime, Lander made a political move that had the potential to render the question moot. Nine months earlier, he had been the most passionate denouncer of the DOE's attempt to collaborate with Celera. Surprisingly, he now reached out with an olive branch of his own—not to Venter directly but to Hunkapiller, who had an interest in the success of both sides and was more likely to be receptive. "The perception of a race going on here is damaging to you, to us, and to science," Lander told him. "If you can convince Craig to sit down and talk, I think I can get Francis to as well."

Hunkapiller conveyed the message to Venter, who reluctantly agreed to a meeting in a Cambridge hotel not far from the Whitehead Institute. He arrived to find Lander, Hunkapiller, and Noubar Afeyan, representing PE Corporation, surrounded by a clutter of half-eaten fruit and Danish, suggesting that they had already been talking for a while. Venter did not trust Lander and the chummy atmosphere made him uneasy.

"Eric has some misconceptions about our business plan," Afeyan said to Venter. "Maybe you can put things in context for him."

"Sure," Venter replied, looking at Lander. "Our business plan is to take your data, combine it with ours, and have the genome done on our own in a year."

Lander's big face reddened with anger. "What you've just said proves you don't give a shit about us," he said. "All you're interested in is winning."

"If you want to talk about a serious collaboration, you have to know where we stand on some fundamental issues," Venter replied. "We're not going to release our data nightly on the web, we're not going to put them into GenBank, and we're not going to refrain from patenting things we think have medical importance."

Lander did not object to any of those conditions. As a former economist and a co-founder of Millennium Pharmaceuticals, which had its own hefty share of patents filed on human genes, Lander was in a position to understand the exigencies of the private sector better than most of his academic colleagues. All he was concerned about was how the data would be released. Venter suggested that in addition to making the genome available on its own web site, Celera would issue a DVD containing the entire code and distribute it free to anyone who wanted to use it, with the condition that the data could not be reprocessed and resold. On this there seemed to be at least some room for further discussion. The next step, obviously, was to bring Collins into the loop and then see if the other key members of the public program could be convinced. Tony White would of course have to be convinced, too, that Celera's business interests were not at risk. A merging of the two efforts still seemed like a long shot, but at least a channel was open.

On the face of it, the Human Genome Project would seem to have had much more to gain from a collaboration than Celera did. As Venter delighted in pointing out, his company already enjoyed a "de facto collaboration" by virtue of its access to the public program's data via GenBank. Breaking through that one-way mirror would give the government-funded scientists not just more data but data of a kind they most desperately needed, especially Celera's mate pair information. The HGP's working-draft strategy called for 5x shotgun coverage of each BAC clone in their clone-by-clone approach. This was enough to combine the shotgunned fragments of each clone into a few large segments. But without the organizing power conferred by the mate pair strategy, it was not enough to arrange those segments in their correct order or to make sure the resulting sequences were pointed in the right direction, not flipped around so that their base pairs read in the reverse of the correct order. If each clone was thought of as a bucket containing a jigsaw puzzle, the working-draft version of the genome would resemble a series of buckets lined up, some with their puzzles mostly put together but others with a jumble of pieces yet to be joined. Finding genes in those unfinished buckets would be much harder than finding those in a true linear sequence. Furthermore, for technical reasons some parts of the genome simply refused to be cloned into BACs. Those places would be represented by gaps in the chain of buckets itself. Celera's random shotgun data could help plug those holes, too.

"I suspect that Eric knows full well that they probably can't win at this point," said Richard Roberts, the chairman of Celera's scientific advisory board, when Venter told him of what had transpired at the meeting. "We're generating loads and loads of data, and have access to theirs, so by definition we have more than they."

Nevertheless, Celera also stood to benefit a great deal by collaborating. Gene Myers's extraordinary prediction in Miami notwithstanding, the success of the whole-genome shotgun method on *Drosophila* was hardly a guarantee that the technique would work on *Homo sapiens.* Not only was the human code thirty times larger, but the percentage of it given over to repeated segments was much greater. There was a good chance that the government's carefully mapped data could plug a lot of holes in those repeat regions. Myers and his assembly team would also benefit from being able to view the raw "trace file" data direct from the HGP's sequencing machines. These provided a more accurate representation of the order of base pairs in the government program's sequences than the computer-processed versions made public in GenBank, and thus would be valuable in resolving discrepancies wherever two sequences did not quite match up. Finally, cooperative interaction with the plethora of experts in the opposite camp would be another huge scientific advantage.

Roberts and the other members of the scientific advisory board, most of whom were academics, were also keenly aware of the political benefits of a truce. They feared that an all-out Celera victory would be an embarrassment to NIH, with ugly repercussions for everyone, possibly affecting NIH's future funding by Congress. It was with an eye toward promoting a private-public collaboration, in fact, that Roberts had joined the Celera board in the first place. Without mentioning Lander's initiative, he called Robert Waterston at Washington University to feel him out. Waterston was skeptical that the two sides could come to terms but did not dismiss the notion out of hand. At the same time, advisory board member Arnold Levine at Rockefeller University broached the subject directly with NIH director Harold Varmus, a personal friend. If Varmus could be brought to the table, Francis Collins would surely follow. Everyone involved agreed on one thing: a genome composed of the combined data and expertise of both sides would be much better, and available to the world sooner, than either side could manage on its own. The question was whether that was a realistic possibility.

"I agree with Eric that it would be in the best interest of science and business if there wasn't all this bullshit going on, so people could focus on the tremendous value the sequence will have," Venter said. "How do I lose by having other people involved? I can only gain by an end to the bullshit."

On October 10, 1999, the Celera scientific advisory board and the company's key scientists—Venter, Ham Smith, Gene Myers, Mark Adams, and Sam Broder—met at the Wye River Conference Center on Maryland's Eastern Shore, where Bill Clinton, Yasser Arafat, and Benjamin Netanyahu had negotiated the Wye River peace accord a year earlier. The scientists were there to discuss Lander's own peace proposal, first among themselves, then with Lander by teleconference. "Do we *want* to share the limelight with these guys?" Gene Myers asked before Lander called in. "It seems to me we don't get anything back."

Lander got on the phone, and after a few pleasantries, Richard Roberts asked him who in the public program was aware of his initiative.

"Francis is open to it, but I'm not authorized to negotiate on his behalf," Lander answered. "I haven't talked yet to Waterston, Sulston, or Gibbs. If I did, this thing would rapidly turn to mush."

"The key issue is data release," said Roberts. "Could Francis live with an agreement where we issue the completed genome on a DVD, like we did with *Drosophila*?"

"Francis likes the idea of instant release, but he knows it's not a religious tenet," answered Lander. "Data release does not seem to be a make-or-break issue, if the principle of fairly accessible release is kept."

As the discussion went on, the issue that generated the most tension was not how the genome would be made public but how credit would be parceled out for doing it. Would there be a joint paper on a single combined genome? Two papers on separate genomes, published at the same time? Much depended, of course, on how closely the two sides collaborated scientifically. But one demand Lander was adamant about. "If Celera is going to make use of public data, then that publication should be a joint paper," he said. "If that can't be agreed on, there is no basis for further discussion."

"One or two papers, it's the same thing," said Venter.

"No, Craig, it is not the same thing," Lander replied. "It's not even *close* to the same thing. You've got twice the data! So how does that look equal? I'm not going to produce my data for your publication."

Venter tried to change the subject but Lander kept veering back to it. If Celera did not include the HGP scientists as co-authors on any paper using publicly available data, it would amount to nothing less than a breach of scientific ethics. His implication did not sit well with the members of Venter's scientific advisory board, all of whom were highly distinguished scientists, with the exception of Arthur Caplan, who was a distinguished bioethicist. Lander kept pressing the point.

"Craig is falling for it," Mark Adams said, passing through the hallway outside the meeting room on his way to the men's room. "He responds to that kind of thing. Luckily, in two weeks he won't remember what he said."

Gene Myers was the next to exit the room, scowling, his black leather jacket already on and his bag packed to leave. "All the guys in that room are older than Eric, a couple of Nobels among them, and *he*'s telling *them* what's the moral thing to do." He shook his head in disgust. "The *real* issue is he wants to see our data because they can't do bupkis with their own." Then he stalked out of the building.

Back in the meeting room, Lander had hung up and the remaining scientists were talking among themselves.

"It's a false assumption to say that just because you use my data, I must be on your paper," said Arthur Caplan.

"I found the p53 gene," Arnie Levine said. "It's part of the genome, so I want to be on the paper, too."

"What about Gregor Mendel?" said Caplan. "He should get co-authorship as well."

The authorship question aside, there did seem to be room under Lander's terms for at least some kind of cooperation, perhaps including an exchange of data that would benefit both sides, and by extension the entire scientific community. The question was whether he could deliver the rest of the public program's leadership.

"Eric says Francis can be persuaded," said Roberts. "That could be used to convince the others. Gibbs would go along, and the DOE would be all for it. Bob Waterston and John Sulston are the make-or-break, and Waterston does what Sulston tells him to. That's been the nature of their relationship all along. So, Sulston's the driving force."

If Lander suspected that the public program could not win the scientific race, Craig Venter was beginning to see signs that Celera was in a similar position in the political one. Two days after the Wye River meeting, the president and first lady held one in a series of intimate Millennium Evenings at the White House, this one given over to the topic of "Informatics Meets Genomics." Vinton Cerf, who had helped father the Internet, was one of the two invited speakers. Eric Lander was the other. Among his plethora of gifts, Lander, a former Rhodes Scholar and MacArthur "genius" award winner, possessed a mesmerizing talent for explaining science with a mixture of mastery and enthusiasm that left his listeners enthralled—including, on this occasion, Bill Clinton, who sat in the front row in the East Room glowing with interest. Lander's central point, that human beings are genetically 99.9 percent the same, would, three months later, became a theme in Clinton's State of the Union address.

Francis Collins was in the audience, too, along with Harold Varmus, Ari Patrinos, Randy Scott, and all the other leaders in the field from both academia and industry. Except for one. A few days before, Venter's office had received a call from the White House asking for the sort of routine security information that usually presages an invitation to an event. But no invitation ever came. As the most prominent figure in genomics in the world, his absence would be conspicuous in any case. But it was even more striking in light of the fact that he was personally acquainted with the Clintons through Tom Schneider, a longtime Clinton friend and political operative. After a previous Millennium Evening, the three men and their wives had retired for drinks in the White House residence, the party lasting until after midnight. Venter had also served on the committee advising the president on biological warfare. In his eager way, Venter even fancied himself something of a "Friend of Bill." But on that evening, he was home watching television. He did not even know the event was taking place until after the fact.

The following day, Paul Gilman tried to find out why his boss had been snubbed. First he called NIH to see if Venter's name had been on the list of invitees to send to the White House. He was assured that it had been. Then he called the White House.

"They told me Craig was on the invite list, but somehow, somewhere, his name accidentally got removed," Gilman related afterward. "And if you believe that, I have the Brooklyn Bridge for sale as well."

While the reason for snubbing Venter was never confirmed, Gilman gave credence to a rumor that had began to circulate widely: the first lady had gotten wind of Venter's remark in Miami about DNA and the Monica Lewinsky affair, and did not think it was funny. "I guess this makes me an FFOB," Venter said in the cafeteria the next day. "A Former Friend of Bill."

Behind the joking, Venter was stung. He was concerned, too, that the lesson had more serious import. Three weeks before, in the midst of the GSAC meeting, the *Guardian* in England had published a story claiming that Prime Minister Tony Blair and President Clinton were negotiating an Anglo-American agreement to eliminate the patenting of human genes. The story clearly identified Celera as the target of the alleged agreement, asserting as a given that Venter's intent was to "patent as many genes as possible." According to the authors, the collaboration between the company and the DOE had been set up by Venter "to protect his investment" but had been scrapped because of pressure from the prime minister's office on the White House. From the frequent mention of the Wellcome Trust in the report, it seemed clear that the writers were getting their information from Michael Morgan or someone else at the trust. Since no one at the White House had been able to corroborate the *Guardian* story and no other papers had picked it up, Venter had decided there was nothing to it. Now he was not so sure. He still considered himself engaged not in a race but in simultaneous chess games, one played against Tony White, the other against Francis Collins and his colleagues in the Human Genome Project. It would not make the latter match any easier if the president of the United States was among the pieces arrayed against him.

At the same time that the rivals for the human genome were tentatively discussing the possibility of cooperating in some yet-to-be-defined way, the collaboration on *Drosophila* between Celera and Gerry Rubin's lab was moving vigorously ahead. Myers's team had succeeded in putting into linear order more than 97 percent of the fly sequence. Plugging the remaining holes would require months of finishing work in Rubin's lab. Equally important, the *meaning* of the sequence had yet to be revealed: how many genes it contained, where they were located, what proteins they directed, what function they served in the life of the fly, and, finally, what answers they might hold to the mysteries of human life and disease. A full understanding of the fly's biology on this level would take decades of experimentation, but a first-pass, computer-based overview of the genome's intimate structure—or what is called "annotation"—would give the research community a much more useful starting place than a numbing progression of As, Ts, Gs, and Cs. For a genome covering some 140 million base pairs, however, even a preliminary survey would take months of work.

"I wonder what could be accomplished if you threw fifty *Drosophila* experts and informatics people together in the same room and divvied up the genome among them," Rubin mused when he was visiting with Venter and Mark Adams one day in September to discuss the publication of the fly code. Venter's face lit up.

"Why don't we find out?" he said. "We could do it right here. We put the genome together all at once. Why not analyze it all at once, too?"

"Cool," said Adams. "An annotation jamboree."

If it was going to work, it had to happen quickly. Rubin contacted his colleague Michael Ashburner at the European Bioinformatics Institute, on the Wellcome Trust Genome Campus next to the Sanger Centre. Ashburner was a pugnacious academic with an ingrained suspicion of Venter and what he stood for. But after a lifetime of work on the genetics of the fly, the prospect of having its complete code opening up before his eyes was too exciting an opportunity to ignore. He and Rubin drew up an invitation list. People canceled vacations and rearranged teaching schedules to be able to participate. Venter agreed to pay all travel expenses, and Adams readied an empty floor of cubicles in Building II. Meanwhile Myers's team completed a final assembly of the fly code using all the data that had run through the machines—in the end, more than 12x coverage. According to the terms of the memorandum of understanding between Celera and Rubin's lab, the company was now obligated to make the information public by uploading it to GenBank. But the business development team raised a protest. "How can we attract customers to pay for subscriptions, if we give it away free to everybody else?" Peter Barrett asked. "And what's to prevent Incyte from downloading our work from GenBank and dishing it up on a platter to its own pharma customers?"

The *Drosophila* MOU was not legally binding. But Venter knew that if he deviated too far from it, he risked alienating the academic community, perhaps even losing Rubin's trust, which he had come to value a great deal. The bearded, easygoing drosophilist was a pure scientist in the best sense: above politics and greatly respected by his peers. He had taken a big risk in collaborating with Celera against the warnings of his colleagues in the public program, and Venter was loath to betray him. But if he did not also consider Celera's bottom line, he would leave himself vulnerable to attack from the opposite flank, even bringing Tony White crashing back into Celera's affairs. After some negotiations with GenBank's administrators at the NIH's National Center for Biotechnology Information, Venter proposed a compromise. Rather than release the fly code into GenBank specifically, Celera would agree to have it placed in a separate NCBI database reserved for unfinished gene sequences where it could not be downloaded in bulk. There, academic researchers and companies alike were welcome to use it any way they wished, including finding and patenting genes—as long as they first clicked their mouse on an agreement not to resell the data itself for commercial purposes.

Barrett and his group said they could live with the arrangement. Venter was not sure the academics would go for it. But in his mind the best way to find out if he was crossing the line was to stick out a toe and see if any alarms went off. Without consulting Rubin, he authorized the circumscribed release of the *Drosophila* genome on November 1, less than a week before the Annotation Jamboree was due to begin. Within an hour, Rubin received an e-mail from Ashburner in England. "WHAT THE FUCK IS GOING ON WITH CELERA?" Ashburner wrote in his subject line. "This is NOT the agreement I thought we had," he continued. "I am very pissed off and am minded (a) to pull out of the Jamboree and (b) e-mail everyone who has been invited telling them that I am and why." Ashburner protested just as passionately to Francis Collins. Soon e-mails were hopping back and forth like fruit flies in a bowl of overripe bananas. Collins wrote back that he shared Ashburner's "deep concern," copying Rubin on his missive. "Gerry, are you reading e-mail?" Collins typed. "This is a MAJOR issue. Failure to put the data in GenBank will be cause for a major protest."

At Celera, Venter was monitoring the reaction with the clinical detachment of a experimenter who had combined some chemicals and was only mildly surprised to find them exploding. Playing simultaneous chess was tricky and not every move was the right one. Rubin called to remind him of his promise.

"To tell you the truth, I wish we hadn't made that promise," Venter said. "There's a lot of value in the *Drosophila* data. But there's more value in us doing what we said we'd do." He called the NCBI and within twenty-four hours the data was moved into GenBank, with no restrictions on its use. Most of the researchers in the fly community were never even aware that there had been a crisis. But it was a sad day for the Celera business development team. "Craig *caved,*" M'Liz Robinson said, resting her forehead on her desk. "He caved! I'd be amazed if Incyte and everybody else isn't sucking this stuff up as we speak." Robert Millman stood in the group's coffee room, his eyes closed, pretending to wish himself back to his old job at Millennium Pharmaceuticals. "There's no place like home, there's no place like home," he chanted, clicking the heels of his red high-top sneakers together.

The following Sunday the academic scientists began arriving in Rockville for the eleven-day Annotation Jamboree. Rubin was there to co-direct the happening with Mark Adams. Celera was providing not only the workspace and equipment but a computer specialist to work hand in hand with each visiting biologist. The first day went badly. The Celera team had not had time to prepare the fly data by processing it through off-the-shelf gene-finding programs, so little work could be done at first. Many of the academic biologists, wary of their host and unaccustomed to working in a corporate setting, were suffering culture shock. "This is the first time I've been in a cubicle in my life," grumbled Mark Fortini, a young drosophilist from the University of Pennsylvania and a former student of Rubin's. "I hope it's the last."

By the next evening, Fortini was so engrossed in what was happening in his cubicle that he could hardly stand to leave it for dinner. With the software running smoothly, the jamboree had turned into a coordinated frenzy of discovery unlike anything anyone there had ever experienced, including Gerry Rubin himself. It was an intellectual Easter egg hunt. Guided by hints from the gene-finding programs and other clues, the biologists were turning up genes under every leaf—adding whole new gene families, spelling out their proteins, visualizing their shapes, tracing the pathways of interacting proteins. There were whistles of amazement, high fives exchanged over cubicle walls. "No human has ever looked at this stuff before," one young scientist effused. "It's like going into outer space or underneath the sea."

Farther down the floor, a gray-haired researcher wearing a white lab coat was on his knees like a transported ecstatic, offering a colleague a printout of what he'd found. It was the fly equivalent of a gene that, when flawed in human beings, causes the folded surface of normal brain cortex to be egg-smooth, resulting in mental retardation. "It's the sort of thing I've been hunting for years!" he said. "There is simply no doubt that it's the same gene in the human and the fly. At this level of exactitude, the three-dimensional structure of the protein must be identical too, and so, therefore, are the connections running to it! So you see, the whole system . . ." He trailed off, hurrying back to his cubicle.

"It's like being a postdoc again," Ashburner said.

Every evening, the scientists left their cubicles and crammed together in a room to report on what they had found that day, share insights, and discuss strategies. Sometimes there would be pizza, sometimes trays of sushi from a local Japanese restaurant. Venter often came up for these brainstorming sessions, offering an occasional idea or wry aside but usually just listening and delighting in the synergy taking place. Whatever notions the academics might have had about him in the abstract, it was hard not to like him in person, or to appreciate the gift he'd provided their community. During one of these twilight sessions a young postdoc named Leslie Vosshall rapturously described how in a few days she had found a whole family of sixty receptor genes guiding *Drosophila*'s olfactory system. She and her colleagues had been searching for them for five years with conventional methods and had found none. When she finished, she looked directly at Venter in the front row. "Thank you," she said. "Thank you for letting me find them."

By the end of the jamboree, the exhausted researchers had maxed out the number of fruit fly genes at 13,600. It was a surprisingly low number, considering that *C. elegans,* an order of magnitude simpler on a cellular level, had 18,000. On the other hand, in the complexity of its genomic landscape *Drosophila* represented a much better model for the human genome. Out of 289 known disease genes in human beings, at least 177 had counterparts in the fly, including the p53 gene involved in many cancers, and others regulating insulin and blood pressure. Mark Fortini, scrolling through his territory of the genome one morning, found a new gene for color vision that scientists had been hunting for for fifteen years. Somehow in that frenetic week and a half, he also found Debbie Morrison of the National Cancer Institute, another young fly researcher, and she found him, and five months later they were married.

The chances of a marriage between Celera and the Human Genome Project were growing more remote. For several weeks after the Wye River meeting in October, Eric Lander and Francis Collins had continued private conversations with Venter about the possibility of cooperating. At the DOE, Ari Patrinos was extremely eager to take an active role in bringing it about. But the British, led by John Sulston, were extremely wary even of sitting down at the same table with Celera. It did not help matters that Sulston learned of the renewal of collaboration talks only when Nicholas Wade called to interview him for a story about it in the *New York Times.* The Whitehead Institute might have surpassed the Sanger Centre as the most productive sequencing operation in the

HGP, but as the only non-American member of the G-5, the British held a powerful political sword over Collins, not to mention $170 million in Wellcome Trust money dedicated to the genome project. Nothing short of a complete giveaway of Celera's data was likely to satisfy Sulston and Michael Morgan at the trust, and no agreement could go through without them.

In the meantime, both sides were sequencing full tilt and sparing no effort to let the world know how fabulously well they were doing. In early October, the Human Genome Project had issued another press release with no apparent purpose but to state that the human sequence was "coming a lot sooner than you thought"—indeed, a billion base pairs would be done in a matter of weeks. Celera promptly announced that it had already delivered over a billion base pairs to customers, in spite of the fact that human sequencing had begun only the month before. Collins and his colleagues then pointed out that for *their* billion base pairs, they counted only those that had been assembled into 2,000-base-pair sequences and placed in GenBank, while Celera was counting every nucleotide that passed through a sequencing machine. The comparison, they said, gave the false impression that Celera was moving much faster than the public program.

To Collins, the most disturbing news was further down in the release, where Celera announced that it was filing preliminary patent applications on 6,500 new human gene fragments. A cry of betrayal rose from the Human Genome Project and its supporters, especially in Britain. Had not Venter explicitly promised to patent only a few hundred genes? The patent filing was not nearly the breach of faith that the public program scientists were making it out to be—certainly not as demonstrative a commitment to wholesale patenting as the 6,700 full applications that Human Genome Sciences claimed to have already filed, or the 173 actual patents that Incyte announced it had been awarded the month before, to not a whisper of protest from the HGP. As Venter tried to point out, sheer attrition in the discovery process made it highly unlikely that 6,500 provisional applications would yield more than the few hundred full patent filings he had always asserted were Celera's IP target. But this reasoning did not satisfy the critics. "There is some logic in what he's saying, but I'm not sure this passes the red-face test," Collins said. "For people who assumed IP was not a major part of Celera's intention, this was a wake-up call."

Indeed, the announcement of the patent filing was meant to be a wake-up call: it just roused the wrong audience. Celera's stock had begun to sink again, and by broadcasting its decision to file the preliminary applications the company intended to send an encouraging message to investors who were worried that the company might be giving away all its assets. But the release had no visible effect on the share price and succeeded only in offending Celera's clients, who wondered if it meant that Celera was going to patent information that its customers were paying millions to mine for IP themselves. With the recent addition of Pfizer to the original three early-access clients, Celera now had four pharmaceutical customers. But Peter Barrett's business team had yet to sign up a single university—a troubling fact, considering that the business model depended on luring thousands of academic customers to pay a reduced rate to drink at the Celera well. Certainly the acceleration of the government program was dampening the universities' enthusiasm for the company's product. But there was at least a little evidence that Collins was taking a more active role in Celera's customer relations. In spite of the unfortunate timing of Vanderbilt's visit to Celera in June, when the water pipe had burst and flooded the floor, officials at the university were working hard on a proposal to become the first academic customer. In October, someone at Vanderbilt leaked a copy of the document to a colleague at Washington University, who passed it on to Waterston, who duly forwarded it to Collins. In a conversation with Lee Limbird, Vanderbilt's associate vice chancellor for research, who was spearheading the project, Collins expressed "serious concerns" about the document.

When an e-mail describing this chain of events was leaked to Venter, he was furious. It was one thing for the government to go after the genome as aggressively as possible, he said, but it was quite another if Collins was trying to deprive Celera of income by intimidating potential customers. Collins indignantly denied any such meddling. But if Vanderbilt or any other university wanted to know how the Human Genome Project was progressing, all they had to do was call his office. "I will certainly not apologize for talking about where we were, and what we could do," he objected.

On November 23, Collins hosted an event to celebrate the deposit of the one-billionth letter in the human sequence into GenBank. A giant screen was broadcasting live feeds from simultaneous celebrations at the G-5 genome centers from Cambridge to California. Collins bounded to the podium with an exuberant jiggle in his rangy stride and a "Billion Base Pair Celebration" T-shirt pulled on over his shirt and tie. "Let's party!" he enjoined the faithful. Secretary of Health and Human Services Donna Shalala and Secretary of Energy Bill Richardson followed him onstage to praise the government program and accept their own Billion Base Pair T-shirts, which they quickly handed off to aides. Senator Tom Harkin (D.-Iowa), a longtime supporter of the Human Genome Project, announced that Congress was increasing Collins's institute's budget by 25 percent, from $269 million to $337 million. Everybody cheered.

"Pretty low, not to invite us too," Venter grumbled afterward, as if he'd momentarily lost track of reality.

Of course, the academics in the Human Genome Project would no sooner have invited Craig Venter to their party than the devil to a christening. Nor did they invite him to their program's next big moment a week later, on December 1, when simultaneous press conferences in Washington, London, and Tokyo trumpeted the finishing of chromosome 22, the first of the twenty-three human chromosomes to be completely sequenced. Two-thirds of the work had been done at the Sanger Centre, with the remainder contributed by Waterston's group in St. Louis, Bruce Roe's at the University of Oklahoma, and a lab at Keio University in Tokyo led by Nobuyoshi Shimizu. To be sure, it was a significant achievement. A number of disease-related genes had already been traced to chromosome 22, including genes implicated in schizophrenia, certain heart diseases, and a form of leukemia. The scientific team, led by Ian Dunham at the Sanger Centre, had managed to close all but eleven gaps on the gene-rich long arm of the chromosome, which they predicted contained a minimum of 545 genes. But there were limitations to the triumph. First, tiny chromosome 22 accounted for less than 2 percent of the entire human genome. The researchers had also ignored the short arm of the chromosome, which consisted almost entirely of repeated sequences unlikely to contain genes and impossible to assemble with the public program's approach. As their own research demonstrated, moreover, there was nothing intrinsically meaningful about the fact that the genes on chromosome 22 were all crowded into the same genetic boat. In a mouse, the analogous genes were scattered over seven different chromosomes.

Nevertheless, the public scientists spared no rhetoric in heralding the milestone. "A new era has dawned," Roe declared at the press conference in Washington. "We have fulfilled the dreams of Mendel, Morgan,

Watson and Crick, and Sanger." His colleagues across the Atlantic were even less restrained. John Sulston called the achievement "as important an accomplishment as discovering that the Earth goes around the Sun, or that we are descended from apes," while Michael Morgan's boss at the Wellcome Trust compared it to the invention of the wheel. Reporters were left to wonder what the completion of the entire genome itself could possibly be compared to when the time came, short of the creation of the universe.

Back in Washington, as Francis Collins left the press conference, a reporter caught up with him and asked if the finishing of the chromosome was more a symbolic than a scientific event. "No, it is a very important scientific milestone, too," Collins answered. "A chromosome is a biological entity." A couple of days later, at Celera's 1999 Christmas party in a Georgetown restaurant, the same reporter asked Venter what he thought of Collins's definition. "A biological entity? Yeah, well, so's a turd," he said. Claire Fraser, standing nearby, put her hand to her forehead. "Are you *ever* going to learn?" she said, leading him away.

The Georgetown party was upbeat. Even Robert Millman was in a good mood. He was dressed in an elegant coal-black double-breasted suit, red suede shoes, and, in place of a tie, a turquoise scorpion, which wagged its sequined-tipped stinger when he turned to face someone. He stood by the bar entertaining a couple with a series of astonishing card tricks. In one, he playfully mimed shuffling motions around an invisible deck in his hands, then asked the woman to pick a card but not let him know what it was. She went along with the mime, whispering to her husband that she had "picked" the Queen of Hearts. Millman reached into his pocket. "Hey look, I seem to have brought some real cards with me," he said, producing a deck still in its cellophane wrapper. He unwrapped it, smiled at his victim, and turned over the top card, the Queen of Hearts.

Ironically, the completion of chromosome 22 by the Human Genome Project had a magical effect on Celera's stock price, sending it up 25 percent in just two days. But this was just a sideshow compared to what happened a couple of weeks later. On December 16, the Motley Fool—an irreverent online market-advice site run by David and Tom Gardner, brothers with a penchant for breaking rules in their stock choices and making tons of money in the process—announced that it was investing $50,000 in Celera. The Gardners were not really fools, and a long analysis, written in their heady vernacular style, detailed the firm's pros and cons, the latter including an assessment of the stock as grossly overvalued, given its market capitalization of over $1.8 billion. "We mean, come on!" read the report. "[Celera] has no profits, no real revenue, and it has no clear business model, just a bunch of promises." Nevertheless, the Motley Fool managers were captivated by the prospect that Venter's enterprise was positioned to become the leading provider of gene information not just to drug companies but to the brave new post-genomic world. "Celera," they told their readers, "may become one of the most important brand names in your life."

In the tinder of the bull market, the report triggered a firestorm. The next day, trading of Celera shares reached an all-time high and continued at such a frenzied pace over the next week that the New York Stock Exchange had to halt trading in the company's stock repeatedly to let things cool down. By the end of the week, a share of Celera was worth $144—up 30 percent in a single day, and the largest gainer for the week on the entire exchange.

Unknown to the market, the negotiations between the company and the Human Genome Project were also coming to a head. Lander and Venter had talked several times on the phone, and while they did not agree on all the issues separating them, they had made remarkable progress, considering the obstacles. John Sulston and his colleagues at the Sanger Centre remained deeply suspicious of Venter but sanctioned a meeting with Celera, provided someone from the Wellcome Trust was included in the negotiations. The two sides arranged to meet at the Hyatt Hotel near Dulles Airport at 10:30 a.m. on December 29.

Leaving aside the inauspicious choice of venue—it had been at Dulles Airport a year and a half earlier that Venter had suggested to Collins that "you can do mouse"—there were signs even before the meeting began that things might not go well. Just before noon on the twenty-eighth, Collins sent Venter a list of "shared principles" to help guide the upcoming discussion, ostensibly reflecting the gist of the conversations with Lander. Venter put on his reading glasses and went down the roster. To his surprise, half the principles listed there Celera had never agreed to at all, including one precluding Celera from publishing any paper using GenBank data without listing Human Genome Project scientists as co-authors. He was worried, too, to learn that Tony White was insisting on flying down to join Arnold Levine, Paul Gilman, and Venter on the Celera negotiating team.

The G-5, meanwhile, had picked Harold Varmus, Robert Waterston, Francis Collins, and Martin Bobrow, a member of the Wellcome Trust's board of governors, to represent the public program. No one had been appointed from the DOE, in spite of—or more accurately, perhaps, because of—the agency's previous attempt to form a collaboration. Far more troubling, Eric Lander had suddenly disappeared from the proceedings.

"Eric felt that he'd done enough for the effort, and it was time to let some others get involved," Francis Collins later explained—which, to anyone who knew Lander, was a little like saying that Napoleon felt he had done enough by conquering Italy and wanted to give other generals a crack at the rest of Europe. The truth was, Lander was not trusted by some of the other powers in the HGP, especially the British. "I had the feeling that they are afraid of Eric," Patrinos told an acquaintance. "He's too powerful. The notion that he would come in on his white horse and get these parties to reconcile was too much." Lander himself admitted that he was not the best person to sell the deal to the other members of the consortium, in particular the Wellcome Trust.

Whatever the reason for Lander's absence, the meeting was a disaster. Collins opened with an update on the status of the public program, demonstrating its now massive sequencing capacity. After some questions back and forth, the HGP representatives began to press home what they would need for a collaboration to happen. They wanted complete access to all of Celera's data, including trace files, as well as use of Gene Myers's algorithms. The merged genome data must be made available not only at Celera's own web portal but through noncommercial ones, such as GenBank. To ensure that the Human Genome Project got equal credit for a joint annotation of the genome, any future Annotation Jamborees should take place at Washington University as well as at Celera.

The Celera team was completely taken aback. Allowing the public program complete perusal of its data and methods was a concession that went well beyond the terms of the *Drosophila* collaboration. They were even more shocked by what the government was willing to offer in return, which was essentially nothing. Celera's chief concern was that the information be protected from commercial reuse by competing database companies, like Incyte. Lander had clearly understood that. But the four people now across the table were opposed to any such restriction. If the public program was to contribute to Celera's effort, the resulting product must be available to the entire world, including Celera's competitors, free of charge or any limitation on its use.

"Why *shouldn't* Incyte be allowed to resell it, if that's what they want to do?" Robert Waterston asked. Arnold Levine, incredulous, rose out of his chair and demanded to know whether Waterston was negotiating for Incyte as well as the Human Genome Project. Tony White's jaw was clenched, his face reddening. He had participated in many business negotiations, but nothing like this: the other party seemed not to accept or understand the fact that Celera was a private enterprise and could not reasonably be expected to give away its primary product to its competitors. Collins, meanwhile, took the high road. "Think what a collaboration like this would mean to the future of science," he said.

"I don't give a shit about that," White interrupted. "If this collaboration is going to make money for Celera, I'm interested. Otherwise you're wasting my time."

He proceeded to lay down terms for an agreement that were as appalling to the other side as theirs had been to him. If Celera's data was to be used in a joint effort, he said, then Celera should be assured a monopoly over the commercial distribution and use of the resulting product for four or five years, including protection of the data, use of information on SNPs, and even on gene chips and other products manufactured using the human genome. Collins said he was prepared to consider such restrictions for six months, or a year at the outside.

"We have the ability to manage the data without you until then," White said. "Why should I make a deal with you, when I don't gain anything in return?"

The meeting broke up with the two sides further apart than they had been when it started. On the way out, Collins took Venter aside. "It's our understanding that whether we collaborate or not, you accept that we should be co-authors on any paper that uses our data, right?" he said.

"Where did you get that idea?" Venter said curtly. Then he turned and walked away.

CHAPTER 24
THINGS BEING WHAT THEY ARE

During the first two months of 2000, Francis Collins tried repeatedly to phone Venter to gauge his interest in continuing the talks. Venter did not return the calls. He was convinced that Celera had been led into a trap. In his view, the Human Genome Project had come into the negotiations with no intention of negotiating at all. Instead, Collins, Waterston, and the others had deliberately provoked Tony White by making demands that no businessman could meet, and the irascible executive had taken the bait. Tony had crossed the line. He had uttered the word "monopoly." In the context of the Book of Life, he might as well have sprouted horns and a forked tail.

"The things they laid on the table were just *shocking,*" Collins said afterward. He gravely shook his head, as if disappointed in a juvenile delinquent's failure to reform. "I got the impression then that there is a public position at Celera, and a real position. And now we were seeing it laid bare."

Up Rockville Pike, nobody was losing sleep over the collapse of the talks. The DNA pipeline was running at full throttle now, and the sequence coming through was of superb quality. In spite of the continued dearth of customers, the stock was performing like an athlete on steroids. On January 3, when Venter appeared prominently in a CNN feature on the genome, it rose thirty-seven points. Meanwhile Gene Myers's team was developing ever more efficient algorithms. At Venter's annual New Year's Eve party, Myers and Hamilton Smith—neither one exactly a party animal—had gone off in a corner and sketched out a new sequencing and assembly strategy on a cocktail napkin. The key was a new way of obtaining 50,000-base-pair clones, devised by Smith and another Celera scientist, Rob Holt. If it worked, Myers's team might be able to put together the genome without using the public program's data at all. To keep the design of the new technique from leaking to the other side, only Venter, Mark Adams, and a couple of others were told about it. Smith gleefully referred to it as "the secret weapon." "Everything is going great," he told an acquaintance. "Everything that we said we would do, we're doing. It's going to be a great year. Craig doesn't just want to be first, he wants to be first *and* have the quality that will last for ages. He wants to stick it to the government program. Make them hurt."

With the collaboration between Celera and the government program seemingly dead, there was no reason to hold back. On Friday, January 7, the company announced that there would be a press briefing the following Monday to reveal a "key milestone." The mere news that there would be news drove the stock up another twenty-five points. Over the weekend the stock chat rooms on Yahoo and the Motley Fool sizzled with speculation about what Venter would divulge.

The press briefing was the first test for Heather Kowalski, Celera's new public relations manager. Determined that everything would go perfectly, she had hardly slept during the weekend, writing and rewriting the press release, compiling charts and graphics, and making sure that all the arrangements were set up for coverage of the event by teleconference and over the web. On Monday morning, piles of shiny press kits covered a table at the door to the Atlantic Room and a string of hookups lay waiting for the expected broadcast crews. Unfortunately, America Online and Time Warner, Inc., chose that morning to announce the largest corporate merger in the history of the planet. With the Washington business press rushing out to AOL's press conference in nearby Virginia, only a handful of reporters showed up at Celera. To make matters worse, the only film crew that did make an appearance was a *Nova* team working on a documentary on the genome race. Visually, Kowalski's big event now consisted of a panel of Venter, Tony White, Mark Adams, and Hamilton Smith blinking into the camera lights in a mostly empty room. Kowalski literally wrung her hands. But Venter seemed almost amused by the circumstances. "Just get some senior people down here to fill these chairs," he told her. "Tell Marshall to bring his inflatable woman."

When a few more people had assembled, Venter walked casually to the podium. Behind it hung an imposing tapestry, borrowed for the day from TIGR, expressing sequences of human DNA in luminous columns of colored wool. Against that backdrop his shining pate took on an aura. Celera's database, he announced, now contained 90 percent of the human code. "This is a monumental moment," he said, "not only in Celera's history, but in the history of medicine."

It was an even greater moment in the history of hyperbole—if not as great as the Human Genome Project's gassy effusion over chromosome 22. Celera had actually sequenced less than 2x coverage of the genome, enough to capture about 80 percent of the sequence. The extra 10 percent had been calculated by adding in the data from GenBank that did not overlap with Celera's own sequence reads. Even with that addition, there was not nearly enough DNA yet for the assembly team to attempt to produce a linear sequence. Still, "90 percent done" had a familiar ring: it was the same standard of completion that the public program had set as the finish line for its own draft, not scheduled to appear until the spring. Judging by the ensuing questions, the handful of reporters in the audience and those teleconferencing in were cautiously impressed.

"I'd like to make it clear, this is not an announcement of the completion of the genome," Venter said in answer to one reporter, who still seemed puzzled over what actually *was* being announced.

"So, is it like you have a few frames of the movie for each gene?" asked another.

"I'm going to turn you over to our Nobel laureate, who will explain the simple math," Venter replied. Hamilton Smith unfolded from his chair and took the microphone. "We've read approximately eighty-one percent of the genome," he said. "Take the public contribution of fifty percent, we overlap that by eighty percent, so twenty percent is not overlapped. Twenty percent is ten percent of fifty percent, so we add that number and get ninety percent."

The room was dead quiet. Smith retreated bashfully to his seat.

"Hopefully that will clarify it," Venter said. "If not, I'll give you Ham's home phone number."

The market did not let a little confusion over the meaning of the announcement dampen its enthusiasm. All day long, Venter kept hopping up from meetings and interviews to check the stock on his computer, like a teenager who could not tear himself away from a video game. By closing time, the share price was up another 55 points and trading at 240. "Congratulations," White said dryly, putting on his coat to fly back to Connecticut. "You're bigger than Motorola."

The scientists of the Human Genome Project were less taken with Celera's "monumental moment." To them, it was nothing more than a statistical shell game meant to delude the press into thinking that Celera was far ahead of the government program. More galling still was how well the ploy seemed to work. One of the reporters covering the briefing was Dick Thompson, a veteran science writer for *Time.* Two weeks later, Collins opened his copy of the magazine to find Venter grinning back at him above a chart of numbers comparing the Human Genome Project and Celera to the proverbial tortoise and hare. Collins composed a long, indignant letter to the writer and his editor, scolding *Time* for its flattering coverage of Celera's "wholly arbitrary" milestones at the expense of the "stellar achievements" of the Human Genome Project. "You owe the hardworking and dedicated public sequencing community a sincere apology, and a commitment to never making these errors again," Collins wrote. Thompson did not respond. But the letter would have repercussions down the road.

At the time, the public program's true totem was neither a hare nor a tortoise but an aroused, frustrated bull. John McPherson at Washington University had finished a clone map of the genome, making it easier to parcel out and sequence the 20,000 BAC clones—the 150,000 base-pair chunks of code—without duplicating effort. The big G-5 labs were spewing out sequenced DNA. Some, like the Sanger Centre, were holding to a high-quality standard, favoring the production of "finished" sequences over amassing raw reads. But Eric Lander had adopted a greedier approach, grabbing more and more genomic territory to feed the superautomated operation at his Whitehead Institute center. What came out the other end was poorer in quality, but there was a lot of it, and Lander waved away apprehensions with the assurance that the data could be cleaned up later. As his stockpile of data grew, so did his political stock within the G-5. When Lander spoke in the weekly teleconferences, Collins deferred. Some of the others might still mutter about his bull-charge approach, but it was getting the job done.

At a G-5 meeting in Walnut Creek, California, on January 13, 2000, the assembled scientists set as their goal announcing that the draft was complete at the annual Cold Spring Harbor gathering in the middle of May. Nobody talked about "winning the race," at least not in public. But nobody had forgotten the wound inflicted at Cold Spring Harbor two years earlier. What could be more exquisite than to celebrate victory in the same place where Venter had predicted their defeat?

Still, a frustrating paradox was embedded in their very determination. There was no getting around the fact that the faster the consortium dumped DNA into GenBank, the faster Celera could scoop it out and add it to its own self-generated lode. The company was like one of those 1950s sci-fi monsters that absorb the energy of the bazooka shells and missiles launched at their flanks, growing ever larger from the attempts to destroy them. The G-5 leaders could take some comfort in the belief that, since Celera was working without a map and had no idea where its masses of random sequence reads belonged in the genome, its scientists would be unable to assemble the repeat sections of code in their proper places. But the public program was facing an assembly problem of its own. In essence, the 20,000 mapped BACs in the consortium's sequencing pipeline represented 20,000 separate puzzles. To be sure, these were little 150,000-base-pair puzzles, as opposed to Celera's one colossal conundrum. But given the mere fivefold coverage of the genome that defined the "working draft," most of the BAC puzzles would be only partly solved, the DNA pieces within them jumbled in the wrong order or—just as bad—oriented in reverse of their true order, like an alphabet reading from Z to A. Finding genes in those unsolved puzzles would be like trying to find whole potato chips after someone has stepped on the bag. And Craig Venter could be counted on to let the world know about it. If the HGP celebrated the draft as complete at Cold Spring Harbor, he would denounce it as an ersatz product cobbled together solely to claim precedence over Celera. All the lofty language Collins could muster would not be able to hide the element of truth in such an attack.

What the public program needed was a quick way of bringing more interior order to the unfinished BACs, ideally by the middle of May, when its scientists would gather once again at Cold Spring Harbor. There was one hope. Most of the HGP's sequences were read off only one end of a DNA clone, but back in October Robert Weiss, a researcher at the University of Utah, had prepared a library of small "plasmid" clones that could be sequenced at both ends, just like Gene Myers's mate pairs. Once their DNA letters were read, these paired ends could function as guideposts to sort out the correct order of DNA fragments within each BAC. Conceivably, the work could be done by May. But there were two hitches. The first was money. Sequencing the paired ends would take at least another $10 million, and Collins did not have it in his budget. And even if he could find the funds, where could the work be done? The genome centers' production capacities were already strained to the limit.

Collins had an idea, but it would require some tact in handling. NHGRI did not have the money for the project, but the SNP Consortium did. The alliance of eleven drug companies and the Wellcome Trust had been set up to look for single-base-pair variations within the human code. The consortium's work was going well, but DNA from a new source would boost its ability to ferret out additional base pair variations. Weiss's plasmid clone library fit the bill perfectly. The scheme was a classic win-win: if the consortium helped fund the project, it would get its SNPs and the G-5 would get the paired ends needed to bring order to the genome draft. Channeling the work through the consortium would also mean that the generated data would not have to be posted to GenBank immediately, according to the Bermuda rules—thus preventing Celera from drawing nourishment for its own assembly from the extra boost of data. Finally, the amount of red tape involved would be far less for a private-sector group like the consortium than for a government agency like NHGRI. With Celera barreling ahead, there was no time for bureaucratic delays.

Collins also had an idea about where the sequencing could take place. He broached the matter with his G-5 colleagues at the Walnut Creek meeting on the thirteenth, just two weeks after the collapse of the collaboration talks. If Celera did not want to cooperate on finishing the genome on the HGP's terms, Collins was pretty sure its chief commercial competitor would. He contacted Randy Scott. Unlike Venter, the Incyte chief was more than happy to take Collins's call.

"Francis came to us and basically said, 'OK, Incyte, make us an offer we can't refuse,'" Scott later said. "We were quite pleased. Our greatest fear was that the HGP wasn't going to step up to the challenge, and Celera would have the genome alone." Incyte quietly undertook a pilot project.

There were some political matters that needed tending to as well. Collins was feeling pressure from the administration, Congress, and the scientific community at large to make a deal with Celera. If that was no longer a possibility, he was intent on making sure that responsibility for the breakdown in the talks rested where he thought it belonged. The question was how to do so without appearing to betray the secrecy of the negotiations. With considerable input from Michael Morgan at the Wellcome Trust and Kathy Hudson, his own policy director, Collins came up with a plan. He drafted a letter to Celera, to be signed by himself and the three other members of the negotiating committee. The letter recounted his unsuccessful attempts to reach Venter after the collapse of the negotiations and in great detail laid forth the company's stringent demands on maintaining commercial control over the merged genome, including the expectation of distribution rights over the product for as much as five years and the requirement that any finishing work done by the Human Genome Project, even after Celera's contribution was over, would still be subject to the same exclusive rights.

"While establishing a monopoly on commercial uses of the human genome sequence may be in Celera's best interests," the letter read, "it is not in the best interests of science or the general public."

The letter ended by setting a deadline of March 6—just a week away—for Venter to reply, after which Collins and his colleagues would assume that Celera had no interest in continuing negotiations. But Venter was not the letter's only, nor even perhaps its primary, intended audience. Tony White had laid bare Celera's "real position" to the HGP, and the HGP was reserving the right to lay it bare to the world.

"We all realized that clearly a letter of this sort would be of interest to a lot of people," Collins later explained. "Things being what they are, the possibility that the letter would ultimately reach public scrutiny had to be considered." He insisted, however, that there never was any explicit discussion among the G-5 over when, or in what manner, the confidential letter was to "reach public scrutiny."

In retrospect, maybe there should have been. The letter reached Celera by fax just after 4:00 p.m. on February 28. At first it caused remarkably little stir. Venter was out of town, but Lynn Holland read the gist of the fax to him over the phone. "More Francis nonsense," he said. If Venter had been less offended by the letter's scolding tone, he might have been more alert to its peculiar content. Why had Collins and his colleagues gone to such trouble to spell out Celera's position for Celera? If the point of the letter was to inquire about the grounds for further talks, why not make an effort to clarify the Human Genome Project's positions instead, perhaps indicating which were nonnegotiable and where some flexibility remained? But if Venter recognized any ulterior motive in the letter, he did nothing to defend himself against it.

Celera was about to make a major new stock offering that week, and Venter was away promoting it in another grueling road show. The secondary stock offering was to finance an expansion from genomics toward the development of therapeutic drugs. The first step was a move into "proteomics." Genes might direct life, but proteins were the actors that made it happen, and a wide-scale understanding of protein form, expression, and patterns of interaction was thought to be as essential to drug development as information on the genome itself. A number of new biotechs were sprouting up in the new field, and most pharmaceutical companies were giving more attention to it as well. But Celera had a particular advantage. PE Biosystems had a new mass spectrometer nearly ready that promised to make it possible to analyze proteins on an industry-size scale, just as the Prism 3700 machines had brought DNA sequencing into a factory setting. With its sister company once again providing the technological backbone, Celera was in a good position to capture the inside track as the world's prime source of protein data. And there was no government-sponsored Human Proteome Project to compete against.

A quick, big move into proteomics would take cash. To finance the enterprise, PE Corporation had announced the sale of nearly 4 million new shares in Celera on February 29. There was a risk in such a secondary offering: in the short term, releasing more shares would dilute the value of existing ones. But Celera's market value could take the hit. Five days earlier, the stock had reached an astonishing $276 per share, even after a two-for-one split the month before. A share bought on the first day of trading less than a year before had increased in value twenty-five-fold. "There should be only one place people go for [genomic and proteomic] information, and it's going to be us," said Charles Poole, PE Corporation's director of investor relations, when he and White came down that day to confer on the progress of the offering. "We're going to be the Big Gorilla."

The venture proved a spectacular success, netting PE Corporation a cash infusion of almost a billion dollars—twice what White had originally hoped for. Venter positively shone with exuberance. His Celera stock alone was now worth almost $700 million. It gave him pleasure to know that Mark Adams, Ham Smith, Gene Myers, and the others who had come with him in the beginning were all rich as well. On paper, even Lynn Holland was a millionaire. Venter made plans to buy a bigger yacht to replace *Sorcerer.*

Collins's letter, meanwhile, was gradually disappearing beneath a stack of other papers in Venter's inbox. With the proteomics initiative occupying everyone's attention, it was hard to hear the sound of a ticking time bomb.

The device went off on Sunday, March 5. Kowalski was at home with her husband in Arlington when she got a call from Paul Jacobs at the *Los Angeles Times.* She was apprehensive as soon as she heard his voice. Jacobs had already written several articles critical of Celera, and for him to be calling on a Sunday morning could not be good news. He told her that he had obtained from the Wellcome Trust a copy of the letter and wanted to know if Celera wished to comment on its implications. Kowalski was stunned. The letter was clearly marked CONFIDENTIAL on every page, and the deadline to respond to it, short as it had been, was still one day away. She told Jacobs she would get back to him and tried to call Venter. But before she could dial his number, the phone rang again. This time it was Justin Gillis of the *Washington Post,* who had also received a copy of the letter. Indeed, according to Gillis, "the Brits were faxing it all over creation."

Their American collaborators were allegedly as shocked and appalled by the leak as Kowalski was. According to one source at NHGRI, the principal architect of the letter gambit at the institute was not Collins but Kathy Hudson. Known within the Human Genome Project for her "kill Celera" attitude, she had conferred directly with Michael Morgan to ensure that the letter would expose Celera's hard-line position. The question was not whether the letter would be leaked to the press but when and how. Nevertheless, like Kowalski, Hudson was rudely awakened by a phone call that Sunday morning. It was Morgan on the line, telling her that the Wellcome Trust had decided not to wait any longer to pull the trigger. "Jesus Christ, Michael, the deadline hasn't even expired yet!" Hudson said. She hung up and tried to reach Collins. He was visiting his father in Stanton, Virginia, and she did not know the phone number or the father's first name. She called directory assistance and one by one began waking up Collins families in the Stanton area. Between calls, her own phone was ringing with reporters. When she finally reached her boss, he quickly packed and rushed back to Washington. Meanwhile Hudson contacted Cathy Yarbrough, NHGRI's press officer, and told her what had happened and what to tell the press. It was imperative now to distance the NIH as far as possible from Morgan's preemptive strike. That was not going to be easy.

"I was ordered to tell reporters that we had nothing to do with the leak," Yarbrough later remembered. "They basically laughed in my face."

Together with Gilman and Arnold Levine, Venter composed a public response to the Collins letter, insisting that it had dramatically misstated Celera's position on intellectual property. White had voiced his demand for a five-year "monopoly" on the data only in response to the HGP's own insistence that its scientists be allowed to see and use Celera's trace files and algorithms. Otherwise, Venter said, Celera's position had not changed at all, the only restriction being that Incyte and other competing database companies not be allowed to upload Celera's data and sell it as their own.

The next day, news stories about the vitriolic breakdown of the negotiations ran on both sides of the Atlantic. The letter gambit had backfired. While it succeeded in outing Tony White's hard-line demands, the premature leak made Celera appear as much a victim as a villain. It looked as though the Human Genome Project had challenged Celera to draw and shoot on the count of three, and then fired its own gun on the count of two. "I'm sort of disgusted that they would send us this threatening 'confidential' letter with a time deadline on it, then fax it to the press [before the deadline]," Venter told Gillis at the *Post.* "I don't even know what to make of it. It's such a low-life thing to do."

"More and more," the reporter wrote, "the Human Genome Project, supposedly one of mankind's noblest undertakings, is resembling a mud-wrestling match."

It was only round one. In an extraordinary coincidence, on the same day the letter went public, Venter found out about Collins's attempt to recruit Incyte into the genome race against him. The G-5's plan to ally with the SNP Consortium to help fund a paired-end sequencing project had first surfaced the previous Friday in the form of a request for applications (RFA) circulated by the consortium—in essence, an invitation to various private and public genome centers, including Incyte, to submit competitive bids to undertake the work. Celera was not on the list, but

Allen Roses of Glaxo Wellcome, a leading force in the SNP Consortium, had made sure that Venter received a copy over the weekend. It was clear from the first paragraph of the RFA that the purpose of the project was not just to generate SNPs for the consortium's own work but to help the government genome project assemble its working draft.

"I took one look at that document and I knew exactly what it was about," Hamilton Smith said, more amused than upset. "There is no reason for anyone needing SNP information to require paired ends. The value of them is in assembling—seeing how you can put things together over a stretch."

Venter was not so amused. Why would the SNP Consortium invest millions in a project so obviously tailored to benefit the public genome program? Clearly Collins had his hand in the business. But it was not until Monday morning, with the leaked letter crisis still at full boil, that Venter learned the full extent of the G-5's involvement and their original intention to source the work directly to Incyte. It was Roses and the SNP Consortium who had insisted that the work should instead be put out for open bidding. "It is strange that Celera is the only one being singled out for being a profit-making corporation while deals are being made with others," he told a reporter from *Science.* "I do not see Incyte making much public."

The revelation of an attempted deal between his government competitor on one side and his chief commercial competitor on the other left Venter feeling utterly disgusted. It seemed as though his enemies were banding together to ensure his demise—including, in his view, using taxpayer funds channeled through a third party to purchase precisely the kind of data the government would have gotten from Celera for free, had the collaboration gone through. At the time, he happened to be reading *Galileo's Daughter,* by Dava Sobel. The great Renaissance scientist suffered much at the hands of rivals seeking to discredit his astronomical theories. "Nothing's changed in four hundred years," he told an acquaintance. "Substitute 'genome' for 'the moons of Jupiter,' and it's the same thing all over again—greed and jealousy, jealousy and greed. Not that I'm likening myself to Galileo. But this is the biggest issue of *our* time, so maybe it's not inappropriate."

Not surprisingly, Venter did not hesitate to share his suspicions with the media. A week later, on March 13, under the headline "Feds May Have Tried to Bend Law for Gene Map." The writer was Tim Friend, the reporter who had published Francis Collins's infamous *Mad Magazine* quote two years before. "Because NIH could not hire Incyte itself without competitive bidding," Friend wrote, citing Roses as his source, "Collins and Michael Morgan of the Wellcome Trust apparently sought out the SNP Consortium to hire Incyte for them."

Collins was furious. He called Friend's story "unforgivable," and adamantly insisted that there had been no plan to single-source the work to Incyte, and certainly no connection between the collapse of the collaboration talks with Celera and the decision to obtain the paired-end sequences somewhere else. "This whole conspiracy thing is bizarre beyond words, and deeply troubling," he told another writer soon after Friend's story appeared. "I'm saying this from the heart. The motivation was to provide a better product for the scientific community. If the collaboration discussion had gone better, then yes, perhaps we would not have needed this. But it didn't, did it?"

Kathy Hudson expressed herself more viscerally. Catching sight of Friend before a White House news conference on March 14, her eyes narrowed to slits. "If he comes over here," she said, "I'm going to punch him in the nose."

Hudson was having a terrible morning. The White House news briefing was the culmination of months of planning in which she had taken a very active role, and it appeared to be on the verge of becoming a wholesale disaster. Earlier that morning, President Clinton and Prime Minister Tony Blair had issued a joint statement pledging to lead an effort to ensure that the "raw, fundamental data of the human genome, including the human DNA sequence and its variations, should be made freely available to scientists everywhere." Within moments, Collins and Neal Lane, the president's science advisor, were to mount the platform in the White House briefing room to clarify the meaning of the statement. That task had suddenly become a major effort in damage control.

The joint statement by the two leaders was a British initiative, begun almost a year before, after conversations between Lane and Sir Robert May, his counterpart in the prime minister's office, and Lord David Sainsbury, the British Parliament's undersecretary of state for science and innovation, at a G-8 world trade meeting in Kyoto. The British ministers were responding in turn to intense lobbying by the Wellcome Trust. The task of drafting the president's remarks had fallen to Hudson at the NHGRI and Rachel Levinson, a colleague of Lane's in the White

House Office of Science and Technology Policy. In light of the threat from Celera, Hudson was anxious that the statement be issued as soon as possible, but it had taken an excruciatingly long time to be reviewed and approved. Once the slew of White House offices and government agencies wanting input had signed off on it, the announcement still had to wait for a suitable context for the president to utter it. The annual Medal of Science and Technology award ceremony that March morning seemed appropriate.

Hudson had gotten the first inkling of trouble while driving into work. She was idly listening to a local radio station when she heard the newscaster mention something to the effect that Clinton was calling for the "elimination of gene patents." The report was so off the mark that at first Hudson wasn't even sure the newscaster was referring to the same announcement: a sentence explicitly supporting patent protection for "gene-based inventions" had been included in the statement she helped to craft. For a few desperate moments she clung to the hope that it was just an isolated case of bad reporting. Then her pager and cell phone began to beep in unison. In announcing the news briefing, Joe Lockhart, the White House press secretary, had made it sound as though the president and the prime minister were advocating a ban on intellectual property protection for *any* genetic discovery.

Hudson met Collins at NIH and they rushed down to the White House together. By the time they arrived, the stocks of Celera and other genomics companies were plummeting, dragging the rest of the biotech sector down with them. Neal Lane had briefed the president on the crisis. "He knows what's happening," Lane told Collins and Hudson when they arrived. "But he's decided not to alter course." The award ceremony and press briefing would go on as planned.

"This agreement says in the strongest possible terms our genome, the book in which all human life is written, belongs to every member of the human race," Clinton said at the ceremony. "Already the Human Genome Project, funded by the United States and the United Kingdom, requires its grant recipients to make the sequences they discover publicly available within twenty-four hours. I urge all other nations, scientists, and corporations to adopt this policy and honor its spirit. We must ensure that the profits of human genome research are measured not in dollars but in the betterment of human life."

Now, with biotech stock prices in a free fall, it was up to Lane and Collins to reassure the press and the market that the president of the

United States was neither voicing his opposition to gene patents nor gunning for a particular private company so recently in the news. Hudson stood against the wall in the briefing room, her arms crossed over her chest, frowning at the crowd of reporters packed into the room. On the podium in front, Collins looked relaxed, even jolly. But Lane stood rigidly, squinting against the glare of the television lights. A mild-mannered physicist who had risen to the summits of public science administration, he looked as if he would be glad right then to be back in the comparative safety of a university lecture hall. "I want to make it absolutely clear that this statement has nothing to do with any ongoing discussions between the private and public sector," he told the press in his opening remarks. ". . . I also want to make it clear that the statement is not about patents or what should or should not be patentable."

"I am happy to be here on what I think is a rather significant day," Francis Collins announced when it was his turn to speak, "where a very important principle about access to the human genome sequence—our common, shared heritage as human beings—is being endorsed by the leaders of the free world."

When he was finished practically every reporter in the room shot up a hand. "We have companies that have applied for thousands of genetic patents. . . ," said one correspondent. "Are they going to be able to receive patents or not?"

Lane repeated his assurance that the joint statement had no effect on patent policy. Then came a volley of questions about the relationship between the statement and the failed talks with Celera. Lane tried valiantly to separate the announcement from the context of the genome race, pointing out that Venter had made clear his own desire to keep the "raw, fundamental data" available. But as hard as he and Collins tried to steer the discussion back to the statement itself, the questions kept coming back to Celera and the effect of the news on the stock market: Was the statement aimed at preventing the company from patenting the human code? To drive it back to the bargaining table? Did Craig Venter know in advance what the president was going to say? If he supported its sentiments, as Lane claimed, why hadn't he been invited to participate in the announcement? Was Dr. Lane aware that Celera's stock had already fallen 15 percent that morning? Was he aware that technology stocks generally were getting hammered because of the statement?

"I have no information on that," Lane answered, clearly uncomfortable. "I see no reason to connect the two."

Perhaps it would not have mattered what he said. By this point, investors were as panicked as a herd of horses fleeing a burning barn. By the end of the day, $40 billion had been drained from the biotech sector, over $2 billion from Celera alone. This was not a mud-wrestling match, it was a bloodbath. The vertiginous momentum carried with it the dot-coms and other tech stocks on the Nasdaq exchange, which suffered its second-highest one-day point loss in history. No one knew it then, of course, but this was the first thunderous rumble of the imploding tech market. The party was over. Soon there would be devastation all around, with smoke blackening the sky, the smell of cordite in the air, cats with their fur blown off staggering around in the streets. In hindsight, the boom was grotesquely overripe. Any number of things could have triggered its collapse. It just happened to be the human genome.

CHAPTER 25
A GARDEN PARTY

Over the next two weeks, Celera's market value continued to plummet, losing another $4 billion by the end of March. There was still the billion dollars in cash from the secondary stock offering to take comfort in. But soon came more bad news. After the crash, a trio of law firms filed class-action lawsuits against Celera on behalf of investors who had purchased the new shares. The suits claimed that PE Corporation had not come clean in its prospectus, having neglected to mention its negotiations with the government genome program, a fact that might have influenced investor enthusiasm.

It seemed ironic, to say the least, that Celera should be sued for failing to disclose a collaboration that did not take place. But the moment was drowning in irony. A presidential utterance having no effect on existing gene patent policy had damaged the value of every private company patenting genes, not to mention hundreds of other companies having nothing to do with genes at all. The company most hurt by the blast of hot air was the one least dependent on patents in its business model and the one most willing to put genetic information in the public domain. And leading the charge against this capitalist menace was the Wellcome Trust—a foundation financed by one of the largest drug conglomerates in the world.

The scientists at Celera did not seem bitter to see their personal wealth evaporate overnight. It was all a paper loss anyway. Still, when one's enemies were recruiting the leaders of the free world to play on their team, it was hard not to feel a little paranoid. No one felt the sting more than Craig Venter. The crash had personally cost him some $300 million. Venter was convinced that all that had occurred—the failed collaboration, the leaked letter, the G-5's secret discussions with Incyte, and the Clinton-Blair sledgehammer—amounted to a single, coordinated effort by Collins and the Wellcome Trust to destroy him and his company so that the Human Genome Project could take the credit for unraveling the code themselves. "Those guys are so tightly linked they don't fart without the other one's permission," he told an acquaintance.

Collins adamantly denied any such conspiracy. He claimed he knew nothing about the letter leak from the Wellcome Trust beforehand. And while Michael Morgan and his boss, Michael Dexter, at the trust had taken an active role in the Clinton-Blair announcement, Collins himself had not. In any case, the event had already been scheduled well before the leaked letter turned the failure of the collaboration into public knowledge. If Venter's company was in trouble, Collins implied, Venter had only himself to blame. "Craig has tried to play both ends," Collins said. "He wants to be the generous scientist and at the same time the clever businessman with an obligation to his shareholders. I think he's made a Faustian bargain and doesn't realize it."

Not even Collins could find much fault with Venter's generosity where *Drosophila* was concerned. The fruit fly genome was published in *Science* on March 24, in a paper co-authored by almost two hundred people, roughly evenly split between Celera employees and the academic scientists led by Gerry Rubin. At the end of the article was a coupon that could be cut out and sent in by anyone who wanted a CD containing the completed, annotated version of the genome, free of charge. At a fly community meeting in Pittsburgh the day before publication, the scientists arrived in the auditorium to find copies of the *Science* article on every chair. Venter received a standing ovation. He was heartened, too, by Rubin's view of the collaboration with Celera, which the drosophilist had made known at the end of his talk a month earlier at the annual meeting of the American Association for the Advancement of Science in Washington. "A lot of my colleagues were not enthusiastic about this collaboration with a private company—or this particular company, or this particular person," Rubin said. "I was given advice that I was going to get in real trouble and be very disappointed. . . . Now that it's done, I can say it's been one of the most pleasurable scientific experiences in my thirty-year career. . . . Maybe it pisses people off that [Venter] says what he thinks. But I think that's a good quality. He always has kept his promises."

The celebration for the publication in the Celera cafeteria, on March 24—right on the heels of the stock crash—was subdued. There was dim sum and Sam Adams beer for the five hundred employees, and CDs containing the fly code and little jigsaw puzzles of a representation of the genome to take home. Robert Millman, in a floppy brown suit, wandered alone through the crowd, shaking his head at the idiocy of so much IP flowing out of the toothpaste tube. A doleful Miles Davis tune issued from a CD player; Marshall Peterson took out the disk and slid in a copy of the *Drosophila* genome in its place, "just to hear what it sounds like." The speaker emitted a low, throaty hum. Venter raised a bottle of beer and called for a toast. "This is *our* moment," he said. "We got our fly on the front page of the *New York Times.* Just like the president's fly. Only our fly will have a lot more lasting impact on history."

Privately Venter sounded much more discouraged. Scientifically, Celera was in fairly good shape—but it might be too late. If Gene Myers's team could work their magic again, Celera would have an assembled, annotated human genome ready for publication by September 2000, a full year ahead of the original schedule. But the Human Genome Project was publicly claiming that their "working draft" would be done in June, three months earlier. Worse, there were rumors that Collins was going to make a surprise announcement at the Cold Spring Harbor meeting in May that it was already finished. That just didn't make sense. A cursory analysis of the data in GenBank was enough to see that the HGP scientists were far short of the 90 percent standard of completion that they had set for the working draft. In fact, they were barely over 60 percent. Even supposing they could sequence another 30 percent in a few weeks, they still would not have the paired-end reads, without which much of the draft would be a mess of unordered fragments, only marginally useful for finding genes. How could they claim they were almost done?

Venter could imagine only two explanations. Either Collins and his colleagues were deliberately withholding human DNA from GenBank to keep it out of Celera's reach, waiting until just before the Cold Spring

Harbor meeting to dump a huge bolus into the database and surprise everyone with a come-from-behind victory. Or else they were going to declare the race over and won well short of reaching the finish line. The first explanation violated the Human Genome Project's own Bermuda rules. The second simply violated the truth. "I don't know how to deal with it," he said, his voice uncharacteristically somber. "I'm just kind of stunned. I can't believe NIH will allow Francis to publish a genome when he hasn't got one."

Fortunately, there was an opportunity coming up to air his feelings in a forum where they might have an impact. After the market swoon, the House Subcommittee on Energy and the Environment had called a hearing for April 6 to review the genome controversy. Venter was determined to make the best possible use of his invitation to testify. In the meantime, Norton Zinder, who had helped found the Human Genome Project but was also a Celera board member, had initiated a meeting between Celera and some top brass at NIH, including acting director Ruth Kirschstein and two institute directors, Richard Klausner of the National Cancer Institute and Anthony Fauci of the National Institute of Allergy and Infectious Diseases. Like Zinder himself, the NIH officials were nervous about the upcoming congressional hearing and wanted to forestall any more public bloodshed. At first Venter refused to attend the NIH meeting if Collins was to be present. Fauci persuaded him. "Don't worry about Francis," he said. "If he tries to submarine a truce, he'll be fired."

The meeting took place at NIH on March 31, a week before the scheduled hearing. Venter and Collins could barely look at each other. Directing his words to the other NIH representatives, Venter launched into a history of the perceived sins committed against Celera by the government program, starting with the *Mad Magazine* comment two years before, moving on to the shutdown of the Celera-DOE deal, the leaked letter, and the devastation wrought by the Clinton-Blair statement. Collins sat rigid and dour. Finally Kirschstein interrupted. "That's enough," she said, sharply. "We have to put all this behind us and move on."

But Venter wasn't finished. If the Human Genome Project declared their working draft done while the BAC fragments were still unordered, he said, it would amount to nothing less than fraud. That order could not be achieved without the kind of paired-end data that only Celera had developed. "How can you call it a sequence when you don't have any order?" he said. "It's like somebody claiming they've conquered Mount Everest after they reach the first base camp."

Thus provoked, Collins spoke up. "But we *are* getting paired-end data, Craig," he said. The critical sequencing project was now to be done, he explained, not by Incyte, whose proposal had been rejected by the SNP Consortium, but by Eric Lander's and Robert Waterston's operations, with funding from both the SNP Consortium and NHGRI.

"Great, Francis," Venter remarked. "I'm glad to hear that the techniques we've developed and you said wouldn't work are proving so useful to you."

Underneath the sardonic retort, he was shaken. Along with Rubin, Adams, and Myers, he had been invited to present a *Drosophila* session at the Cold Spring Harbor meeting in May—and now it looked even more possible that the HGP scientists would be greeting him there with the announcement that their draft of the human code was completed. When he and the others returned to Celera, he immediately called Gene Myers and Mark Adams down to his office. Hamilton Smith and Paul Gilman were already there. "I need an assembled genome in three weeks," Venter said.

"But we don't even have all the data yet," said Smith.

"Then we assemble what we have, and add the rest as we go," Venter replied. "Gene, how soon could you have something ready to announce?"

Myers was slouched back in his chair, wearing his black leather jacket. He took a moment to answer, and when he did a little grin played on his face. "How about at Cold Spring Harbor?" he said.

"You mean we reveal that we've finished the human genome during the *Drosophila* session?" asked Adams, incredulously.

"Why not?" Myers replied. "A little surprise for them. And guess what. Francis isn't speaking until two days later."

Hamilton Smith let out a big laugh. The prospect of beating the public program to the punch by a couple of days, and on their own sacred ground, was too outrageous an idea to take seriously. Or was it?

"We'd better bring some bodyguards," said Venter. "Or we'll never get out of there alive."

While nothing substantial had been resolved at the NIH meeting, the two sides agreed to issue a joint statement before the congressional hearing took place. Through the following week, drafts of the statement kited back and forth between Norton Zinder, Celera, and the NIH directors, emphasizing the "complementary" nature of the genome projects and their agreement on gene patent policy. If there could not be a collaboration, at least it was beginning to look as if there might be a public easing of tension. Zinder was deeply engaged. Since his retirement, he had been longing for an opportunity to be useful, and he was glad to play a role in negotiating such an important truce. But the détente fell apart abruptly on April 5, when Ruth Kirschstein received a draft of the written testimony that Venter was to present the next day to the congressional subcommittee. In spite of the conciliatory language of the statement in the works, the testimony laid out before the lawmakers all of Celera's grievances against NIH in shockingly blunt language. "I find myself in the peculiar position of warning you," Venter wrote, "that in the race to complete a draft human sequence, the publicly funded human genome program may be at a stage where quality and scientific standards are sacrificed for credit." He even implied that the slow pace of the HGP before Celera's appearance may have cost thousands of lives lost to cancer and other diseases.

Kirschstein phoned Venter, so angry she could barely speak, and canceled the joint statement. After the call, he wandered into Paul Gilman's office wearing the contumacious look of an unjustly reprimanded child. "Maybe she thought that if NIH played nicey-nice for one day, I'd just roll over and let my business get ruined," he said.

"They need this more than we do," Gilman said. "Let it go."

As Gilman knew, Celera had a surprise in store. At 7:30 the next morning, the company issued a press release: "Celera Genomics, a PE Corporation business, announced today that it has completed the sequencing of one person's genome and will now begin to assemble the sequenced fragments of the genome into their proper order based on computational advances." The release, timed as a prelude to Venter's appearance a couple of hours later before the House subcommittee, also announced that the company was moving on next to tackle the mouse genome, which it said was as important to biomedicine as the human code. Two hours later, Venter entered the hearing room in the Rayburn House Office Building, radiant in a dark blue suit. Tony White and a train of other Celera and PE Corporation people took seats in the row immediately behind the witness table. Also testifying were Gerry Rubin,

Neal Lane, and, representing the Human Genome Project, Robert Waterston. But it was Venter's show. Building on the news in the release, he announced that Celera's assembly of the sequenced pieces would be ready in "three to six weeks," a prediction that shocked even some of his own staff. Then, instead of reading from his inflammatory written statement, he ad-libbed a more temperate, upbeat oral testimony, stressing Celera's open-research business plan, the unmitigated success of the *Drosophila* collaboration with Rubin, and his own vigorous opposition to companies who would file patents on mere "gene-like" sequences. The questions that followed were partisan, but even some of the Democratic members of the panel seemed won over by his artlessness. The Republicans, meanwhile, lobbed softballs. "[Celera's business model] sounds very altruistic," said one. "Why do people object to it? What do people fear? Are they afraid you will get to the mountaintop sooner?"

As the hearing ended, Venter was mobbed by reporters. Unnoticed, Robert Millman walked quickly out of the building and rushed back to Celera. He had left instructions for his assistant to print out a new batch of patent applications to be delivered to the Patent Office. He arrived just in time to delete the frequent phrase "gene-like" from the applications before they went out the door.

Wall Street loved Venter's testimony. "This is huge. This is wonderful," said one analyst. "[Celera is finishing] three or four months before expectations." The stock price, which had already regained some of its lost altitude the day before when President Clinton clarified that his previous statement did not mean he was opposed to gene patents, ticked up another twenty-five points. But the share value did not stay high for long. Francis Collins had been warned from higher up not to make provocative statements about the genome controversy. But in his opinion, Celera's press release "hit a new low in communication clarity." Celera's alleged milestone amounted to nothing more than the admission that it had abandoned the sequencing of human DNA after having accumulated only three- or fourfold coverage of the genome, instead of its originally intended tenfold coverage. That this technicality would be lost on the average science reporter, Collins believed, was part of Celera's public relations plan. On April 10 he could not help voicing his opinion to a reporter.

"You should not take at face value any claim by any group for at least two years that says we have finished sequencing a human genome sequence," he said, before addressing a genetics conference in Vancouver. "It will not be true." The reporter published the remark, the wire services picked it up, and down went Celera's stock again, relinquishing all the territory it had recovered. The company thus attained the notable distinction of being the biggest one-day percentage gainer on the New York Stock Exchange and the biggest one-day loser, in the span of less than a week. This was more than irony; it seemed a kind of madness. Word came down to Collins from the Department of Health and Human Services to keep his mouth shut or else.

At this point, with the rancor of the race threatening to dominate the national economic scene, Norton Zinder decided to give diplomacy one more try. A decade before, Zinder had helped his old friend James Watson get the nascent Human Genome Project on its feet. He still thought of the program as a child that he and Watson had nursed through its infancy, and he had joined the Celera board only because he had lost faith in NIH's resolve to get the genome done while he and Watson were still alive to see it. In January 2000, Zinder suffered a stroke. It was not serious, but it left him with an unshakable anxiety that a bigger one could strike at any time, killing him or leaving his mind in shambles. Since the stroke, he had witnessed the hostility between his two allegiances in a growing state of agitation and despair. Bitterness, ego, and misunderstanding were destroying everything he and Watson had worked for. Zinder might not have the political power he once possessed, but he had time, passion, and deep connections in both camps. Someone had to do something to save the dignity and integrity of the genome project, and it might as well be him.

At seventy-two, Zinder was compact, excitable, and still handsome, with swept-back curly hair and a gnomic twinkle in his eye. He talked eagerly and often, his speech broken up by little yelps of laughter and jolts of profanity that paid witness to the stress of fifty years of dealing with the egos and petty jealousies of his peers. Ruth Kirschstein, as acting director of NIH, had the authority to intervene and force a rapprochement, but in Zinder's assessment she didn't have "the *cojones*" to use it. He could see nothing ahead but woe. If Celera completed its assembly first—as seemed inevitable, given that it had twice the data to work with—it would leave NIH scarred in reputation and hobbled in its future funding appeals to Congress. But if Collins tried to fend off defeat by declaring the Human Genome Project finished while its draft was still a chaos of disordered fragments, the long-term embarrassment would be even worse. Something had to be done to bring about a reconciliation while there was still time. Zinder could think of nothing else, day and night. He was terrified that his anxiety was going to well over into another stroke, and that thought made him still more distraught. "I'm living on Valium," he confided to an acquaintance. "I can't sleep. I can't think."

In the middle of one restless night, Zinder hatched a new plan to bring the two sides together. The scheme centered on the National Academy of Sciences' annual garden party at its headquarters in Washington, scheduled for April 28. The event was an elegant affair for members and invited guests, including the brightest luminaries in American science. Ruth Kirschstein's attendance was assured. Collins, Anthony Fauci, Rick Klausner, and the other NIH institute directors would no doubt be there as well, along with Neal Lane and Rita Colwell, who was the director of the National Science Foundation. Eric Lander was a member of the academy and could be approached to make sure he attended. Zinder would also make sure that Watson was there; his presence was critical to the plan. Venter, in spite of his accomplishments, was not an academy member, but it would be easy to arrange an invitation for him.

Zinder's idea was to take advantage of this transitory massing of scientific clout to crush any opposition to a new agreement carefully worked out in advance. On the tap of a shoulder, the key players would "spontaneously" absent themselves from the outdoor party and regroup in an office upstairs, where paper, food, drink, and a laptop would be waiting. With Kirschstein presiding and Bruce Alberts, the NAS president, acting as host, they would formalize a new accord between NIH and Celera, without the knowledge of either Tony White or the Wellcome Trust. The deal could be signed right on the spot. With the doyens of American biology pressuring him to cooperate, Collins would have to go along, whether he liked it or not. Everybody would shake hands. Perhaps they would open some champagne and turn to Zinder to make the toast.

He approached Kirschstein, Fauci, Klausner, and Venter with his plan. The terms of his proposed truce were substantial. To begin with, the HGP would agree to drop its plans to announce the completion of the working draft in mid-May. If the two sides were to work together, this demand was nonnegotiable. Over the summer, the groups would merge their data and convene an "Annotation Jamboree" like the one held for *Drosophila.* They could then decide whether to publish one genome together or two genomes separately in the same journal issue. The main obstacle to previous collaboration attempts—how to protect Celera's intellectual property while still keeping the public program's data freely available—might be worked out through a copyrighting scheme being proposed by David Lipman, the head of the National Center for Biotechnology Information, which ran GenBank. If that did not work, then Celera should be granted a period of time, perhaps three years, in which it could retain some commercial rights over the genome.

In Zinder's mind, the scheme depended on getting the right people on board beforehand and holding them there until the garden party took place. The key players were Venter, Lander, and Watson. These were volatile, powerful men, any one of whom could abort the plan in an outburst of ego or temper. To make sure that didn't happen, Zinder assigned chaperones to monitor each one. He enlisted Richard Roberts, the chairman of Celera's scientific advisory board, to keep Venter in line. "Rich's job is to watch your penis, and my job is to watch Jim's," Zinder told Venter. "Rich will be on your fucking tail. Rick Klausner watches Eric. Meanwhile, the Wellcome Trust is not told shit. The party is on Friday. On Monday morning, if the deal goes through, Mr. Clinton calls Mr. Blair and says, 'Put up or shut up.'"

It is a tribute to Zinder's once potent stature in science and his sheer passion that such an improbable scheme gathered a little momentum, however briefly. Venter, at least, seemed willing to entertain the notion. "In theory, I'm supposed to bump into Ruth at the party, then we all go up to the president's room and something will get signed," Venter told Gilman. "Francis has met with Tony Fauci. Obviously *something* will happen there. Maybe a fist fight."

Gilman did not want to disillusion his boss, but he thought the scheme had as much chance as solving the Middle East crisis by getting Yasser Arafat and Ehud Barak together over beers at a baseball game. "Norton wants to do good," he told an acquaintance. "But he's living in a different universe."

A week before the event, Kirschstein told Zinder that while she very much wanted to continue talks with Celera, rushing into things would be a mistake. "Let's not mix a wonderful party with business," she said. But Zinder would not let go. He forwarded Kirschstein's e-mail to Fauci and Klausner, desperately urging them to change her mind. "Prop her up!!" he wrote. "It must all be over after the garden party or never. . . . We've history in our hands. Let's not fumble the ball." But Fauci and Klausner also cautioned against "rushing in somewhat of a frenetic manner" to force an agreement through at the party. Meanwhile, whatever enthusiasm for the idea that there had been at Celera had cooled, especially after Robert Millman pointed out that there was no way that Lipman's copyright plan could protect the company's raw sequence.

When the day of the party arrived, there were no taps on the shoulder, no handshakes in an upstairs room, nothing for history to remember. The event went on pretty much as it had in previous years. In the morning, scientists gave talks on bioterrorism and other pressing topics. In the afternoon, senators mingled with Nobelists in the academy's elegant garden. Young women in white jackets passed through the crowd bearing trays of Chardonnay and salmon hors d'oeuvres. The NIH stars stood along the gravel path leading to the big tent on the lawn, where they would be sure to be noticed. Beneath the tent, Hamilton Smith wandered alone from table to table, checking out the food.

CHAPTER 26
END GAME

Zinder's scheme may have been improbable, but his diagnosis of the situation was disturbingly accurate. History *was* watching, and it was not a pretty sight. Both sides in the race were now suspicious that the other was about to make a preemptive announcement, and the mutual distrust only increased the likelihood that one of them would. Rumors of a showdown at Cold Spring Harbor had the press clamoring for admittance to the private meeting. The editors of *Science* were appealing to Collins and Venter to find a way to resolve the conflict before "an irreversible collapse of collegiality" damaged the scientific community. John Sulston meanwhile was telling the BBC that Celera's genome was "a con job." Celera's stock was losing value like a burst barrel leaking oil. Zinder was right: someone had to do something. But for all his heartsick hope, he was not the right someone.

At Celera, everything depended now on Myers's team, just as it had at the end of the *Drosophila* project. Since the completion of the fly genome, its algorithm had been rewritten to run faster and more efficiently, and the computer power at its disposal had been vastly beefed up. The team itself had moved into more spacious quarters, with a lounge area, dozens more cubicles, and a larger, custom-designed "war room." At the entrance, Myers's black-and-white scarf, neatly folded and framed like some archaeological artifact, hung beside the cover of the *Drosophila* issue of *Science.* Except for a pair of windows, all four walls of the room were composed entirely of whiteboard, covered now by a multi-colored Jackson Pollockian din of sketchy lines, crenellations, equations, graphs, and interconnected canister-shaped objects. High in one corner of one wall, Myers had scrawled something in English: *"Assembly will take* 3–6 weeks": *JCV, 4/6/00.*

"Order in three to six," he told the team. "That should be your mantra." There was a glum silence when he said it, the kind made heavy by too many eyes staring down at a table. Perhaps everyone was practicing the new mantra.

Three of those weeks had now passed. The Cold Spring Harbor meeting was set to begin on May 10, a week away. Myers sat in his corner cubicle facing away from the door, staring into his laptop. His leather jacket hung loosely around his hunched shoulders, and with his wild black hair obscuring his face, the effect was that of a human black hole. Apparently things were not going too well. "When Craig asked me if I could have an assembly in three weeks, I said yeah, we could," he explained, once he noticed the visitor in the doorway who had been trying to get his attention. "And we did. We've done the assembly. The problem is, a whole lot of the genome just didn't come together. Like, forty percent. I'm bummed. Really pissed. I mean, where are all these *expectations* coming from?"

The original expectation had been to assemble the human sequence using a whole-genome shotgun algorithm, working on enough raw data to cover the code ten times over—10x. Back in September, the team had proved that a complex organism's genome could be put into linear sequence with as little as sixfold coverage, albeit with plenty of small gaps that would need to be filled later. But that was the fruit fly. In spite of Myers's optimism back then, the human genome was proving to be far more repetitive, with as much as 45 percent now estimated to consist of repeats, as opposed to only 8 percent in the fly. He still believed that with the human genome covered seven or eight times over using Celera data, the team could put together a product that met the standards they had set with *Drosophila.* The problem was, Celera was never going to produce that much human DNA. Venter had made the tactical decision to halt sequencing human DNA after only 4x coverage. Myers could hardly argue anymore with the business logic: the public program's data was available to use instead, and stopping at 4x not only saved $100 million but allowed Celera's sequencing machines to be turned over to the mouse genome project. Getting the mouse genome done would be a huge draw for customers. In many ways the mouse genome was more useful for finding drugs than the human, and—for a couple of years, at least—Celera would be the only place to get the information.

On the other hand, the decision to abort human sequencing meant that using the public data in the assembly was no longer just an option; it was a necessity. And Myers and his team were beginning to realize what a load of woe that meant for them. They had designed the Celera algorithm to run on the kind of highly accurate, paired-end libraries that only a master molecular craftsman like Hamilton Smith could produce. Feeding in the public data instead was like filling the gas tank of a race car with regular unleaded. The team was finding a great deal more contamination from foreign DNA in the sequences downloaded from GenBank, mostly from the viral vectors used to hold the human inserts in place in the cloning process. There were also more chimeras—instances in which two bits of DNA from different parts of the genome had been mistakenly welded into one. And of course, most of the available public data consisted of single reads instead of mated pairs. Overall the data quality in GenBank might be good enough for the Human Genome Project's working draft, which by definition presupposed a lot of tidying up down the road. But to make it work on the Celera algorithm would require a lot of hard work and hope.

Under the circumstances, the best strategy amounted to a kind of wholesale deconstruction of everything the government's team had put together. In order to perform a true whole-genome shotgun assembly, Myers needed random five-hundred-letter DNA fragments covering the whole genome. But the publicly available human DNA data was not random at all: it was mapped to specific places on each chromosome and was already partly assembled into two-thousand-letter segments before going into GenBank. The only way to see the raw, unassembled five-hundred-letter "traces" off the consortium's machines was to collaborate. But with the two projects at war, that was not an option. Myers's only choice was to computationally rip apart the Human Genome Project's partly assembled edifice, shredding the BACs into five-hundred-letter bits and treating them as if they were actual random fragments spat from a sequencing machine. Gone would be any information on where they belonged in the genome and what other five-hundred-letter bits they had once been joined to in GenBank. That was very important, because all too often the Celera scientists were finding that the public data were improperly joined. In effect, it was like smashing a hastily half-built model airplane to splinters with a hammer. The splinters—"faux reads," Myers called them—would then be mixed in with the 20 million fragments off Celera's own machines, and the assembly algorithm would attempt to put the whole pile together according to its own calculus, blind to what was once part of a wing, a tail, or a fuselage.

Myers had most of his team working on this whole-genome shotgun assembly method. They called it "Grande," not because of the enormous scale of the compute involved, but because Myers drank a lot of grande cappuccinos. Under pressure from Venter, however, he had also directed two of his team members to develop an alternative assembly method. Nicknamed "the Overlayer," the program made full use of the Human Genome Project's information on where each piece of its genome belonged. It sorted Celera's millions of random reads into twenty thousand separate piles corresponding to the public project's BACs, and then treated each pile as a separate assembly problem. In effect, it was a mutual backscratch: the public data pinned Celera's random reads down to a particular region of the genome, while Celera's data, with its mate pair information, could be used to bestow the correct order and orientation of the pieces within each of those regions.

The Overlayer represented the quickest way to deliver an ordered sequence. Venter was urging Myers to get it done first and worry about the whole-genome shotgun version later, if only to get a product out to Celera's handful of data-hungry customers. But it was hardly a creation Myers could unveil with pride—especially since it wasn't working. The two data sets weren't meshing well. While the reason was hard to pin down, it appeared that flaws in the public data were preventing the algorithm from making firm matches between fragments, like air bubbles weakening an adhesive bond. A computer crash over the first weekend of May only made things bleaker. The following Monday, Granger Sutton went downstairs to give Venter and the others a report. Myers was away in Europe lecturing on *Drosophila* but joined in by speakerphone. *"Comment allez-vous?"* he said, calling from Denmark. "How is everybody?" Sutton gave him the news. Myers gave a huge sigh, his mouth too close to the mike. It sounded like someone opening the door of a blast furnace.

"I thought you said you guys could have this done in three weeks," Venter said.

"We said we could have it on time if everything went right," Myers replied. "It sounds like this weekend everything didn't go right."

"Did anything go right?" Venter asked.

Everybody waited for Sutton to reply. "The weather was good," he said.

"We need Grande," said Mark Adams. "If there are problems with the way they've put things together, breaking their contigs apart will help." But both programs could not be run at once on the supercomputer.

"We can't accelerate Grande until after we get out from under the yoke of Overlay," Myers said.

"Which doesn't look like anytime soon," Sutton added. Nobody spoke for a while.

"If we go up to Cold Spring Harbor with nothing to say," Myers said, breaking the silence, "we're going to get crucified."

"If we can't say anything, fine," Venter replied. "Our goal is to publish the genome when it gets to the quality of *Drosophila,* and not before. Cold Spring Harbor is an artifact. If we wanted to make a major announcement, I don't think we'd choose that as a venue." It was as if there had never been mention of doing just that.

"I certainly agree with that reprioritization," Myers said.

Venter was running out of moves. Throughout the race he had counted on his nimbleness in reconstituting his aims to adjust to circumstance. When the public program had sped up ferociously to challenge him, he had deftly shifted his strategy to use their charge against them. When the sequencing machines would not work, he had used Tony White to trump Michael Hunkapiller, and when White and Robert Millman had threatened his vision of an open genome, he had used the PE board to buy some time. But now his space to maneuver was limited. Unless Myers's team had a breakthrough soon, one afternoon in June a beaming Francis Collins would be standing in front of the cameras in the Rose Garden, with the president beside him wearing that earnest clenched smile of pride that wrinkled up his chin. Maybe the prime minister would join in by video, or drop down by parachute. A quartet of air force jets, one for each letter in the code, could fly by and dip their wings. It would not matter to Wall Street or to the man on the street that the Book of Life being revealed was a half-assembled approximation that was not even useful for finding genes. The perception would be that Celera had lost.

For once, Venter wasn't sure what to do. Attempting a first-strike announcement of his own was not an option. He had publicly stated already at the congressional hearing that the assembly of the code was the next step, and with the whole-genome shotgun method there was no fudging on what that meant. You either had a linear sequence or you didn't, and Myers and Sutton had made it very clear where things stood. Briefly, Venter conceived what he called his "evil plan": let the Human Genome Project publish its draft in June, but schedule a human Annotation Jamboree to take place the same day at Celera, inviting every reporter in the vicinity to take a seat on the sidelines. Thus the public program's one big news day would be swamped by the flood of revelations on the meaning of the code coming out of the jamboree. But there was no assurance that there would be a Celera sequence to annotate by then, and the idea was too contrived in any case. The best they could do was to continue as fast as possible toward a *Drosophila*-quality genome, standing quietly aside while the public program published its semi-ordered draft, and let the difference speak for itself when the time came. In any case, there was not much now to do but wait. To fend off depression, he kept occupied with side projects, like getting a scrap of Einstein's brain to sequence. Venter was giddy about the prospect of the Einstein genome. But the project fell through.

On the evening of May 4, Venter got a call at home from Ari Patrinos of the Department of Energy. It was not unusual for Patrinos to telephone him at home. The two men had been friends ever since the DOE had become a major supporter of Venter at TIGR, and they had made it a regular tradition to talk by phone on Sunday afternoon and catch up on things. Patrinos was also a friend and neighbor of Francis Collins, and was always careful not to betray either's confidence. Lately Venter and Patrinos had been discussing the possibility of renewing a collaboration between Celera and the DOE. But this was not the reason for Patrinos's call that evening.

"What would your response be if I were to invite you to come by my house this weekend?" Patrinos said.

"I guess I'd ask if your intentions were honorable," Venter said.

"What about if Francis just happened to come by, too?" said Patrinos.

"Now you're getting kinky."

"I'm serious," said Patrinos. "I'll make sure my family is out. It would just be the three of us."

"I guess it depends on the purpose of the meeting."

"Just to talk."

"The last time I sat down 'just to talk' with Francis, I got a class-action suit filed against me," Venter said.

"If nothing comes of it, then it never happened," Patrinos said. "Francis lives just down the street. If he agrees, too, I'll arrange it so he just happens to drop by at the same time. That way we all have plausible deniability."

Venter was skeptical, but he told his friend he'd think about it. He hung up and reported the conversation to his wife.

"Maybe you should wear a wire," she said.

Venter was still pessimistic when he rang the doorbell of Patrinos's townhouse in Rockville the following Sunday afternoon. He was late, and Collins was already there. Patrinos invited Venter in and offered him a beer. There were some chips and a plateful of cheese and crackers on the table in the dining room, which was strewn about with toys and colorful drawings by Patrinos's two young daughters. His wife had taken the girls out shopping. In one corner was a parakeet in a cage. The three men sat down around the table. Patrinos had never seen either one of them so tense. For a while they danced around the issues, but at least the conversation was polite. Collins went out of his way to voice his admiration for what Celera had achieved in so short a time, and Venter returned the compliment. Venter assured Collins that Celera was not going to make a surprise announcement at the Cold Spring Harbor meeting, and Collins said that the Human Genome Project had no plans to do so either. The two scientists found some common ground in their mutual resentment of the feeding frenzy in the press over the race, especially Nicholas Wade of the *New York Times,* who each man felt had favored the other's side. When they got down to business, however, it was clear that they were still miles apart. At one point Collins suggested that the Human Genome Project publish its working draft and then the two sides could sit down together and work together on finishing the genome.

"Francis, there is one thing that is certain here," said Venter. "That won't happen."

"Why not?" Collins asked. "It's just a *draft.* We've made it very clear all along, this is not a finished product that directly competes with yours."

"Whatever you say, whatever you think, the whole world is looking on it that way," Venter said. "If you do a rough draft, there won't be a collaboration. Period. It's been extremely frustrating, feeling that you're rushing ahead to publish, which puts pressure on us to do the same before we've got the high-quality product we want."

"That pretty much sums up the way *we've* felt all along," said Collins.

Up until then, Patrinos had mostly just listened, getting up to refresh the snacks and speaking only to steer his two friends away from their perennial points of conflict. But now he spoke up. "You're both nearly done with your respective projects," he said. "What if you forget about trying to collaborate and simply agree to coordinate the timing of your announcements?"

"You mean, we declare a tie," said Venter.

"Call it what you like," said Patrinos. "But the point would be to say, so everyone can hear, 'We have our differences, but we're done, they're done, hooray for everybody, now let's see what we can do to work together to figure out what the genome means.' "

"We'd have to coordinate the timing of the publications to follow as well," said Collins.

"Of course."

"It's something to think about," Venter said. "But for obvious reasons, we have to be able to distinguish Celera's quality from what else is out there, if we want to stay in business. I don't mean to be critical, Francis. Half of your BACs are in good shape. But the other half are terrible."

Collins merely shrugged, not wanting to get into an argument about it. About Patrinos's proposal, he was noncommittal. The three men chatted a little more, and after agreeing to return for another private talk in a couple of weeks, Collins left. Patrinos was pleased, but Venter was still pessimistic. The two talked a little more on the sidewalk.

"Whatever Francis agrees to, there is no possibility of cooperation if he can't control his center directors and the Wellcome Trust," Venter said. "They do the dirty work for him."

"Let's just take one step at a time," Patrinos replied.

Considering the residue of gall left behind after he last set foot in Cold Spring Harbor, Craig Venter's return there two years later was remarkably uneventful. The day before the meeting began, NHGRI issued a press release announcing that chromosome 21 was completely finished and the working draft itself almost done. But there were no surprise announcements, no genome showdowns, and when the visitors from Celera arrived and stood in a wary clutch in the corner of the patio behind the dining hall, no rotten fruit was lobbed their way. Venter himself seemed determined not to stir up any trouble, presiding over the *Drosophila* session on the opening morning with all the cheek and swagger of a retiree on Ritalin. "I'd like to thank all our speakers for their great presentations" were about the most provocative words out of his mouth, and except for a brief, chilly exchange in the question-and-answer period between Gene Myers on the podium and Philip Green, his nemesis, in the audience, the session went off without incident. Venter shook hands with James Watson, standing against the wall in a green raincoat, and the Celera contingent withdrew. Two days later, Collins gave the keynote address with his usual sturdy vigor. "You'll tell your grandchildren you were here in Grace Auditorium, in the presence of the giants of genomics, including Watson himself. This is the moment!" he said, referring, evidently, to the moment of almost-doneness.

Collins had not mentioned his secret meeting with Venter to anyone, not even his own staff. On May 24, ten days after Cold Spring Harbor, the two men met again at Patrinos's house. Both arrived precisely on time, which Patrinos took as an encouraging sign. They agreed to the concept of coordinating announcements on the same day, but perhaps across town from each other. Patrinos carefully led them toward more difficult issues. A merging of the data sets was no longer on the table, for all the old reasons. But they might still publish separate papers at the same time. Donald Kennedy, a former president of Stanford University and soon to become the new editor of *Science,* was eager to get both papers in one mega-genome issue of the journal.

The sticking point once again was public access to the data. It is a deeply held principle of scientific publication that anyone should be able to access the data used to support a paper's conclusions. In genetic experiments, which since the early 1980s often involved data sets too large to fit between the covers of a journal, this had come to mean placing the relevant information in GenBank. Venter had already made it clear that Celera was *not* putting its human genome in GenBank but would make it available on the company's web site, where it could be protected against data piracy by other firms. He told Collins that Kennedy had indicated that *Science* would agree to some such arrangement, provided researchers could examine the data and download it for their own analyses for free and publish their results without interference from Celera. Collins was doubtful that *Science* would really agree to such a break in tradition. But at least they were making additional progress. The three men had begun a discussion of some technical problems in assembly when Venter's beeper went off. It was his wife, telling him he was way overdue for a dinner engagement.

A week later Patrinos had his strange bedfellows over again, this time ordering pizza. Collins arrived with some good news. He had taken the step of informing the genome center directors, as well as Neal Lane at the White House, about the secret meetings and the concept of a joint announcement. All were pleased. Most important, Michael Morgan of the Wellcome Trust had given the plan his blessing. There were still many details to be worked out, but having the British on board was a great relief to everyone. One problem concerned the timing of publication: Collins wanted to publish in *Science* as early as July, but Venter was opposed to rushing out something less than the best that his team could achieve, arguing for publication at the end of the year. Gently, Patrinos suggested September as a compromise.

Up to this point, Venter had told only two colleagues about the meetings: his wife and Mark Adams. The next day he called in his senior staff, divulged what had been going on, and laid out the notion of a coordinated announcement and simultaneous publication. Everyone hated it. Myers scowled, Heather Kowalski left the room in disgust, then angrily swept back in, and even Hamilton Smith shook his great white head in disbelief. It was abundantly clear to them all that Celera was far closer to getting an assembly than the public program was—in fact, the only obstacle seemed to be the exasperating quality of the public data that Myers's team was trying to wrestle into submission. Under the circumstances, conceding a tie seemed insane.

"Why are you *trusting* these people?" said Kowalski. "Do you really think they are all of a sudden going to play nice with you, after everything else they've done?"

"Francis says he won't make an announcement until we do," Venter said.

"And you believe him?" asked Mark Adams.

"He's under pressure from the administration to work with us, and getting pressure from his own people not to announce anything when their data are still bad," Venter replied.

"There has to be some way of differentiating our product from theirs," said Gilman. "Why else would anyone pay for ours?"

"The rules are, we talk about our genome, they talk about theirs," Venter replied. "We don't diss their data, and they don't diss ours."

"*Please* don't do this," Kowalski pleaded. "If you do, it's going to have a huge negative impact. It's going to look like we couldn't deliver what we said we could, and had to come together with them to finish." Venter gazed at her sadly. He understood what she was feeling, having just been through it himself.

"Is it better to say we finished second?" he said, looking from one face to another around the table. "Because that's how it will be perceived if they make their announcement without us."

"But it seems to me that we're ahead in the press now," said Myers.

"Do you think it will be the same after the president gets involved?" said Venter. "We have to decide whether we want to be part of this or get hit by it. The last time we got hit, we lost six billion in market cap. This time they'll have not just the president and the prime minister, but Germany, Japan, every little group that wants to stand up and take some of the credit for uncovering the Book of Life. If we try to combat that by dissing the quality of their product, it's going to look like sour grapes."

"Is it dissing them to say that we have an ordered genome and they don't?" Gilman asked.

"To be fair, they do have partial order," said Venter.

"And the other part is shit," said Myers.

Venter sighed. "Look," he said. "Everybody wants a clear-cut win. But we ain't gonna get it. We've got three choices. Either we ignore them and risk getting slaughtered, or we continue to treat it as a race and try to rush something together and publish before they do. But that means abandoning our own standards, and in the meantime we get crushed by their announcement anyway. You know what the third choice is."

"So the least damage is in holding hands and crossing the line together," said Myers.

"There's no damage," Venter said. "I don't see how we lose this way."

Everybody was quiet, taking it in.

"I like the idea of peace," Myers said, breaking the silence. "But I'm worried about the constraints."

"What constraints?" Venter asked. "That we can't diss them properly?"

"Yeah, actually."

"Look, if we have ninety percent of the genome in scaffolds, we win," said Venter. "If we have only fifty percent scaffolded, frankly, I'd rather have their data than ours. It's localized to the chromosomes. Not just a bunch of fragments."

"You've been spending too much time with Francis," said Kowalski.

Barring some last-minute catastrophe, one of the most contentious episodes in the history of science was about to conclude with the peaceful pomp of a double wedding at the end of a Shakespearean comedy. The press got its first inkling that a truce was in the works when Venter and Collins appeared together at an NIH conference on June 6, shaking hands and trading compliments in front of a gaggle of reporters. Investigative minds began to whir and plumb their channels, and despite a news blackout on all fronts, within a week the *Los Angeles Times* was reporting that a compromise of some sort was in the offing. The *Wall Street Journal,* the *Washington Post,* and *USA Today* quickly followed with their own stories on the rumors of a truce. By Thursday, June 22, the *New York Times* was citing an unnamed source at the Wellcome Trust saying that a major event was scheduled for the following Monday, June 26.

In the meantime, two potentially divisive articles appeared in the press. On June 12, a profile of Venter by Richard Preston in *The New Yorker* opened with a quote from an unnamed senior scientist in the public program: "Craig Venter is an asshole." On the same day, Venter appeared on the cover of *Business Week,* his bright gray eyes and sardonic smile emerging from behind a veil of As, Ts, Gs, and Cs. The article declared Celera about to announce the genome "finished" within two weeks, and quoted Venter describing Collins's behavior as "despicable." Both these stories had been in the works for weeks; they were like land mines left behind in a war after armistice has been declared, waiting for someone to step on them.

Throughout June, Dick Thompson at *Time,* whose last article on the genome war had inspired the written rebuke from Collins, had been gathering material for his own big "end of the race" story. Like everyone else in the press, Thompson was unaware of the get-togethers in Patrinos's townhouse. For him and his editors, the rumor that a big announcement was scheduled for Monday was bad news. *Time* publishes its print version on Mondays, with an online version appearing the evening before. Thompson could hardly write a major news feature on an event that had yet to take place or an agreement whose terms were not disclosed, and had thus resigned himself to being scooped by a full week by every major daily newspaper in the world. The best he could do was write a background story for the following Monday on "What's next?" now that the genome was largely done. He interviewed Collins and Venter and let them know the slant he was taking. Neither, of course, divulged anything about their agreement. Collins, in fact, was reluctant to talk to Thompson at all. "I don't know why I'm doing this," he told the writer. "You always stab me in the back."

When Thompson got back to his own office, he found the situation radically changed. His editors had read the explosive *New Yorker* and *Business Week* profiles of Venter and decided that whatever détente was being planned for Monday, Venter was still the man of the hour. "This guy has done all these incredible things," said Philip Elmer-DeWitt, *Time*'s science editor and Thompson's immediate superior. "Now he wins this race, and people are *still* calling him an asshole. Why? *That's* the hook. And we should publish it *this* Monday, the day of the announcement. With Venter on the cover."

Thompson was happy to rewrite his comparatively bland "What's next?" story to fit under a "Celera Wins!" headline, not just because it was more sensational but because he believed it was the truth. But he also had to cope with a pang of conscience. He knew that when Francis Collins saw the pro-Celera story he would feel terribly betrayed. Under the circumstances, Thompson felt he should at least try to soften the blow by letting Collins know beforehand what kind of story *Time* was planning. "[Otherwise] he's going to think I intentionally misled him," he told Elmer-DeWitt. "There will be hell to pay no matter what, but I'd rather he wasn't surprised Monday."

"Definitely tell him before Monday," Elmer-DeWitt replied. "But if you tell him too soon, he'll think he's got time to turn the ship around with heavy lobbying. I'd rather be spared that. How about Friday evening?"

Thompson's conscience would not let him wait. He called Collins on Wednesday, June 21, and informed him of the change. Collins did not have time to discuss the matter right away. His sister-in-law was gravely ill with breast cancer and not expected to live more than another day or two. He also had an appointment to meet with Venter and Patrinos one more time, with some of their staff in tow this time to help work out the details of the agreement.

Since his wife and daughters were home, Patrinos led his visitors down to the family room in the basement. Once again he served pizza and beers, followed by Greek brandy. Venter sat next to Kathy Hudson, the NHGRI policy director, and every time the two sides reached agreement on some point, they clinked glasses. The parakeet had gotten loose from its cage and flew about the room squawking. But Collins could not enjoy the moment. He was thinking about his sister-in-law, and about what Thompson had told him.

After the meeting, Collins called the reporter back at his home. Thompson had underestimated the amount of hell there was to pay. Collins called him a liar for misleading him about the story and accused him of being a "journalistic prostitute" for pandering to the public with a "*People* magazine piece." He said that Thompson must own stock in Celera to be so consistently biased, a charge Thompson, who for professional reasons owned no biotech stocks at all, heatedly denied. "You have to admit you've played a role in making this a contentious, highly personal battle, which only heightened people's interest in Venter," he told Collins. "Moreover, with the *New Yorker* piece and the *Business Week* cover, my editors want to know who this 'asshole' is."

"Editors are fools," Collins replied.

Before nine the next morning, Thompson got a call from Eric Lander's publicity person at the Whitehead Institute. "You're writing the wrong story!" she said. "This event is about people coming together, not racing with each other!" She told him a little bit about the secret meetings that had been going on and hinted there was a lot more to tell. "If we give you the details, will you put Francis on the cover, too?"

"I don't have any influence over the cover," Thompson said. "But if you give us an exclusive, it could have impact on the direction of the story." A few minutes later, Lander himself called. He said he could persuade Collins to offer *Time* the exclusive. He told Thompson the terms of the détente, and how it had all come about through Ari Patrinos's "pizza diplomacy." He even leaked a spicy bit of news from the genome itself: Early analysis from both groups indicated that the total number of human genes might be less than 50,000. If that was true, then Incyte and Human Genome Sciences, which had both filed patents on an excess of 130,000 genes, were staking claims to a lot of junk DNA.

"I would need confirmation of an exclusive from Venter and the White House," Thompson said. "There's another hitch. I'd need to interview Francis, and he won't talk to me anymore."

"I will get him to talk to you," Lander said. "What about the cover?"

"Not my decision," said Thompson.

He then telephoned Collins. Kathy Hudson intercepted the call. She told Thompson that Collins's sister-in-law had died during the night, and though he had come in to work, he was obviously shaken. "Francis agrees to the deal," Hudson said. "*Time* gets an exclusive for a weekly."

"No way," Thompson said. "Exclusive means *exclusive.* I don't want to be an echo of Nicholas Wade in the Sunday *Times.*"

"OK," said Hudson. "But only quid pro quo. Put Francis on the cover, too."

A few minutes later Collins himself called. It was not yet ten o'clock. There was no jauntiness in his voice. Thompson offered his condolences. Collins thanked him and turned to the matter at hand. He wanted reassurance about the cover. Thompson sighed. "All I can say is that having the exclusive is very attractive," he said.

The next time the phone rang, it was the White House. Jeff Smith, a staffer in Neal Lane's office, gave Thompson the scoop on Monday's event: President Clinton would announce the completion of the human genome in a ceremony in the East Room, with Collins and Venter by his side and Tony Blair joining in by video simulcast. He also read Thompson a memo that the president, angry about the public battling over the genome, had sent to Lane. "Fix this," the memo commanded. "Get these guys together." Smith offered to do what he could to get Thompson an interview with the president, but he wanted to know something first: "What about the cover?" Smith asked.

In an e-mail to his editor, Thompson summarized everything he'd learned so far. "This [story is] ours alone, until Monday morning," he wrote. Then he called Venter, who agreed to give *Time* the exclusive. It was still before noon. "I hope the trade-off isn't putting us together on the cover," Venter told the reporter. "That would annoy the hell out of me."

A few minutes later, Collins called Venter, who was conferring with Kowalski and Gilman. He told Venter about his sister-in-law, then broached the *Time* matter. "I wonder what they're doing with the cover?" he said. "If we're putting to bed the rancor, to have just the public program or just the private sector alone on the cover would send the wrong message, don't you think?"

"They are being very noncommittal with me," Venter said, giving Kowalski a look. "I'm not in the loop."

Collins persisted. "I don't want to put you in a corner," he said. "Obviously Celera on the cover would be in your best interests. But this is delicate."

"If you're asking whether I'm opposed to sharing the cover with you, the answer is no, not at all," Venter said. "We want to emphasize burying the hatchet."

Kowalski's jaw dropped. She gave Venter a fierce look and shook her head. "What are you *doing?*" she said, when he had hung up. Venter shrugged.

"I'm trying not to justify the first line of *The New Yorker,*" he replied. "Francis is an emotional wreck. He just lost his sister-in-law. I'm not going to manipulate behind the scenes to keep him off the cover of *Time.*"

It seemed that the race was over. That same day the protocols for the White House announcement were set in motion, arrangements for a joint press conference began to take shape, and a *Time* photographer was rushing down from New York for a midnight shoot of Venter and Collins, together. In four days, the president would tell the world that the human genome had been brought to light independently by two competing but now happily reconciled enterprises.

There was only one problem: neither Celera nor the Human Genome Project actually had a genome to announce. The evening before, while everybody was clinking champagne glasses in Ari Patrinos's basement, Gene Myers, Granger Sutton, and the rest of the team were back at Celera, watching the Overlayer program sputter, gag, and hawk up another mediocre assembly.

"Fuck this overlay shit," Myers said. "It sounded like a great plan, but it depends on there being a certain level of quality to the data underneath. The public centers don't measure up."

"It's not fair to judge them by this standard," said Sutton. "They're just throwing stuff out there. But you can see from chromosomes twenty-one and twenty-two that they can fix it all. They just haven't done it yet."

"Yeah, well we don't have the time to wait around for them," Myers replied.

The public genome was in even worse shape than Celera's. Not only was it 100 million base pairs short of the 90 percent coverage promised for the working draft but there was no overarching order to it at all. While some of the BACs were correctly aligned, over half still were not, and even the arrangement among the BACs on the chromosomes was full of holes. It would take a miracle to make any kind of coherent sense of it before Monday's announcement. Several computational specialists in the public program had already tried, including David Haussler at the University of California, Santa Cruz. "The only way we could have gotten an assembly in time was if some genius came along who could do it all by himself," Haussler later said. "What were the chances of that?"

The miracle took the form of Jim Kent, Haussler's own graduate student. Kent had a background in computer animation and the bearded burliness of a lumberjack. The previous December, Eric Lander had asked Haussler and some others if they could devise a program that could look for genes in the public genome under construction. At this point, it consisted of some 400,000 pieces from 25,000 separate BACs. Haussler quickly realized that before he or anyone else could effectively search for genes, he would have to put the 400,000 pieces in some kind of order. He persuaded his university to advance enough money to purchase a network of a hundred desktop computers to run the program and got to work. Without sufficient mate pair information, however, the task seemed next to impossible. By early May, neither Haussler nor anyone else attempting to solve the problem had gotten close.

Kent had just passed his Ph.D. oral qualifying exam and was looking for something to do to fill time. As he later put it, "It ended up being a bigger project than I had thought." With the paucity of mate pair information at his disposal, Kent threw together whatever other kinds of information he could scrounge and hoped that collectively it would all work. He began with known "exons"—the portions of genes transcribed into RNA—using them in a manner similar to the way Celera was using paired reads. If two or more exons existed from the same gene, then their known order and distance apart could be used as reference points to correctly align the corresponding portion of a BAC and orient its fragments in the right direction. He gradually added additional clues to the mix, eventually devising a program that made use of thirteen different kinds of information. Kent completed the first version of his program in four weeks, doing it all on his own, because that was the only way to keep everything straight, and working so hard at his computer that he frequently had to stop to ice down his wrists. The program ran for the first time on Thursday, June 22.

On that Thursday, Myers's team at Celera completed a second run of the Overlayer program, this time leaving out the least-finished, most error-prone portion of the public data. This worked, more or less, but the point of the great experiment all along—especially for Myers and his group—was to show that the whole-genome shotgun technique could put the human code together, and the Overlayer was most emphatically not a whole-genome shotgun approach. But the Grande program was. Over the previous week, the team members had poured all their efforts into getting Grande ready for its first full run, which would finish on Friday, June 23. For fodder, the program had 27 million paired-end reads from Celera's own sequencing machines to digest, plus another 14 million "faux reads" of shredded-up public data. Combined, it came to around sevenfold coverage of the genome, roughly equal to the amount of coverage the team had had available for the first successful assembly of *Drosophila* back in September. In addition, there was Hamilton Smith's "secret weapon" to throw into the mix, providing an extra, order-enhancing set of mate pairs a full 50,000 base pairs apart. The sheer scale of the problem justified Grande's name, whatever size coffee Myers happened to prefer. In fact, when the program ran, it would be one of the largest single computes ever attempted. To Myers and the others, Overlayer was merely an afterthought.

That Friday, Overlayer suddenly took on greater importance. In the morning, a glitch in Grande sent the program into a core dump, adding another twenty-four to thirty-six hours to the run—not a serious problem under normal circumstances but a little worrisome when there were only seventy-two hours to go before the president of the United States said you were done, whether you were or not. In a pinch, the results of the Overlayer run could be used to keep the president honest. In the meantime, the team members would take turns monitoring the Grande run around the clock in case it needed nursing through any more glitches. A final meeting in the war room that afternoon left behind this message on the whiteboard:

WHAT THE FUCK CAN WE SAY ABOUT OVERLAY AND GRANDE ON MONDAY?

(A take home assignment for the weekend)

1. We're pretty sure we'll be done in a few weeks.
2. We'd have been done a lot sooner if the public data wasn't so crappy.
3. We think Gene can whip Francis at ping pong.
4. What's the hurry, we don't think they're done either.

On Saturday, Venter was just getting off the phone with Collins and Patrinos when Myers came into his office. He was yawning.

"Up late?" Venter said.

"I yawn when I'm stressed," Myers said. "It's a nervous reaction. We're having a bad Grande moment."

The program had choked again. Only 55 percent of the genome was coming together. Now there were only forty-eight hours to go. Several more hours of soul-searching and bug-searching elapsed before Myers came up with an idea—a possible fix to the program and a personal revelation, wrapped into one. Was he asking for too much perfection? In joining the DNA fragments into bigger ones, he had set the program to require sixty base pairs of overlap to confirm a match. But perhaps that was too stringent a requirement. What would happen if only forty base pairs were needed to call a match instead? This simple adjustment might allow more of the genome to fall together, with plenty of confidence left that the program was not sewing pieces together that were not in fact adjacent. In other words, maybe Grande, like its chief architect, just needed to relax a little.

Granger Sutton made the adjustment, and while everyone waited to see what effect it would have on the run, Myers did two very uncharacteristic things. First, without any reason beyond blind faith, he trusted that the fix would work. Second, he went shopping. There was a White House ceremony and press conference coming up, after all, and even though he and his wife had made the fortunate decision to sell a million dollars' worth of Celera stock just before the crash, Myers still did not own a suit. He and M'Liz were not getting along very well, and having no experience in shopping for respectable clothes on his own, he enlisted the services of a personal shopper at Nordstrom, who brought armloads of suits, ties, shirts, belts, shoes, and socks up to the dressing room for him to try on.

"This belt doesn't feel right," he said, standing in front of a three-sided mirror while a tailor fussed around him.

"You missed a loop," said his personal shopper. "You don't want to miss a loop when you go to the White House." After an hour they had narrowed the suit choices down to three. "I'll take them all," Myers said, then selected half a dozen shirts, as many ties, three pairs of shoes, and a couple of belts. The bill amounted to far more than he had spent on clothes since he'd arrived at Celera, but he was enjoying himself for once, and when he took out his battered nylon wallet to pay and the personal shopper mentioned that Nordstrom also carried an abundant selection of fine leather wallets, he bought two—one for regular use and a slim, elegant number for special occasions. While everything was being bundled up, his cell phone rang. It was Sutton. The genome was coming together. The program would not finish its run until the next morning, but it was already clear that they were going to triumph.

Out in Santa Cruz, meanwhile, Jim Kent had already run his "Gigassembler" program two days before. With only a little over 80 percent of the genome sequence to work with, the program necessarily made a lot of mistakes and left many large holes in the sequence. But the following Monday morning, when the Human Genome Project issued its press release announcing that it had assembled a working draft of the human genome, there was at least some measure of truth to the operative word.

Just before ten o'clock on that Monday, Gene Myers filed up the stairs leading to the East Room at the White House with the other invited guests, looking proud and elegant in the sharpest of his three new suits. Hamilton Smith, Mark Adams, Granger Sutton, and the others who had led the effort at Celera were there, too, along with their counterparts on the public side, mingling with an assortment of ambassadors, cabinet secretaries, and senators. Dapper, diminutive Ari Patrinos was wearing a red tie with a bright yellow sunburst on it—a very happy tie. James Watson smiled and squinted at the flashbulbs aimed his way. Norton Zinder chatted and chortled. A few people had copies of *Time* under their arms, with Venter and Collins side by side on the cover. Michael Hunkapiller, whose machines had made both genomes possible, was home with chicken pox. But there was a silent tribute to him on the cover of *Time:* behind the two scientists in their white lab coats were rainbow columns of red, blue, green, and yellow bars—a readout from one of Hunkapiller's automated sequencers.

Eric Lander stood at the door of the East Room, grinning and greeting people like a host until a marine guard politely asked him to move inside so others could get through. Inside, the chandeliers glistened, the TV crews bustled, and a string quartet played Mozart beneath a cheerful murmur of anticipation. The doors in the front of the room swung open, a band struck up "Hail to the Chief," and Bill Clinton appeared, flanked by the two genome leaders.

"We are here to celebrate the completion of the first survey of the entire human genome," the president said. "Without a doubt, this is the most important, most wondrous map ever produced by humankind." He acknowledged "the robust and healthy competition" that had led to the achievement and noted with earnest satisfaction that the public and private efforts had agreed to publish their data simultaneously. When Clinton finished his remarks, Tony Blair's face appeared on a large-screen monitor. After he had offered his own tribute to the greatness of the moment, first Collins, then Venter had their chance to speak. Neither there in the East Room nor in the giant press conference that followed in the Washington Capitol Hilton did either man allude to the contentiousness that had marked the previous two years or to the competing interests that had made a true collaboration impossible.

"I'm happy that today the only race we are talking about is the human race," Collins concluded, to much applause. He then introduced Venter, whose talk was surprisingly modest, stressing not so much the glory of the moment as the commonality of humankind revealed by the genome and the continuity it manifests between our species and all others. "Some have said to me that sequencing the human genome will diminish humanity by taking the mystery out of life," Venter said at the end of his talk. "Nothing could be further from the truth. The complexities and wonder of how the inanimate chemicals that are our genetic code give rise to the imponderables of the human spirit should keep poets and philosophers inspired for the millenniums."

Clinton then returned to the podium for the last word. "I suppose, in closing, the most important thing I could do is to associate myself with Dr. Venter's last statement," he said. "When we get this all worked out and we're all living to be one hundred and fifty, young people will still fall in love, old people will still fight about things that should have been resolved fifty years ago, we will all, on occasion, do stupid things, and we will all see the unbelievable capacity of humanity to be noble. This is a great day."

Collins was happy, Venter was happy, the president was happy, everybody was happy. Or almost everybody. At the Hilton, a stocky, red-faced, grim-looking man was standing in the hallway by himself when Venter and Collins passed through on their way to the press conference, a crowd of photographers and cameramen surrounding them, eager for a shot of the two men together. Somehow in the crush the red-faced man was pushed against the wall and nearly knocked over. It was Tony White.

EPILOGUE
A BEAUTIFUL MOMENT

Once, when Craig Venter was serving as a medic in Vietnam, he decided to kill himself. He had come to see that the war was without purpose. The daily strain of tending to the mangled bodies of other young men left him feeling that beyond the specific meaningless of the war was a larger, more awful meaningless to life itself. So one afternoon after his shift was finished, he put on his bathing suit and walked into the ocean. The plan was to swim out until he was exhausted, then let go and sink down into the darkness. His steady, practiced strokes soon took him well out to sea. The shoreline was nearly out of sight when it occurred to him that the sun was going to set soon, the water he was swimming in was infested with sharks, barracuda, and poisonous sea snakes, and he was hungry. *What the fuck am I doing?* he thought. Then he turned around and swam back again.

Venter related this incident to me one evening in February 2002. I was surprised that he had never told me about it before, since he described it as a turning point in his life. We were sitting on the deck of his new, grander yacht, *Sorcerer II,* anchored just outside the mouth of Gustavia harbor on St. Barts, in the Caribbean. The sun was setting then as well, turning the water briefly to gold. There were no sharks or sea snakes, but occasionally a sea turtle would appear near the boat, holding its head up above the waves before diving out of sight. Venter pointed with pleasure at the turtle every time it resurfaced. The suicide attempt, he explained, had been a kind of experiment. You can choose to kill yourself or choose not to. If you choose to live, you have to embrace life totally, and achieve as much as you possibly can. But even if you take that course, the other door is always beckoning. "I still think there is at least as good a chance that I'll commit suicide as die from some disease," he said, in the flat, matter-of-fact way one might weigh the advantages of two job offers.

Even knowing his penchant for the melodramatic (especially when I was around taking notes), it seemed an extraordinary admission. By every standard measure, Venter was a very successful man. He was no longer an "almost-billionaire" but was still rich enough to have spent the day shopping on the island for a villa in the $5 million range. Since the White House announcement in June 2000, his contribution to the discovery of the human genetic code had been recognized by a torrent of prizes and awards. Among them were the Paul Ehrlich and Ludwig Darmstaedter Prize, which he eagerly referred to as "the German Nobel," the Takeda Award ("the Japanese Nobel!"), and the Gardiner Award ("the Canadian Nobel!"). After his lectures, people lined up for his autograph, and strangers approached him in restaurants wanting to shake his hand. On the other hand, he had just been fired, and his academic detractors were repudiating his accomplishments more than ever. Apparently the plan to break down the wall between academic science and business had not worked out as planned.

"My greatest success," he said, in the same deadpan tone, "was I managed to get hated by both worlds."

The cordiality between Celera and the HGP at the White House announcement in June 2000 had lasted about as long as it takes a television crew to pack up its gear. The truce negotiated by Ari Patrinos that spring had two major components: the two sides would declare that they were finished at the same time, and they would subsequently publish papers explaining their methodology and giving some preliminary analysis of the code simultaneously in the same journal, presumably *Science.* Negotiations with the journal quickly broke down over the stubborn issue of data release. The actual sequence of the human code was far too long, of course, to publish between the covers of a magazine; it would be released instead through a publicly accessible web site. *Science*'s previous policy was to require that genetic data substantiating a paper be made public through GenBank. But Celera was unwilling to follow that precedent, because that would amount to giving away the information to its commercial competitors.

After much negotiation, Donald Kennedy and Barbara Jasny, his deputy editor, worked out a compromise. The company would be allowed to publish in *Science* and still make the genome available only at its own web site, so long as certain key conditions were met. Academic, nonprofit researchers must be allowed free access, and be allowed to patent at will any discoveries they made using Celera's information without any commercial obligation to Celera. Academics could download one million base pairs of DNA from the site per week simply by mouse-clicking their agreement not to redistribute the data for commercial purposes. If they wanted complete access to the genome, they had to submit a letter signed by an institution official agreeing to the same terms. The rules for private-sector users, however, were much more restrictive, and in effect prevented any commercial use without compensation to Celera. The compromise was not exactly the "unrestricted access to all" that had been promised in the beginning. But to the editors at *Science,* having such precious, possibly lifesaving information freely available only to academic researchers seemed a whole lot better than having it available only to those who could pay.

"We thought we were doing a *good* thing," said Jasny. "The alternative seemed crazy."

The decision nonetheless outraged the academics. Harold Varmus, by then president of the Memorial Sloan-Kettering Cancer Center, wrote a letter to Kennedy co-signed by a host of heavyweights warning the editor that he was setting a dangerous precedent by allowing the company to dictate the terms by which its published data could be used. The British *Drosophila* expert Michael Ashburner kited a series of explosive missives to Kennedy announcing his intention to cut all ties with *Science,* and to advise all his colleagues to do the same.

"You have lowered a proud journal to the level of a newspaper Sunday supplement, accepting paid advertisement in the guise of a scientific paper . . ." Ashburner wrote. "The problem comes, of course, because Celera, particularly in the form of Craig, want[s] the best of both worlds. They want the commercial advantage of having done a whole genome shotgun sequence and they (or at least Craig) want the academic kudos which goes with it."

When Kennedy stood his ground, the public program withdrew its paper from consideration at *Science* and submitted it to the British journal *Nature* instead. Francis Collins, who through the latest wrangling had been doing his best to maintain friendly relations with Venter, called him at his home to tell him the news. Venter took it in stride. "That's too bad, Francis," he said. "But this way, at least we both get our own covers."

The simultaneous publication of the two epochal papers was heralded at another joint press conference on February 12, 2001. Once again, everybody dressed up nicely, the various leaders of both efforts smiling together before the cameras and hordes of reporters, with Ari Patrinos, the quiet diplomat, gracefully presiding. Venter quipped, Collins waxed, and Lander clapped and nodded cartoonishly at their salient points, impatient for his own turn to speak. Gene Myers would also have a chance to describe his assembly strategy to the world. He sat at the far end of the platform, leaning back in his chair, a sweater tossed casually across his shoulders and a roguish grin raising one corner of his mouth, as if he were posing for a fashion shoot. Since his Nordstrom makeover he had taken to dressing sharp. His wife was in the audience. They were going through a divorce, but as a gesture of support to him in his moment of glory, she had put her wedding ring back on and driven down for the event. When she caught sight of his new girlfriend sitting in the front row, she went out to the women's room, pulled off the ring, and washed her hands.

The star of the show was the genome itself. Both teams had included preliminary analyses of its main features in their publications. They agreed on the surprisingly low number of genes that it takes to operate a human being—probably somewhere around 35,000, roughly only twice the number required to operate a fruit fly or a little soil worm. But the early analyses indicated that human genes were much more "modular" than those in the *Drosophila* and *C. elegans* genomes—that is, they were able to produce multiple proteins from the same gene by varying which parts of the gene get expressed and shuffling those parts around in novel arrangements. Thus at life's fundamental level, complexity was a matter not of size but of versatility. Both papers also revealed that the genes were not sprinkled evenly along the chromosomes but clustered into densely packed regions separated by vast genetic deserts of noncoding DNA. Such revelations were just the most tentative of beginnings. It would take decades or even centuries to completely understand the language of the code—how the tens of thousands of genes and their proteins interacted to create the biological symphony of a human being. But at least there was general agreement between the two initial attempts to divine its rough structure.

After the press conference, the two sides celebrated their respective achievements at separate parties that seemed to caricature their ideological frames of reference. The public program's fest, convened in the hangar-like main hall of the National Building Museum in Washington, was a gigantic hootenanny, with little children bouncing on their fathers' shoulders and the kindly eyes of Gregor Mendel looking down on the gathering from an immense screen. Francis Collins strapped on his electric guitar and from the stage he and the other members of his middle-aged rock band sang and played golden oldies with genetically engineered lyrics. To the tune of "This Land Is Your Land," they belted out:

This draft is your draft, this draft is my draft,
And it's a free draft, no charge to see draft.
It's our instruction book, so come on, have a look,
This draft was made for you and me!

A couple of weeks later, Celera held a swankly catered celebration at the San Francisco Design Center during the 2001 meeting of the American Association for the Advancement of Science. Columns of liquid light threw colored shadows on the gowns and tuxes, making everybody look more beautiful and exotic than they actually were. People picked at an array of caviar, oysters, and tenderloin and drank vodka served in ice-lined glasses, while an orchestra played and a dance company dressed in silver and black body suits performed onstage what appeared to be a mating ritual of nematodes. Then a master of ceremonies in leopard-accented tails took the microphone. "Ladies and gentlemen!" he roared. "How about a big welcome for your host . . . the man who made the human genome . . . CRAIG VENTER!"—and in came Venter from backstage, escorted by Heather Kowalski on one arm and an equally radiant blonde on the other, his broad bald pate outshining them both. He seemed to be enjoying both the moment and its silliness. He seemed to be enjoying everything.

Many of the other Celera scientists in the crowd, especially Gene Myers, were less adept at forgetting their troubles. Since the publication of the genome papers two weeks earlier, Eric Lander had been waging a ferocious campaign in lectures and through the press to let the world know that Celera's human genome was a flop. Myers and the others had taken to calling him "Eric Slander." In one regard, the company scientists had left themselves wide open to the attack. In their *Science* paper, they had presented both methods for assembling the sequence: the Grande whole-genome shotgun version, and the alternative Overlayer assembly, which explicitly exploited the public program's mapping information to locate sequence fragments on specific places along the chromosomes. In their discussion of the annotation and analysis of the genome, however, they had opted to use only the second version, because it was 2 percent more complete. It was a costly strategic mistake. Having touted the virtues of random shotgun assembly since the beginning, Celera now appeared to be relying on the public program's mapping information after all.

For Lander, this criticism was just a warm-up. Even by exploiting the public program's mapping information and having twice as much raw data to work with, he claimed, the Celera product was still only marginally better than the public program's draft version. Indeed, a quarter of Celera's genome was composed of more than 100,000 isolated pieces, most of which were too small to be pinned down to any one location. This was not a sequence, Lander said. It was "genome tossed salad." He saved his most serious criticism for the whole-genome shotgun assembly itself. According to Lander, when Myers's team shredded the public program's joined sequences into bits to mimic random fragments covering the whole genome, they had done so in a way that preserved the underlying mapping information that had allowed the public program to put the sequences together in the first place. In other words, he said, Celera had not even *attempted* a whole-genome shotgun assembly, much less proved that it could work.

"I must say, it has caused quite a stir to discover that the whole-genome shotgun assembly was an utter failure," Lander told me, "—even worse than Maynard Olson had predicted. As one wag put it, 'the emperor is stark naked.'" Worst of all, according to Lander, this crucial fact had been deliberately obscured in Celera's paper. It wasn't just that the emperor had no clothes. Lander was saying that the emperor *knew* he was naked and had used an elaborate computational fan dance to hide that fact from the world. His claim came very close to accusing the Celera team of wholesale fraud.

Venter responded to Lander's accusations with a kind of elemental weariness, as if he'd resigned himself to the fact that nothing he and his team could do or say would ever satisfy his critics. But the other Celera scientists, especially Gene Myers, were furious and deeply hurt. Certainly, the company's human genome was not as complete as it would have been if they had used 10x or more of their own data, as they had with *Drosophila.* Myers conceded that there were over 116,000 pieces in the Celera genome that were "free-floating." But these, he said, were "mere detritus"—tiny flecks of repeated sections of no consequence to the way the genome was really put together or to the maintenance of the organism. What mattered was that 95 percent of the Celera genome *did* fit together into large hunks, making it far more useful for finding genes than the Human Genome Project's draft version. Myers was even more appalled by Lander's assault on his whole-genome shotgun assembly. Far from using the public program's information on where pieces matched up as a secret crutch, the Celera algorithm had in fact deliberately *excluded* that information when assembling the genome, because the public program's information was so unreliable. "Eric is a fucking liar," Myers told me. "He's smart enough to read this the right way. He *knows* what we did."

Lander, however, insisted that his contentions were purely scientific and got to work drafting a paper for *Nature* that would set the record straight. At the same time, he was pumping e-mails to Nicholas Wade and other reporters and attacking the Celera genome with such tireless gusto that even some of his own colleagues were taken aback. "I don't understand why Eric has launched this jihad," Ari Patrinos told me. "But it's going to be a very big embarrassment if he is wrong." For Patrinos, the litmus test would be the mouse genome. "If Celera can do a whole-genome shotgun assembly on the mouse," he said, "then Eric is wrong. If not, then he's right."

There were no such "ifs" for Lander. "They didn't sequence the genome," he told me. "End of story."

By the middle of March 2001, Celera was on the verge of completing the first assembly of the mouse. Venter, Myers, and their colleagues were so confident that the whole-genome shotgun technique would indeed work that they had elected to use only their own data in the assembly, ignoring the mouse data available in GenBank. They had less than a sixfold coverage of the genome to work with, and it might not work. But at least no one could accuse them of cheating. The prospect of vindication aroused even the normally imperturbable Granger Sutton. "We have to get a really great mouse assembly out there," he said, "so we can shove it up Eric's ass."

Venter announced the results of the pure whole-genome shotgun mouse code on April 27. It was not only as good as their published human code—it was better. Without the sequencing errors and incorrect matches in the public data to trip it up, the assembly algorithm had put more pieces together into a linear sequence, with far fewer gaps. The Celera team followed this triumph by reassembling its human DNA data, this time stripped of the sequences they had borrowed from GenBank. Like the mouse genome, the new human assembly was significantly better than the original one published in *Science,* though it made use of only half as much data.

"I hope we've reached a turning point, as we did with bacterial genomes in 1995, where people can see that whole-genome shotgun works as a strategy," said Sutton in early June, when he revealed the results of the new human assembly at a workshop held at the Howard Hughes Medical Institute in Bethesda. Scientists from both Celera and the public program were attending. If anyone doubted it, Sutton said, there was "an open invitation for anyone to come out to Celera and analyze the quality of the data for themselves."

Not one person from the public program took him up on the offer. I later asked Francis Collins why Sutton's evidence that the strategy really worked had aroused so little response. "Once again we are in the position of hearing verbal positive reports from Celera about their achievements, without being able to see the data," he told me. "Can you blame us for being skeptical? We've been to this movie. Show me the data."

Presumably, Collins could have gone out to Celera to judge the quality of the assembly for himself. But this was not what he meant by "show me the data." In academia, only publication would satisfy that imperative, and Celera could not make public any more genomes on its web site without betraying the customers who had paid for their privileged access to the information. It did not matter that a year later, when Lander's damning critique was published, it was accompanied by a credible rebuttal by the Celera team. It did not matter that eventually Celera, to demonstrate how well the assembly had worked, published one complete chromosome of the mouse genome in *Science,* thus placing the data in the public domain and risking its subscribers' wrath. It did not even seem to matter that soon afterward, when the public program finally published its own mouse genome, the technique used to assemble the draft was revealed to be . . . whole-genome shotgun. It probably will not matter that when other complex genomes are sequenced—chimp, cow, dog, corn, and others useful for biomedical and agricultural research—they, too, will almost certainly be done using the method pioneered at Celera.

"They will never be satisfied," Venter told me aboard *Sorcerer II* off St. Barts. "Nothing will satiate them but to see us destroyed and humiliated."

On April 14, 2003, the Human Genome Project announced the completion of its "finished" version of the human code, two years ahead of the original projected finish in 2005. Timed to coincide with the fiftieth anniversary of the discovery of the DNA structure by Watson and Crick, the event was celebrated with a two-day orgy of symposia, speeches, and festivities in Washington, with Francis Collins presiding and Watson himself the star of every gathering. In spite of Ari Patrinos's efforts to persuade Collins to include Venter in the program, he was not invited to participate. In the talks, his name was scarcely mentioned, and then only in derisive asides.

If the HGP scientists seemed intent on erasing Celera from history, Tony White seemed equally determined to erase Venter from Celera. While he was standing beside the president on that June day in the White House in 2000, White, whose corporate decision making had godfathered and financed the enterprise, was sitting unnoticed in a side row in the audience. In all the speeches that day, nobody so much as mentioned his name. In Venter's view, White never forgave this slight, and the subsequent deterioration of their already frayed relationship was caused solely by the CEO's intense personal jealousy. White certainly resented being ignored. But when I talked to him following the press conference that day, his primary concern seemed to be not about himself but how the news would be perceived by his shareholders. The declared tie between the two sides notwithstanding, the Celera genome was more complete and usable than the government's offering. Venter could not say this, because to do so would violate the agreement with Collins not to "diss each other's genomes." But *not* to say anything left Wall Street wondering what the hell was going on. While Venter was being herded through interviews that day with Ted Koppel and Charlie Rose, White was left to plead Celera's business case on RadioWallStreet.com.

If anyone was listening to him, it did not register in Celera's stock price, which by the end of the day had dropped by 11 percent. In the long run, however, it probably would not have mattered what Venter did or did not say at the White House that morning. Celera's bid to become "the world's definitive source of genomic information" required the company to be years ahead of anyone else in sequencing. It was doomed as soon as the Human Genome Project showed the will and ability to stay just a few months off the company's pace. In a commercial sense, it might have been better if Venter had downplayed his ambitions from the start and never roused the government program to such fervent determination.

"He got what *he* wanted," Robert Millman said at the Celera staff party after the White House announcement, gesturing with a flick of his chin toward his beaming boss across the room. "It's great for *him.* But it's shit for the company. All we have now is cash. Our best bet is to give it all back to the shareholders and go home."

Ironically, shortly after this, Celera's information business finally came to life. In July 2000, the government of Australia signed up all its research institutions, followed shortly by Harvard University, the second university subscriber. A flurry of other universities soon piled on, including the entire University of California system. American Home Products became the next big pharmaceutical company to subscribe, and in September the mighty Howard Hughes Medical Institute, the largest biomedical research philanthropy in the United States, signed up on behalf of all of its grantees. The main attraction was not Celera's version of the human code but the imminent prospect of the mouse genome, which the government would not have for at least a year, even in draft form. In June 2001, not wanting to penalize its researchers by depriving them of the advantages the mouse genome held for making biomedical breakthroughs, NIH itself took out a Celera subscription, through the National Cancer Institute. The Wellcome Trust was of a different mind, specifically prohibiting its grantees from using its funds to buy a Celera subscription, on the grounds that there was nothing available from Celera that could not be gotten for free in the public domain. In the trust's view, apparently, the complete, assembled genome of the most widely used laboratory animal in the world did not count.

Within a year, subscriptions to the Celera database were drawing $100 million in revenues. It was not enough to sustain the company in the long run, but at least the information business was more or less paying for itself. Then there was that billion dollars in cash from the secondary stock offering to play with. White and Venter had agreed that Celera would be moving into proteomics. The next logical step was to commit the company to using its own enormous gene database and computing power to seek out and develop drugs of its own. At first, Venter threw himself wholeheartedly into the new direction. "What better act to follow the sequencing of the human genome," he told me late in 2000, "than to discover the cure for cancer?" He was instrumental in the acquisition of Axys Pharmaceuticals, which specialized in the discovery of small-molecule therapeutics and already had some drugs nearing clinical trials. Together, he declared, Celera and Axys would transform the way pharmaceuticals were developed. But screening molecules for possible drug targets is a slow, rarely rewarded process, absurdly unsuited for a man who had named his company after the Latin word for speed and whose greatest fear was that he would die before he achieved something memorable. By the fall of 2001, Venter's enthusiasm for leading the enterprise had dwindled. He was restless, bored, and frustrated by his frequent run-ins with "corporate incompetence and greed."

Tony White, meanwhile, was determined to impress upon Venter that Celera's days of creating controversy and grabbing headlines were over. "We are a drug company now," he said, shortly after the Axys acquisition. "We are trying to produce products. We are going to cool it, live on our accomplishments, and not on what we boast we'll do."

In retrospect, given these tensions, Venter's departure was only a matter of time. Its abruptness still came as a shock. On January 22, 2002, a press release from Celera chimed into my e-mailbox. J. Craig Venter, it said, was stepping down as president. According to the release, now that the company had turned to drug discovery, the board of the parent company (now called Applera, an amalgam of Applied Biosystems and Celera), Tony White, and Venter himself all agreed that it would be better to make room at the top for an executive skilled in pharmaceutical development. While the search for Venter's replacement was under way, White himself would take over the day-to-day operations of the company.

"We are now at a critical juncture where my best contributions can be made in a scientific advisory role," Venter was quoted in the release, "allowing the rest of the organization to continue Celera's progress toward becoming a successful pharmaceutical business."

The words sounded like they were coming from a ventriloquist's dummy. No warning about the "resignation" had been given to Wall Street, and no successor had been named: typical signs of a forced ouster. I drove out to Celera to see whether I could learn anything more and ran into Heather Kowalski in the hallway. She was not permitted to say anything about what had happened and rushed past, looking pale and shell-shocked. Venter, she said, had "gone sailing," and was not available for comment. Her vacated look echoed the emptiness in the hallway itself: all the framed articles, magazine covers, and other Venterabilia had been stripped from the walls. I continued on into Venter's office. It, too, had been cleared out. Gone were the photos from Vietnam, the awards, the trophies, the framed letters from the president, and the pictures of Claire Fraser, Hamilton Smith, and *Sorcerer* that had sat on the credenza behind the desk. There was nothing left but bare furniture. Like Kowalski, Lynn Holland and Chris Wood were not allowed to discuss what had occurred. Both looked like they had been crying.

I eventually learned that the day before, the board had voted unanimously to ask for Venter's resignation, effective immediately. A termination agreement was worked out that same day, including a provision that prohibited Venter from saying anything at all about the matter, including the fact that there was a termination agreement in the first place. Bill Sawch, the Applera chief corporate lawyer, strongly advised him not to attempt to return to the building to say goodbye to his employees. Holland and Wood packed up his things, and Venter himself crept back in after midnight to collect them. By the time I arrived later that morning, there was not a trace of him left. Outside his empty office, I ran into Tony White, dressed in a casual brown shirt and brown slacks. He gave me his beefy hand to shake. "This should be good for your book," he said.

A month later, down on St. Barts, Venter consoled himself with the conviction that White had forced him out only because he knew he was about to quit anyway, and wanted to strike first. "Ever since the publication of the genome, I wanted to move on," he said. The "ultimate turning point" had come at an investors' conference in San Francisco a couple of weeks before he was let go. Talking up investors was what White wanted him to do most, and what he enjoyed the least. "I felt soiled as a person and as a scientist to have to do that," Venter said. "Tony and Dennis Winger [the chief financial officer of the parent company] were sitting in the front row, scowling up at me. Both of them were particularly fat at the time. They seemed to represent in physical form the grossness of what I was doing. I realized then that I had made a Faustian bargain."

After the investors' conference Venter drove across the Golden Gate Bridge to Sausalito. On the spur of the moment, he dropped in to visit the nonprofit clinic of his friend Dean Ornish, the physician and advocate of lifestyle changes to prevent heart disease. Ornish invited him to join a group of his prostate cancer patients for a low-fat, Ornish-style dinner. Among the other guests were a priest and some professors. For Venter, the contrast of the gathering with the company he'd just been keeping clinched his decision.

"If you were going to leave anyway, then why are you so depressed?" I asked.

He thought for a moment. "It's like a divorce," he said. "Even if you're sure it's the right thing to do, it still kills you."

A few weeks after returning from the Caribbean, I drove out to the building off Rockville Pike one last time. Only a few spaces in the visitors' parking lot were occupied. Inside, the place looked the same, but its spirit was gone. It was no longer a "great experiment" outside the normal course of commerce and science, but just another drug development company going about its business. I did not recognize most of the people I passed in the hallway, and they did not recognize me. I went downstairs and found Hamilton Smith in the mostly deserted basement cafeteria, where that first all-hands meeting had convened three and a half years earlier and people predicted that someday they would be telling their grandchildren about what would happen here.

"The tragedy is that business concerns overruled science—that we abandoned the original 10x plan," said Smith. "If we'd been at TIGR, we would have done it right and blasted [the HGP] to hell. But of course, we wouldn't have had the money to do it in the first place." In his view, Celera had sequenced the genome too fast. But there hadn't been any choice; the pace was dictated by the competition from the public program. "The bastards!" He laughed. "But I can see it from their point of view. It all goes back to that first insult. Those four words Craig never should have said. 'You can do mouse.'"

By this time, many of the characters in this story besides Venter had also departed. Robert Millman had quit to become patent attorney for a new biotech in Cambridge—co-founded, ironically, by Eric Lander. Paul Gilman had returned to government service, to head the EPA's Office of Research and Development. Marshall Peterson had been let go, and more than a hundred others would soon be dismissed on a single day of downsizing. Mark Adams was still playing an active role in the company, and Gene Myers was still there, though he was spending most of his time giving lectures around the country and checking out job offers to return to academia. He and Smith had developed a new cloning technique that, theoretically, could assemble a genome without any gaps. But the innovation wasn't being put into development, because there were no more genomes at Celera to assemble. Smith himself was just biding his time there until his stock options matured. But he already knew where he was going next.

Craig Venter, meanwhile, had pulled out of his Caribbean funk and resurfaced, issuing a double-barreled, part outrageous, part spectacular, wholly Venterian notice that he was back in the game. The first shot was the admission to the world, by way of Nicholas Wade in the *Times,* that the Celera human genome was largely composed of his own DNA. This news had long been an open secret to anyone close to the Celera project, and hardly came as a surprise to those who were acquainted with the limits to his self-expression, or lack of them. ("Sounds like Craig," said James Watson, on hearing the news.) Venter explained that he had volunteered his blood and sperm out of scientific curiosity—"What scientist *wouldn't* take this opportunity, if he got the chance?" he asked me—as well as to set an example as we all begin to balance the advantages of genetic self-knowledge against the risks that it might be misused. He had learned that he had inherited from one parent the "bad" form of the gene called APOE ε4, which meant that he had a greater than normal risk of developing heart disease and Alzheimer's disease. He could not do much about Alzheimer's, but he had cut back on his calories and begun taking cholesterol-reducing drugs to reduce his chances of a heart attack. So there was justification not just for being the primary donor to his own project but for going public with that information.

At the same time, there was something crudely indulgent, even a little obscene, about the decision, as if every copy of a new version of Windows were to come with a screen-saver of Bill Gates's nose installed. "Any genome intended to be a landmark should be kept anonymous," said Arthur Caplan, the biomedical ethicist and Celera science board member who had helped oversee the exacting protocols for taking DNA donations for the project. "It should be a map of all of us, not of one, and I am disappointed if it is linked to a person." But self-fulness—as opposed to selfishness—had always been the working edge of Venter's talent, the tool he used to pry open the box and get outside it. Surely this was one of the reasons the academic genome scientists despised him. It was all right for a scientist to have a "healthy ego"—in fact, it was a necessity. But you were supposed to be discreet in the way you promoted your own achievements, not to flaunt them like a toddler delighting in its own bodily functions. And when it came to something as noble and grand as the human genome, of course, you were supposed to swaddle your ego in altruism. What Francis Collins and the others could not understand was that for Venter, as for Faust himself, altruism and the fullest possible gratification of self weren't at opposite ends of a spectrum but were one and the same. At a press conference after the genome was announced where both Venter and Francis Collins were present, a reporter asked whether the achievement deserved a Nobel Prize.

"Well, you would have to give it to three or four thousand people!" Collins selflessly replied. Venter did not answer the question. But among his friends and colleagues he did not bother to mask his conviction that if anyone in genomics deserved the Nobel Prize, he did. He even chatted about it with officials at the Swedish Royal Academy of Sciences—a shocking audacity and probably counterproductive. One does not lobby for the prize, or even speak its name and one's own in the same breath before it is in one's grasp, any more than a pitcher muses aloud about his chances for a no-hitter after seven perfect innings. Venter must have known this. He just couldn't help himself. Whatever his critics say, if he does win the prize, it will come in spite of his self-promotion, not because of it.

Venter's second, more substantive news in the spring of 2002 was about his next endeavor. He had considered a return to his beloved TIGR. But his wife was now happily ensconced in his old job there, and they had different, perhaps conflicting, agendas. Claire Fraser wanted to carry on directing the successful research institute she had built up, while Venter told me he "was looking for something else to do to change the world." He decided to set up an affiliate of TIGR called the J. Craig Venter Foundation, a nonprofit endowed with $100 million from his stock holdings. The foundation would offer a platform to change the world through genomics. One of the first efforts mounted under the foundation's umbrella was The Institute for Biological Energy Alternatives—a deceptively modest name, considering its ultimate aim was to solve the global energy crisis by inventing a new life-form. The idea was to use the tiny genome of the bacterium *Mycoplasma genitalium* as a model to string together a synthetic life code consisting of the bare necessities for growth and reproduction. Researchers at the institute would insert this ersatz genome into a *Mycoplasma* nucleus from which the natural genetic material had been removed. If the organism functioned, then additional genes might be spliced into its code to orchestrate a biochemical pathway using sunlight and the carbon from greenhouse gases to harvest hydrogen atoms from water—a clean, bountiful new source of energy, derived from the waste products of burning fossil fuels. Venter gathered a new scientific team to carry out what he was calling "the Genesis Project." It was to be led by his old friend Hamilton Smith.

"I can't resist!" Smith told me, holding up his big liver-spotted hands in mock protest. "I'm drawn like a magnet."

Smith would be the first to admit that solving the world's energy crisis by conjuring up new life-forms that feed on pollution was not an idea he would have come up with on his own. That takes a different kind of genius. Faust himself would have been envious of an alchemy so sublime and self-elevating all at once. "By his noble deeds a man is deified," Goethe's hero declares, and it is this ambition that prompts him to seek out the devil. In fact, the devil's claim on Faust's soul comes due only if Faust stops striving and finds contentment in what he has achieved. He tells Mephistopheles:

> *If ever to the moment I shall say:*
> *Beautiful moment, do not pass away!*
> *Then you may forge your chains to bind me,*
> *Then I will put my life behind me*

By this measure, Venter seems in no danger of being swept off to hell anytime soon. In his view, heaven is not an option either.

"As a biologist, I firmly believe that when you're dead, you're dead," he told me off St. Barts, the night he was talking about suicide. "Except for what you leave behind in history. That's the only afterlife."

A NOTE ON SOURCES

In May 1998, shortly after Craig Venter announced his intention to sequence the human genome, he agreed to allow me exclusive access to the inner workings of his enterprise as it went forward. The one condition was that I sign an agreement with his company (later named Celera Genomics) not to disclose proprietary information for a period of three years. Francis Collins, the head of the competing Human Genome Project, declined my request for similar access to the HGP's operation, half-jokingly citing the quantum mechanical law known as "the Heisenberg Principle": my presence as an outsider at private meetings of the HGP might influence the behavior of the participants, much as the measurement of quantum mechanical events changes the events being measured.

While I was initially disappointed by Collins's refusal, I later came to appreciate his point of view. Nevertheless, his reluctance to throw open the doors to the HGP left me with far more access to one genome project than the other, and ultimately led to my decision to locate the narrative primarily at Celera. Until the White House announcement in June 2000, I spent hundreds of hours shadowing Venter and the other key players, attending scientific and business policy meetings, and recording the numerous small agonies and ecstasies that led to the final resolution of the conflict. It soon became apparent that the competing ambitions within Celera—and even within Venter himself—were just as compelling as the conflict between the company and the government program that was receiving so much media attention.

Conveying the excitement of what happens over such a long period requires some distilling of reality. I have occasionally collapsed what was said in two meetings into a single scene or left out inconsequential intervening dialogue. Unless otherwise noted, when someone in the narrative is characterized as speaking to a "visitor," "writer," or "acquaintance," the person referred to is me. Of course, many important discussions and developments took place in my absence, especially in the HGP. I have reconstructed these scenes, including dialogue, through interviews with multiple participants. When two or more recollections contradicted one another, I followed the account which I believed to be the more trustworthy.

When possible I confirmed the accuracy of my sources' memories with documented records, such as minutes and summaries of meetings. In this regard, the degree of cooperation from the government was disappointing. From its inception in the spring of 1999, the senior leadership of the Human Genome Project—the so-called G-5 group—held weekly meetings in person or by teleconference. Their discussions were later summarized in writing. I appealed to Dr. Collins and others at the NHGRI to provide me with these summaries, without success. In June 2001, I submitted a Freedom of Information Act request to obtain the summaries along with some other documents. After nearly a year and a great deal of negotiation I received some of the summaries, but many of the discussions, and virtually all of the information revealing the actual progress of the sequencing at the public genome centers, had been blanked out. When I appealed for the reason for the deletions, I was told by the head of NIH's FOIA office that they were permitted under Exemption 4 to the FOIA, which prevents the release of "commercial and financial information that is privileged and confidential." Considering the concerted efforts the HGP leaders made during the race to distinguish their totally free, totally public version of the genome from Celera's commerical one, the explanation sounds oddly discordant. To this day, I remain perplexed why an enterprise that prided itself on global access to its genomic treasures should be so secretive about how those treasures were obtained.

In a few years, the biomedical knowledge gained by the struggle depicted in these pages will begin to bear fruit, and such matters will seem trivial. When the benefit of knowing the human code is measured in the thousands of human lives saved, no one is going to care who won this race; we will only be grateful that the prize was achieved.

NOTES

PROLOGUE

6 "Craig likes to do high dives": Bruce Cameron, interview by the author, December 8, 1998.

CHAPTER ONE: MAY 1998: "YOU CAN DO MOUSE"

13 introduced to Tony White: The account of the meeting is based on interviews by the author of Mike Hunkapiller, Craig Venter, Nicholas Wade, Tony White, and Tony Russo of Noonan/Russo, Perkin Elmer's public relations consultant.

17 "There is only one human genome program": Robert Cook-Deegan, *The Gene Wars: Science, Politics, and the Human Genome* (New York: W. W. Norton, 1994), p. 341.

18 tone began to get nasty: Richard Gibbs, interview by the author, April 8, 2000. The description of this meeting of the HGP in December 1997 is based as well on the recollections of Mark Adams, Francis Collins, David Cox, and Eric Lander.

21 "Craig's *mystery* guest": Francis Collins, interview by the author, May 21, 1999.

23 response attributed to Francis Collins: According to Wade (interview by the author, May 21, 1999), when he interviewed Collins and Varmus that weekend he found both men "in a mood of stark defeat." In his book *Life Script,* published in 2001 by Simon & Schuster, Wade quotes Collins as saying NIH could "pursue other model organisms like the mouse" (p. 45). Collins denies he ever considered such a course and told me the allusion to it in Wade's original story was "an unfortunate misinterpretation on Nick's part."

25 *Wall Street Journal* and other major dailies published their own stories: Bill Richards, "Perkin-Elmer Will Join Venture to Decode Genes," *Wall Street Journal,* May 13, 1998, p. CA2.

26 journal accepted the paper, but published it with a damning rebuttal: J. L. Weber and E. W. Myers, "Human Whole-Genome Shotgun Sequencing,"

Genome Research 7, no. 5 (May 1997): 401–9. P. Green, "Against a Whole-Genome Shotgun," *Genome Research*, op. cit., 410–17.

CHAPTER TWO: THE SECRET OF LIFE

27 the case of Gregor Mendel: My discussion of Mendel is based largely on Robin Henig's excellent biography, *The Monk in the Garden: The Lost and Found Genius of Gregor Mendel, the Father of Genetics* (Boston: Houghton Mifflin Company, 2000).

31 some enzyme . . . wasn't doing its job: Matt Ridley, *Genome: The Autobiography of a Species in 23 Chapters* (New York: HarperCollins, 1999), p. 39.

32 Morgan was the original drosophilist: My discussion of the work of T. H. Morgan and his students on the genetics of *Drosophila* is based on Jonathan Weiner's superb book, *Time, Love, Memory: A Great Biologist and His Quest for the Origins of Behavior* (New York: Alfred A. Knopf, 1999).

36 "nuclein": Much later in his career, Miescher speculated, in a letter to his uncle, that certain proteins composed of repeating, but not quite exact, chemical subunits would be capable of carrying the hereditary message through generations, "just as the words and concepts of all languages can find expression in twenty-four to thirty letters of the alphabet." If only he had known that the real molecule of heredity was not a protein, but the new substance he had himself discovered.

37 The result of their meeting was the discovery of the double helical structure of DNA: The definitive accounts of the discovery of the structure of DNA and the "golden age" of molecular biology that followed are still Horace Freeland Judson's majestic *The Eighth Day of Creation* (originally published by Simon & Schuster in 1979, and reissued in an expanded edition in 1996 by Cold Spring Harbor Laboratory Press) and James Watson's autobiographical account, *The Double Helix* (New York: Random House, 1967).

38 he and Watson had found "the secret of life": James Watson, op. cit., p. 167.

CHAPTER THREE: DOWN BUNGTOWN ROAD

40 proposal by the Department of Energy to uncover the entire sequence of the human genome: Discussion of the origin of the Human Genome Project and its early years is based largely on Cook-Deegan, *The Gene Wars*, and Jerry E. Bishop and Michael Waldholz, *Genome: The Story of the Most Astonishing Scientific Adventure of Our Time—the Attempt to Map All the Genes in the Human Body* (New York: Simon & Schuster, 1990).

41 "indenture all of us": Cook-Deegan 1994, p. 111.

42 "I do not have *a* Nobel Prize": Richard Roberts, interview by the author, October 12, 1999; confirmed by Hamilton Smith.

44 heeded an urgent call from the community: Maynard Olson, "A Time to Sequence," *Science* 270 (October 6, 1995): 394.

46 "The establishment of this principle": F. Collins, *Genome Research* 11, no. 5 (May 2001): 641.

49 Venter entered the blond-paneled conference room: Reconstruction of this meeting as based on the recollections of numerous participants, including Mark Adams, Ari Patrinos, David Cox, Francis Kalush, Francis Collins, Richard Gibbs, and Craig Venter. Kalush, Gibbs, and Patrinos all mentioned that Waterston was particularly concerned about data release.

50 "you've completely misapplied Lander-Waterman!": Eric Lander, interview by the author, May 21, 1999.

51 There would be widely divergent opinions: Interviews by the author with David Cox, October 12, 1999, and Francis Collins, May 21, 1999.

53 "If you join Craig's board": David Cox, interview by the author, October 12, 1999.

53 "I understand the fruit fly is going to be Poland": Gerry Rubin, interview by the author, January 27, 1999.

CHAPTER FOUR: GENESIS

60 it could sequence more DNA in a day: Leslie Roberts, *Science* 238, no. 4825 (October 16, 1987): 271.

63 "If I dreamed it, too bad for them": Tony White, interview by the author, September 8, 1999.

67 "someone who recognizes a good idea": Hamilton Smith, conversation with the author, November 12, 1999.

CHAPTER FIVE: THE CODE BREAKER

69 "My mother would ask him, 'Why are you such a failure?'": Keith Venter, interview by the author, October 21, 1999.

75 he wore bell bottoms: Claire Fraser, conversation with the author, December 3, 2000.

CHAPTER SIX: THIS GUY CAN GET SEQUENCERS TO WORK

79 "Jim could see we were way ahead of everybody else": Richard McCombie, interview by the author, November 23, 1999.

79 "The Human Genome Project is going to succeed": Gerry Rubin, interview by the author, February 1, 2000.

83 "When Craig first started talking about ESTs": Richard Roberts, interview by the author, October 12, 1999.

83 "a bargain in comparison to the genome project": Leslie Roberts, *Science* 252 (June 21, 1991): 1618–19.

85 "You could see the dagger go in": Robert Cook-Deegan, interview by the author, August 8, 2000.

85 "too hard on Craig": Ibid.

86 "It's really remarkable": S. Sugawara, "A Healthy Vision," *Washington Post,* November 16, 1992.

87 Haseltine would call up his competitors at three o'clock in the morning: Interview by the author, July 13, 2000. The source asked not to be identified.

87 a scientist at Incyte Pharmaceuticals in California: Jeff Seilhamer, interview by the author at Incyte Pharmaceuticals, September 10, 1998.

87 "I was running the company de facto from day one": William Haseltine, interview by the author, November 11, 1999.

88 "When he went off with Bill Haseltine": Norton Zinder, interview by the author, October 25, 1999.

88 "An academic can use the data for anything he wants": William Haseltine, interview by the author, November 11, 1999.

CHAPTER SEVEN: THE QUIETER WORLD

94 over 100,000 serious gaps in the assembled sequence: Maynard Olson, testimony before the House Committee on Science, Subcommittee on Energy and the Environment, June 17, 1998.

95 "I love bench work": Hamilton Smith, interview by the author, December 22, 1998.

95 Ham Smith had grown up with a precociously brilliant brother: Material on Hamilton Smith's background based on interviews by the author and on Douglas Birch's superb three-part series on Smith in the *Baltimore Sun,* April 11–13, 1999.

95 "We did a lot of incredibly dangerous things": Hamilton Smith, interview by the author, October 21, 2000.

99 a letter full of science news and gossip: Letter from Dan Nathans to Hamilton Smith, June 17, 1969.

CHAPTER EIGHT: H FLU

107 "Look Craig, we are in a race!": William Haseltine, interview by the author, November 11, 1999.

107 "I'm going to get you": Author's interview with John Coleman, TIGR board member, September 19, 1999.

108 "We were damned if we were going to pay ten million a year": Haseltine interview, op. cit.

110 "a great moment in the history of science": Nicholas Wade, "First Sequencing of Cell's DNA Defines Basis of Life," *New York Times,* August 1, 1995.

110 "I sat there looking at this piece of paper in my hands": Bill Efcovitch, interview by the author, September 11, 1998.

110 the most cited article in all of biology: Jeremy Cherfas, "Complete Genome Sequences Yield Answers to Fundamental Questions," *Science Watch* 8, 5 (September/October 1997): 8.

CHAPTER NINE: A HUNDRED MILLION CUSTOMERS

117 "But I can see this guy is a winner": Tony White, interview by the author, September 8, 1999.

121 An estimated 106,000 people die from drug side effects a year: Nicholas Wade, "Tailoring Drugs to Fit Genes," *New York Times,* April 22, 1999.

CHAPTER TEN: THE GENE HUNTER

123 Collins had Hunkapiller's assurance: Francis Collins, interview by the author, March 20, 2000.

124 the five men discussed their options: The description of this dinner meeting is based on recollections of Eric Lander, Francis Collins, and Richard Gibbs.

125 "Let me assure you, we will work together": Testimony of Francis Collins at U.S. House of Representatives Committee on Science meeting, June 17, 1998.

126 "the *Mad Magazine* version of the genome": The account of this phone conversation is based on an e-mail to the author from Tim Friend, March 28, 2001. Francis Collins did not recall the conversation.

127 Francis Collins's upbringing: Description of Collins's background is from an interview by the author, December 7, 1999.

134 "I was getting a tiny glimpse into God's mind": Francis Collins, interview by the author, April 20, 2000.

134 a key role in the discovery of more genes: Leslie Roberts, *Science* 249 (July 20, 1990): 236.

135 White, forgotten in Utah, remained bitter: Ray White, interview by the author, August 14, 2000.

CHAPTER ELEVEN: ALL HANDS

148 "Nobody can say for sure": Hamilton Smith, conversation with the author, December 22, 1998.

CHAPTER TWELVE: DEAD ON ARRIVAL

156 "I went there with the best of intentions": Francis Collins, e-mail to the author. The description of this meeting is reconstructed also from the recollections of Sam Broder, Paul Gilman, and Craig Venter.

158 a copy of the *Science* article they'd published the previous June: J. C. Venter et. al., "Shotgun sequencing of the human genome," *Science* 280, no. 5369 (June 5, 1998): 1540–42.

160 Frazier saw the developing agreement between Celera and the DOE as a win-win: Marvin Frazier, interview by the author, December 2, 1999.

161 "I ran into Sir Robert May": Paul Gilman, interview by the author, November 30, 1998. Rita Colwell remembered the conversation with May but could not confirm exactly what he said.

161 "it was to let me save face": Ari Patrinos, interview by the author, November 10, 1999.

161 "everybody was very upset": Frazier interview, op. cit. The description of the meeting to discuss the Celera-DOE collaboration is based as well on recollections of Elbert Branscomb, Francis Collins, and Eric Lander.

162 "what he'd done was outrageous": Interview by the author, September 25, 2000. The source asked not to be identified.

163 "These were extremely smart people with big egos and a lot at stake": Patrinos interview, op. cit.

CHAPTER THIRTEEN: VENTER UNITS

168 "If Venter wins the genome race": Dick Thompson, "Gene Maverick," *Time,* January 11, 1999.

174 "the choicest job there is in biotech": Robert Millman, conversation with the author, February 8, 1999.

CHAPTER FOURTEEN: WAR

182 "Will this MAVERICK unlock the greatest scientific discovery of his age?" Justin Gillis, "Copernicus, Newton, Einstein and VENTER?" *USA Weekend,* January 29–31, 1999, p. 4.

182 "like Dorothy at the door of the great and powerful Oz": Lewis Ricki and Barry A. Palevita, "Sequencing Stakes: Celera Genomics Carves Its Niche." *The Scientist,* July 19, 1999.

185 "that visit wasn't just for a little chat": Interview by the author, July 13, 2000. The source asked not to be identified.

186 key strategists in the public genome program: Francis Collins, interview by the author, May 30, 2001. The depiction of this meeting is also based on interviews with Richard Gibbs, Eric Lander, and Rick Wilson and e-mail from Robert Waterston.

186 "I'm Bruce Roe, the grandfather of sequencers": Bruce Roe, conversation with the author, May 21, 1999.

187 "Eric's mind is simply overwhelming": Conversation with the author, June 18, 2001. The source asked not to be identified.

187 Collins and his colleagues now had more than enough cash: Budget figures are from Antonio Regaldo, "Whitehead's Lander Kept Public Sector in the Gene Race," *Wall Street Journal,* February 12, 2001, p. B12.

188 "Jesus, did you do the arithmetic"?: Elbert Branscomb, interview by the author, June 18, 2001.

192 "Private sector companies . . . claim to be able to sequence the human genome": Hearings from a subcommittee of the Committee on Appropriations, House of Representatives, 106th Congress, February 25, 1999, p. 855.

193 "You can imagine that is a substantially more difficult process": Quote taken from the original hearing transcript. Witnesses at congressional hearings are given a chance to copyedit their testimony before it is printed. In the printed record of Collins's testimony, the word "undoubtedly" has been edited out.

CHAPTER FIFTEEN: THE IDES OF MARCH

196 a review of ABI's new capillary machine in *Science:* J. C. Mullikin and A. A. McMurray, "Sequencing the Genome, Fast," *Science* 280, no. 5409 (March 19, 1999): 1867–68.

201 The completion of the genome of the tiny roundworm *C. elegans:* The *C. elegans* Sequencing Consortium, "Genome Sequence of the Nematode *C. elegans:* A Platform for Investigating Biology," *Science* 282, no. 5396 (December 11, 1998): 1945.

CHAPTER SEVENTEEN: THE HAND OF MAN

226 "told me how everybody was going to spend a dollar a day": Robert Millman, conversation with the author, July 6, 2000.

227 The same four basic guidelines that Hopkins had to fulfill: Until 1880, inventors were also required to submit a model of the invention with their application. For another half century they were free to do so if they chose, but in 1925 the Patent Office, its attics stuffed, expressly forbade any more models. The bulk of the collection of 200,000 was sold at auction. The winning bidder was Henry Wellcome, the founder of Glaxo Wellcome and the benefactor of the Wellcome Trust. When he died the collection was sold off and scattered, and some of it eventually found its way through flea markets into Millman's office and living room.

235 "Nobody here understands the land grab like me": Robert Millman, conversation with the author, October 6, 1999.

CHAPTER EIGHTEEN: EVIL BOY

236 "Craig should not be portrayed as a maverick": Bruce Rose, conversation with the author, op. cit.

CHAPTER NINETEEN: CHESS GAMES

245 "Six weeks later, he was reporting to me": Tony White, interview by the author, September 8, 1999.

246 The newspaper published Wade's story on May 18: Nicholas Wade, "The Genome's Combative Entrepreneur," *New York Times,* May 18, 1999, Section F, p. 1. The publication of the story just a day before the Cold Spring Harbor meeting was viewed by both sides as more than a coincidence. Wade was at the meeting, and many of the scientists there believed that he'd come to harvest the resentment to Venter's remarks in the story. Paul Gilman agreed. "It's the most manipulative thing I've ever seen the press do," he said.

250 "How dare you diss my people like that!": The account of the offending sign incident is based on interviews with Michael Hunkapiller, Robert Thompson, Craig Venter, and Tony White.

251 "I consider this to be very important": Tony White, letter to Craig Venter, July 12, 1999.

252 "You've got to stop the bleed": Robert Millman, conversation with the author.

CHAPTER TWENTY: HOW TO ASSEMBLE A FLY

257 "That man doesn't look so good": M'Liz Robinson, conversation with the author, May 15, 1999.

260 unitigs: Mathematically, "a unitig is a maximal interval subgraph of the graph of all fragment overlaps for which there are no conflicting overlaps to an interior vertex" (Myers et al., *Science* 287 [March 24, 2000]: 2196).

264 They set the figure at almost 50 percent above Incyte's trading value: Neither Incyte nor PE Corporation officials would divulge the actual asking price. The

figure of $700 million for the market value of Incyte at the time was based on approximately 28 million shares of the company outstanding, then trading at about $25 per share.

264 "He more or less told them, 'We will bury you'": The meeting description had been reconstructed from interviews with Randy Scott, Paul Gilman, Peter Barrett, Craig Venter, and Tony White.

266 "The public program has played right into our hands": Craig Venter, Celera senior staff meeting, May 8, 1999.

CHAPTER TWENTY-TWO: DANCING IN MIAMI

280 "the human organism is considerably more complex": Nicholas Wade, "Count of Human Genes Is Put at 140,000, a Significant Increase," *New York Times,* September 23, 1999, p. A19.

286 "We could have been gods": M'Liz Robinson, conversation with the author, September 25, 1999.

CHAPTER TWENTY-THREE: GETTING TO NO

290 Eric Lander had emerged: Eric Lander, interview with the author, July 13, 2000.

291 "Eric did a very Eric-ky thing": Elbert Branscomb, conversation with the author, June 21, 2001.

297 *Guardian* had published a story: David Hencke, Rob Evans, and Tim Radford, "Blair and Clinton Push to Stop Gene Patents," *Guardian,* September 20, 1999.

300 Rubin received an e-mail from Ashburner in England: E-mail from Michael Ashburner to Gerry Rubin, November 1, 1999, supplied to the author by Ashburner along with responses from Collins and others.

302 Sulston learned of the renewal of collaboration talks: John Sulston and Georgina Ferry, *The Common Thread: A Story of Science, Politics, Ethics, and the Human Genome* (London: Bantam Press, 2002), p. 207.

303 The patent filing: Maggie Fox, "Celera Files Preliminary Patents on DNA," Reuters report, September 27, 1999.

303 "I'm not sure this passes the red-face test": Interview with the author, November 16, 1999.

304 "serious concerns": Collins's "serious concerns" are mentioned in an e-mail from Lee Limbird, associate vice chancellor for research at Vanderbilt University, to George Stadler, another Vanderbilt official, dated October 16, 1999.

304 "I will certainly not apologize for talking about where we were": Francis Collins, interview by the author, March 8, 2002.

305 they predicted contained a minimum of 545 genes: I. Dunham, et al., "The DNA Sequence of Human Chromosome 22," *Nature* 402 (December 2, 1999): 489–494.

305 "A new era has dawned": Nicholas Wade, "After Ten Years' Effort, Genome Mapping Team Achieves Sequence of a Human Chromosome," *New York Times,* December 2, 1999. Sulston's quote and comment by Michael Dexter of the Wellcome Trust are from Kevin Davies, *Cracking the Genome: Inside the Race to Unlock Human DNA* (New York: The Free Press, 2001), p. 194.

306 "A chromosome is a biological entity": Francis Collins, conversation with the author, December 1, 1999.

307 "We mean, come on!": "Rule Breaker to Buy PE Celera Genomics Corp.," at www.fool.com, December 16, 1999.

307 largest gainer for the week on the entire exchange: "Celera Genomics Leads Charge in Biotech Sector," at www.thestreet.com, December 23, 1999.

309 White's utterance was reported by Collins and confirmed by Venter. Reconstruction of this scene based also on interviews with Paul Gilman, Robert Waterston, and Tony White.

CHAPTER TWENTY-FOUR: THINGS BEING WHAT THEY ARE

310 "The things they laid on the table were just *shocking*": Francis Collins, interview by the author, March 20, 2000.

313 Collins opened his copy of the magazine to find Venter grinning back at him: Dick Thompson, "The Gene Machine," *Time*, January 4, 2000, p. 58.

313 "You owe the hardworking and dedicated public sequencing community": Francis Collins to Dick Thompson and Philip Elmer-DeWitt, February 8, 2000.

313 What came out the other end was poorer in quality: Characterization of Whitehead's sequence reads as "poorer in quality" is based on several sources within the HGP.

315 He broached the matter with his G-5 colleagues: Summary notes of G-5 meetings in early 2000 were obtained by the author in heavily edited form under the Freedom of Information Act. In some cases the complete notes were obtained through other sources.

315 "Francis came to us and basically said, 'OK, Incyte, make us an offer'": Randy Scott, interview by the author, March 12, 2002.

316 "While establishing a monopoly on commercial uses": Francis Collins et al. to Craig Venter et al., February 28, 2000.

316 "We all realized that clearly a letter of this sort would be of interest to a lot of people": E-mail from Francis Collins to the author, April 4, 2002. "I figured that the letter would probably leak out eventually, but I had no plans to do this intentionally . . . I am aware of no explicit discussions amongst our group about how the letter might ultimately become public—just a general expectation that, given the number of individuals who would see it, it would probably not remain confidential forever." The fact that the letter was intended to be leaked is supported by Sulston and Ferry (op. cit., p. 217): "Everyone involved understood that the contents of the letter were likely to be made public at some point, although exactly how and when this should happen was left vague at first."

317 "There should be only one place people go for [genomic and proteomic] information": Charles Poole, conversation with the author, February 1, 2000.

318 "the Brits were faxing it all over creation": E-mail from Justin Gillis to the author, April 4, 2002.

319 "the deadline hasn't even expired yet!": Kathy Hudson, interview by the author, June 30, 2000.

319 "I was ordered to tell reporters that we had nothing to do with the leak": Cathy Yarbrough, phone interview with the author, April 4, 2002.

319 "I'm sort of disgusted": Justin Gillis, "Gene-Mapping Controversy Escalates," *Washington Post,* March 7, 2000, p. E1.

320 "Celera is the only one being singled out": Roses expressed his position on the paired-end sequencing project in an e-mail to Eliot Marshal at *Science* magazine on March 6, 2000. "The TSC (The SNP Consortium) is involved is [*sic*] progressing a request from Francis et al. that involved and named Incyte for plasmid sequencing," he wrote. "The TSC is moving an RFA forward for open bids, not wanting to 'pick sides.'"

320 "Nothing's changed in four hundred years": Craig Venter, phone conversation with the author, March 6, 2000.

321 Collins's infamous *Mad Magazine* quote: Tim Friend, "The Book of Life: Twin Efforts Will Attempt to Write It," *USA Today,* June 9, 1998, p. D6.

321 He called Friend's story "unforgivable": According to the summary notes from the G-5 meeting held in Walnut Creek, California, on January 13, 2000, obtained through the Freedom of Information Act, "a few suggestions were made about how to get and to pay for additional, external sequencing capacity. *Francis agreed to follow up on the one most likely to materialize.*" According to Randy Scott, Incyte was the firm being referred to. The matter was discussed again in a G-5 conference call on February 11. According to the summary notes, "The situation with the potential subcontractor still seems promising in terms of their ability to produce these results."

321 "This whole conspiracy thing is bizarre beyond words": Francis Collins, interview by the author, March 20, 2000.

321 Kathy Hudson expressed herself more viscerally: Kathy Hudson, conversation with the author, March 14, 2000.

322 "He knows what's happening": Kathy Hudson, phone interview by the author, May 8, 2002.

CHAPTER TWENTY-FIVE: A GARDEN PARTY

326 Collins adamantly denied any such conspiracy: How closely NHGRI and the Wellcome Trust coordinated their activities during this period is difficult to ascertain. In response to a Freedom of Information Act request to NHGRI for e-mails and other correspondence between Collins and Michael Morgan, I was informed that no such documents existed. When I asked Collins directly why there was no record of his correspondence with Morgan, he told me that he had deleted the e-mails to save space on the server.

326 "Craig has tried to play both ends": Francis Collins, interview with the author, March 20, 2000.

326 The fruit fly genome was published in *Science:* M. D. Adams et al., "The Genome Sequence of *Drosophila melanogaster,*" *Science* 287, no. 5461 (March 24, 2000): 2185–95.

328 "I don't know how to deal with it": Craig Venter, conversation with the author, March 28, 2000.

328 "Don't worry about Francis": According to Craig Venter, op. cit. Dr. Fauci did not respond to requests for an interview.

331 "This is huge. This is wonderful": Eric Schmidt of SG Cowen, quoted in Stephanie O'Brien, "Celera Finishes Genome Sequencing," CBS MarketWatch Biotech Report, April 6, 2000.

331 "hit a new low in communication clarity": Francis Collins, e-mail to the author, April 10, 2000.

331 On April 10 he could not help voicing his opinion to a reporter: Allan Dowd, "Expert Urges Caution on Genome Discovery Claims," Reuters news wire, April 9, 2000.

333 "I'm living on Valium": Norton Zinder, conversation with the author, April 21, 2000.

333 The terms of his proposed truce were substantial: Zinder, op. cit., and e-mails between Zinder and Klausner, Fauci and Kirschstein.

334 "Norton wants to do good": Paul Gilman, conversation with the author, April 26, 2000.

CHAPTER TWENTY-SIX: END GAME

336 John Sulston meanwhile was telling the BBC: David Whitehouse, "Gene Firm Labelled a 'Con Job,'" BBC News Online, March 6, 2000.

342 Collins was already there: Depictions of this meeting and the subsequent ones at Patrinos's house are based on interviews with the principal participants.

347 Investigative minds began to whir and plumb their channels: Peter G. Gosselin and Paul Jacobs, "Rivals in Gene Mapping Seek to Tie Race," *Los Angeles Times,* June 14, 2000, p. 1; Scott Hensley, "Celera, U.S. Group Discuss Publishing Findings," *Wall Street Journal,* June 15, 2000, p. A12; Tim Friend, "Competing Genome Camps Could Share Glory," *USA Today,* June 15, 2000, p. 10D; Nicholas Wade, "Rivals in the Race to Decode Human DNA Agree to Cooperate," *New York Times,* June 22, 2000, p. A20.

347 In the meantime, two potentially divisive articles appeared in the press: Richard Preston, "The Genome Warrior," *The New Yorker,* June 12, 2000, and John Carey, "The Genome Gold Rush," *Business Week,* June 12, 2000.

348 "I don't know why I'm doing this": Dick Thompson, interview by the author, June 23, 2000.

348 "he's going to think I intentionally misled him": E-mail from Dick Thompson to Philip Elmer-DeWitt, June 21, 2000. Elmer-DeWitt's response came minutes later.

349 Thompson had underestimated the amount of hell there was to pay: Dick Thompson, interview by the author, op. cit. Thompson also reported the gist of this conversation with Collins in an e-mail to Philip Elmer-DeWitt at 8:51 a.m. on June 22, 2000. Collins, in an e-mail to the author on July 27, 2003, said he "spoke to [Thompson] and tried to convince him to shift the focus to reflect the actual facts."

350 Thompson summarized everything he'd learned so far: E-mail from Dick Thompson to Philip Elmer-DeWitt, 11:40 a.m., June 22, 2000.

355 A few people had copies of *Time* under their arms: Fred Golden and Michael Lemonick, "The Race Is Over," *Time,* July 3, 2000, p. 18. While July 3 is the date of record, the issue was published on June 26, 2000.

EPILOGUE: A BEAUTIFUL MOMENT

361 "You have lowered a proud journal": E-mail from Michael Ashburner to Donald Kennedy and Barbara Jasny, December 1, 2000.

362 simultaneous publication of the two epochal papers: J. C. Venter et al., "The Sequence of the Human Genome," *Science* 291, no. 5507 (February 16, 2001): 1304; The International Human Genome Sequencing Consortium, "Initial Sequencing and Analysis of the Human Genome," *Nature* 409, no. 6822 (February 15, 2001): 860.

364 "the whole-genome shotgun assembly was an utter failure": Eric Lander, e-mail to the author, February 5, 2001.

366 when Lander's damning critique was published: The critique was eventually published not in *Nature* but in *Proceedings of the National Academy of Science,* and with Robert Waterston and John Sulston joining Lander as co-authors (Waterston et al., vol. 99, no. 6, 3712–16, March 19, 2002). The main contention of the paper vis-à-vis the validity of Celera's whole-genome shotgun method was that the public data shredded into "faux reads" by Myers's team implicitly contained the information on overlaps and location in the genome needed to put them back together. Waterston et al. attempted to demonstrate this was the case by similarly shredding up the published public human chromosome 22 and showing that it reassembled perfectly. In their rebuttal, the Celera team (Myers et al., pp. 4145–46) pointed out that not only did chromosome 22 represent a problem one hundred times smaller than the whole genome but it was a finished sequence to begin with, so it was bound to came back together in the correct order. When applied to the unfinished genome as a whole—as Celera had done in its published paper—the simulation by Waterston et al. failed to make any sense of the sequence at all.

In a subsequent issue of *PNAS* (100, 6 [March 18, 2003]: 3022–26), the HGP team rebutted the Celera team's rebuttal, and the Celera team rebutted the rebuttal of the rebuttal. The arguments were highly technical, and, needless to say, nobody's mind was changed. The question of how well Celera could have assembled the human genome without the public program's data can only be put to rest by the publication of a genome composed entirely of Celera sequence reads—a possibility made more likely in 2003 when Craig Venter obtained from Celera the right to make public the sequence reads from his own genome.

367 his name was scarcely mentioned: Ari Patrinos, e-mail to the author, July 1, 2003.

367 the Celera genome was more complete and usable: "We just put together what we did have and wrapped it up in a nice way, and said it was done . . ." John Sulston later said about the public program genome at the time of the White House announcement. "Yes, we were just a bunch of phoneys!" (Sulston and Ferry, op. cit., p. 224). He blamed "Washington politics" for forcing the government pro-

gram to inflate its results. By the time the two genomes were published, seven months after the White House celebration, the HGP's version had improved, though it was still not the product some members of the program had hoped for. ("I was ashamed to be associated with it," the DOE's Elbert Branscomb later said.) It continued to improve as additional data was sequenced, but as late as June 2002, the *New York Times* was reporting that an Icelandic genetics company had revealed that more than a hundred large-scale corrections were still needed. By the time the HGP celebrated its "finished" version of the human genome in April 2003, the euchromatic portion of the code was more than 99 percent complete, with the sequence fragments in the correct order and orientation (Nicholas Wade, "Once Again, Scientists Say Human Genome Is Complete," *New York Times,* April 15, 2003, p. F1).

369 "We are a drug company now": Tony White, interview by the author, September 7, 2001.

372 checking out job offers to return to academia: In November 2002, Myers accepted a faculty position at UC Berkeley. Mark Adams later left for a position at Case Western Reserve University.

372 "Sounds like Craig": Nicholas Wade, "Scientist Reveals Genome Secret: It's Him," *New York Times,* April 27, 2002.

372 "Any genome intended to be a landmark should be kept anonymous": Ibid.

374 "If ever to the moment I shall say": Johann Wolfgang Goethe, *Faust, Part One,* trans. David Luke (New York: Oxford University Press, 1987), lines 1699–1702.

ACKNOWLEDGMENTS

I would first like to thank The Alfred P. Sloan Foundation's Public Understanding of Science and Technology Program and The Alicia Patterson Foundation for their generous support in the research and writing of this book. I am especially grateful to Doron Weber and Margaret Engel for their guidance, faith, and insight when it was just an idea.

I owe a profound double debt to the former scientists and staff of Celera Genomics, who served as both sources and subject matter for much of this account. I am particularly grateful to Craig Venter for allowing me virtually unrestricted access to his company from before it had a building, staff, or name. Venter was naturally aware of the historical importance of Celera's attempt to sequence the human genome, and of the advantage of having an outsider there to witness and record it. But he also had the courage to grant me complete independence, relinquishing any control over what I wrote, including the right to review the manuscript. Such an arrangement took a great deal of trust on his part, and while I expect that he will not agree with all of my interpretations and depictions of events, I hope that I have returned his trust by writing an account that is objective and fair to all parties. I also want to thank him and his wife, Claire Fraser, for their graciousness in inviting me into their homes, on land and at sea.

If one were to judge by some published accounts of the genome race, Celera in those years was a sinister warren populated by sell-out scientists and greedy capitalists interested mainly in maximizing their personal gain at direct cost to the Public Good. This is untrue. The scientists and businesspeople I came to know at the company were just as infused with a sense of higher purpose in their mission as those I interacted with in the rival government program, and judging by their willingness to let me witness what went on, a great deal more confident that they had nothing to hide. In addition to Venter, I am particularly in debt to Mark Adams, Gene Myers, Hamilton Smith, and Granger Sutton for serving as my principal guides in a complex scientific world, and for reviewing my attempts to make that landscape navigable to a layperson. I am of course wholly responsible for any errors and omissions. Paul Gilman spent many hours briefing me on policy matters and on events that had transpired at Celera in my absence, always with patience and equanimity, often with memorable humor. I thank Robert Millman for his equal patience with my questions about patent policy and strategy, and for being such a gift of a character. I will always be grateful to Lynn Holland, Heather Kowalski, and Chris Wood for making my working life a lot easier, and a lot more enjoyable. Many others at Celera also generously gave me their time, most of all Peter Barrett, Vivien Bonazzi, Sam Broder, Barbara Culliton, Anne Deslattes Mays, Rob Holt, Mike Knapp, Marshall Peterson, Cindy Pfannkoch, M'Liz Robinson, Bob Thompson, and the stalwart Nerf warriors of the Assembly Team—Art Delcher, Ian Dew, Saul Kravitz, Clark Mobarry, and Karin Remington.

At Celera's parent and sister companies, Tony White and Mike Hunkapiller were always forthright and forthcoming, and I am much in their debt. Rich Roberts and the other members of the Celera Scientific Advisory Board were equally obliging, and I am especially grateful to Norton Zinder. I would also like to thank Bill Efcovitch and Susan Eddins at ABI for helping me understand the intricacies of DNA sequencing technology, Alex Lipe at Morgan Stanley for guiding me through the landscape of corporate financial structuring, and Keith Venter for helping me understand his big brother. At other companies, Randy Scott of Incyte and Bill Haseltine of Human Genome Sciences were frank and cordial sources.

I am extremely grateful for the cooperation of the scientists and staff of the public Human Genome Project, without which my account of the HGP's own dramatic story would have been much the poorer. Most of all I would like to thank Francis Collins, for his unflagging readiness to respond to a never-ending stream of e-mails and requests for interviews, and Ari Patrinos, for his wisdom, perspective, and sheer kindness. I am indebted as well to Michael Ashburner, Elbert Branscomb, David Cox, Marvin Frazier, Richard Gibbs, Mark Guyer, Kathy Hudson, Jim Kent, Rick Klausner, Neal Lane, Lauren Linton, Dick McCombie, John McPherson, Michael Morgan, Bruce Roe, John Sulston, Harold Varmus, Bob Waterston, and Kathy Yarbrough. Among my own colleagues, I would like to thank Jonathan Harr and Barry Werth for advice early on, and Bob Cook-Deegan, Tim Friend, Dick Thompson, and Nicholas Wade for their coverage of the genome wars and for enlightening my own.

For giving me their support and encouragement, especially when the going was hard, I am forever thankful to Elizabeth Shreeve, Walton Shreeve, Missy Vineyard, and Tom McKain. Missy, Steve Braun, Peter Gould, and Lori Cuthbert read all or parts of the manuscript. I am grateful for their comments, and most especially to Robin Henig for her careful, caring reading at the end of a very long haul. Sara Lippincott subsequently gave the manuscript her keen professional eye, which greatly improved the result. At Knopf, Vicky Wilson was patient, astute, and right more often than I may have wanted to admit. Susanna Sturgis's copyediting was superb. I am immensely grateful to my agent, Charlotte Sheedy, for her unwavering faith in this project, and for her magical ability to transform seemingly terminal catastrophes into mere bumps along the road. If I had known that at the end of that road Jessica Falcon Shreeve would come into my life, I would have run the whole way.

INDEX

Note: HGP refers to the Human Genome Project

PHOTO: JOYCE RAVID

JAMES SHREEVE is the author of *The Neandertal Enigma: Solving the Mystery of Modern Human Origins,* and co-author of *Lucy's Child: The Discovery of a Human Ancestor.* His articles have appeared in *The Atlantic Monthly, Discover, National Geographic, Science, Smithsonian,* and other publications. He has been a fellow of the Alfred P. Sloan Foundation and of the Alicia Patterson Foundation. Shreeve lives in South Orange, New Jersey.

Printed in the United States
by Baker & Taylor Publisher Services